W9-ADM-461

Calcium Channel Modulators in Heart and Smooth Muscle

Balaban Publishers 

© VCH Verlagsgesellschaft mbH, D-6940 Weinheim (Federal Republic of Germany), 1990

Distribution

VCH, P. O. Box 101161, D-6940 Weinheim (Federal Republic of Germany)

Switzerland: VCH, P. O. Box, CH-4020 Basel (Switzerland)

Great Britain and Ireland: VCH (UK) Ltd., 8 Wellington Court, Cambridge CB1 1HZ (England)

USA and Canada: VCH, Suite 909, 220 East 23rd Street, New York, NY 10010–4606 (USA)

ISBN 3-527-28251-3 (VCH, Weinheim) ISBN 1-56081-123-4 (VCH, New York)

Calcium Channel Modulators in Heart and Smooth Muscle

Basic Mechanisms and Pharmacological Aspects

Proceedings of the 33rd Oholo Conference
Eilat, Israel 1989

Edited by
S. Abraham and G. Amitai

Balaban Publishers

Editors:
Dr. Shlomo Abraham
Dr. Gabriel Amitai
Department of Pharmacology
Israel Institute for Biological Research
P. O. Box 19
Ness Ziona 70450
Israel

This book was carefully produced. Nevertheless, editors, authors and publisher do not warrant the information contained therein to be free of errors. Readers are advised to keep in mind that statements, data, illustrations, procedural details or other items may inadvertently be inaccurate.

Published jointly by
VCH Verlagsgesellschaft mbH, Weinheim (Federal Republic of Germany)
VCH Publishers, Inc., New York, NY (USA)
Balaban Publishers, Rehovot (Israel)

Editorial director: Miriam Balaban and Nurit Katzir

VANDERBILT UNIVERSITY
MEDICAL CENTER LIBRARY
OCT 25 1991
NASHVILLE, TENNESSEE
37232

Library of Congress Card No. applied for

British Library Cataloguing in Publication Data:
OHOLO Biological Conference: 33rd (1989; Eilat, Israel)
Calcium channel modulators in heart and smooth muscle.
1. Man. Smooth muscles. Calcium. Antagonists. Action
I. Title II. Abraham, S. (Shlomo) III. Amitai, G. (Gabriel)
615.77
ISBN 1-56081-123-4

Deutsche Bibliothek Cataloguing-in-Publication Data:
Calcium channel modulators in heart and smooth muscle/
[OHOLO Biological Conferences. Organized by Israel Institute
for Biological Research. Scientific organizing committee
(1989 conference) co-chairmen: Shlomo Abraham and Gabriel Amitai]. –
[New York]: Balaban Publ.; Weinheim; Basel; Cambridge;
New York, NY: VCH, 1990
ISBN 3-527-28251-3 (VCH, Weinheim ...)
ISBN 1-56081-123-4 (VCH, New York ...)
NE: Avrāhām, Šelomo [Hrsg.]; OHOLO Biological Conference
<33, 1989, 'Elat>; ham Mākôn le-Meḥqār Biyyôlôgî be-Yiśrā'ēl <Nēs Ṣiyyônā>

© VCH Verlagsgesellschaft mbH, D-6940 Weinheim (Federal Republic of Germany), 1990

Printed on acid-free paper.

All rights reserved (including those of translation into other languages). No part of this book may be reproduced in any form – by photoprinting, microfilm, or any other means – nor transmitted or translated into a machine language without written permission from the publishers. Registered names, trademarks, etc. used in this book, even when not specifically marked as such, are not to be considered unprotected by law.
Printing: betz-druck GmbH, D-6100 Darmstadt
Bookbinding: Großbuchbinderei Josef Spinner, D-7583 Ottersweier
Printed in the Federal Republic of Germany

Oholo Biological Conferences

Organized by

Israel Institute for Biological Research

Permanent Committee

S. Cohen, M. Feldman, A. Golombek, N. Grossowicz,
I. Hertman, A. Keynan, A. Kohn, M. Sela

Scientific Organizing Committee (1989 Conference)

Co-Chairmen: S. Abraham and G. Amitai

Members:
D. Atlas, The Hebrew University of Jerusalem
Y. Barak, Ichilov Medical Center, Tel-Aviv
B. Vidne, Ichilov Medical Center, Tel-Aviv
B. A. Weissman, Israel Institute for Biological Research, Tel-Aviv
R. Zimlichman, Sheba Medical Center, Tel-Aviv

Technical Management: R. Pniel
Secretary: N. Ben-David

Acknowledgements

The Organizing Committee of the 1989 OHOLO Conference gratefully acknowledge the generous support of the following organizations (alphabetical order):

Abic Ltd., Netanya, Israel

American Cyanamid Co., Pearl River, N.Y., USA

Fidia Research Laboratories, Abana Terme, Italy

ICI Pharmaceuticals Group, Wilmington, Delaware, USA

Israel Institute for Biological Research, Ness-Ziona, Israel

Israel Academy of Sciences and Humanities, Jerusalem, Israel

Medifisher Pharmaceutical Industries, Bnei Brak, Israel

Merck Sharp & Dohme Inc., Rahway, N.J., USA

Ministry of Tourism, Jerusalem, Israel

National Council for R & D, Jerusalem, Israel

Rafah Laboratories, Jerusalem, Israel

Research Biochemicals Inc., Natick, MA, USA

Rorer Central Research, Horsham, PA, USA

Sandoz AG, Basel, Switzerland

Siev, Dan Ltd., Tel-Aviv, Israel

Squibb, E. R. & Sons, Princeton, N.J., USA

Teva Pharmaceuticals Ltd., Petach Tikva, Israel

Wyeth-Ayerst Research, Philadelphia, PA, USA

Preface

This book has emerged from the 33rd Oholo conference on "Calcium Channel Modulators in Ischemic Heart Disease and other Cardiovascular Disorders: Basic and Therapeutic Approach" held 1989 in Eilat, Israel. The main objective of this meeting was to create the environment for the encounter of researchers who explore more basic aspects of calcium control and homeostasis in cells with those who study the pharmacological mechanisms which underlie cardiovascular diseases.

The introductory review "Calcium: The Control of a Charismatic Cation" by D. J. Triggle summarizes the different calcium control mechanisms and their involvement in various cellular events. This review is indeed the professional introduction of this book and provides the reader with the current notion about the wide spectrum of cellular pathways which involve calcium. The book is devided into five parts concerning the following: receptor-operated and voltage-dependent channels and signal transduction, excitation-contraction coupling in smooth muscle and heart, pharmacodynamic effects of calcium channel modulators and related endogenous factors, model systems for myocardial ischemia and calcium channels and new approaches in drug design of calcium channel agonists and antagonists.

The scientific data compiled in this book may be useful for the novice as well as for the more established researcher. We do hope that these proceedings will eventually contribute to the common effort of studying the various mechanisms underlying cardiovascular diseases which are the primary cause of mortality in the world.

The editors would like to express their gratitude to the members of the organizing committee: D. Atlas, R. Zimlichman, B. Vidna, Y. Barak and B. A. Weissman for their help in the preparations of the 33rd Oholo meeting. The excellent technical assistance of the late Roni Pniel, who regrettably passed away after the meeting, is appreciated. We would like to acknowledge the skillful assistance of Nita Ben-David in typing the book of abstracts and managing the correspondence before and after the meeting. Finally we thank Nurit Katzir and Miriam Balaban of Balaban Publishers for their most efficient production and their rapid response during the editing process.

Israel Institute for Biological Research
September 1990

Shlomo Abraham
Gabriel Amitai

Contens

Introductory paper

Receptor-operated and voltage-dependent mechanisms

Drug design and new approaches

Introductory paper

Calcium Channel Modulators in Heart and Smooth Muscle: Basic Mechanisms and Pharmacological Aspects
S. Abraham and G. Amitai, editors
© 1990, VCH, Weinheim/Deerfield Beach, FL and Balaban, Rehovot/Philadelphia

Calcium: The control of a charismatic cation

DAVID J. TRIGGLE

The School of Pharmacy, State University of New York, Buffalo, NY, USA 14260

INTRODUCTION

A role for calcium in the control of cellular function was early described by Sidney Ringer [Ringer, 1883]. Subsequently, calcium has been shown to play multiple roles in the control of excitable cell function [Campbell, 1983]. These roles are not limited to the support of excitation-contraction and stimulus-secretion coupling processes alone, but rather extend to numerous other events from the fertilization response to the control of cellular integrity and the finality of cell death from Ca^{2+} overload. Ca^{2+} thus plays a dominant role as a cellular messenger. However, it is important to note that the excess availability of Ca^{2+} constitutes a lethal signal.

Such Ca^{2+}-regulated processes have their origin in a decision, presumably made early in cellular development, to exclude ionized Ca^{2+} from the cell interior [**Figure 1**]. This decision, with its depicted theologic overtones, is encoded in a pattern of rules or dogma [Kretsinger, 1976]:

1. The free ionized Ca^{2+} concentration in the cell cytosol is maintained at approximately 5 x 10^{-8}M or less during the resting state.

2. During excitation the cytosolic concentration of Ca^{2+} rises to approximately 5 x 10^{-7}M to 10^{-6}M.

3. The function of cytosolic Ca^{2+} elevated subsequent to cell stimulation is to transmit information from the exterior to the interior of the cell.

4. This physiologic information is coupled to response through a set of homologous Ca^{2+} binding proteins, including the ubiquitous calmodulin, that serve as cellular Ca^{2+} receptors.

This dogma is described in the appropriate literature [Hausman, 1982].

Figure 1. The expulsion of Ca^{2+} from the interior of the cell.

The messenger function of Ca^{2+}, made possible by its highly asymmetric distribution across the plasma membrane of the cell, has two components. The inward movement of Ca^{2+} serves a current carrying [depolarizing] function and simultaneously fulfills a coupling function by interaction with the Ca^{2+} binding proteins. This duality of function is unique to Ca^{2+}. The chemistry of Ca^{2+}, notably its ability to form complexes with polyanionic ligands, its flexible ligand coordination distances, its variable coordination number and its rapid complexation rates, distinguish it from Mg^{2+}, the other physiologically available divalent cation [Levine and Williams, 1983].

It is thus clear that Ca^{2+}, to fulfill the multiple roles described, must be a tightly regulated species. The several regulatory processes are outlined in **Figure 2.** Not all of these are of equal significance to every cell type, to every stimulus or to all times. In principle, each of the processes of **Figure 2** should be susceptible to control by specific drug action. These agents should act at the following control points:

1. Potential-dependent Ca^{2+} channels

2. Receptor-operated Ca^{2+} channels

3. Plasmalemmal Na+:Ca^{2+} exchange

4. Plasmalemmal Ca^{2+}-ATPase

5. Plasmalemmal Ca^{2+} binding [intra- and extracellular

6. Mitochondrial Ca^{2+} uptake and release

7. Sarcoplasmic reticulum Ca^{2+}-ATPase

8. Sarcoplasmic reticulum Ca^{2+} release channel

9. "Undefined" Ca^{2+} leak pathways.

In practice, however, the existence of specific drugs has thus far been confined to the voltage-dependent Ca^{2+} channel. It is likely that other classes of specific drugs will be developed.

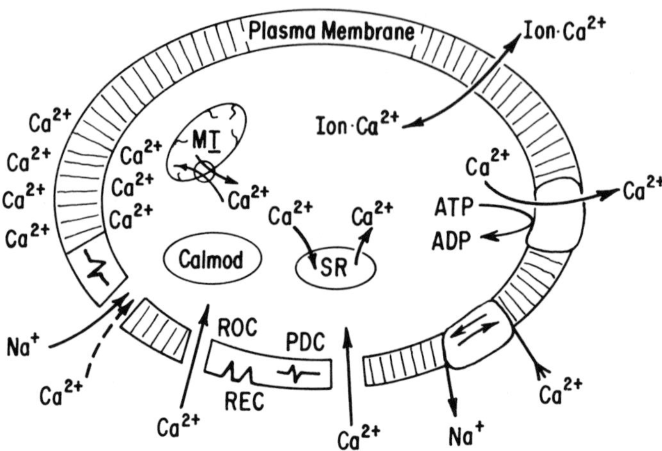

Figure 2. Schematic representation of the control of cellular Ca^{2+} by Ca^{2+} channels [ROC, PDC], Na+:Ca^{2+} exchange, Ca^{2+}-ATPase and Ca^{2+} sequestration and release from mitochondria and sarcoplasmic/endoplasmic reticulum.

THE CALCIUM CHANNEL ANTAGONISTS

The calcium channel antagonists include the clinically available verapamil, nifedipine and diltiazem and have important cardiovascular applications from hypertension and peripheral vascular disorders through the several types of angina and some cardiac arrhythmias. They are a chemically heterogeneous group of agents interacting with a set of allosterically linked binding sites on a major protein of the L class of voltage-dependent Ca^{2+} channel.
The characterization of such binding sites by structure-activity relationships, the existence of potent activators at the 1,4-dihydropyridine site and the association of site occupancy and response suggests that the Ca^{2+} channel be regarded as a pharmacologic receptor. Accordingly, the existence of endogenous regulatory species has been proposed, although not proven, for the Ca^{2+} channel [Triggle, 1988].

It is likely that other drug binding sites exist that accommodate other structural classes of agent [Janis and Triggle, 1984]. The ongoing search for new agents may serve both to facilitate characterization of the channel and its subclasses and to provide new therapeutic agents with different profiles of selectivity. Projected sales in North America of the Ca^{2+} channel antagonist as antihypertensive agents indicate the continued importance of this class of agents [**Figure 3**].

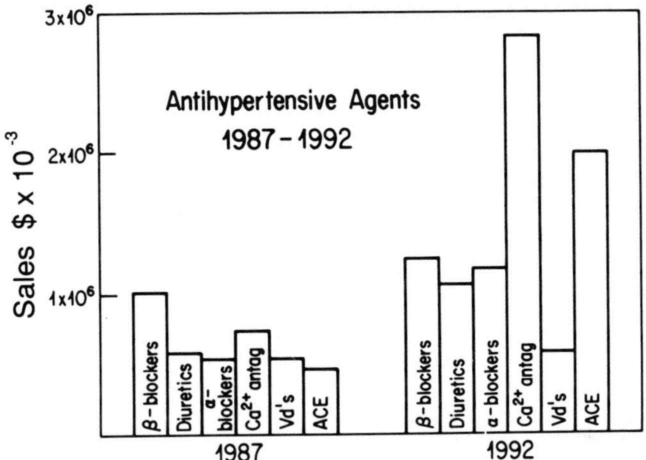

Figure 3. Current and projected sales of antihypertensive agents in North America for the periods indicated. Data taken from C. E. Stinson, Chem. Eng. News, pp. 35-70, Oct. 3, 1987.

Since the L channel is but one of at least three major categories, L, T and N, of voltage-dependent Ca^{2+} channel each characterized by electrophysiologic and pharmacologic criteria and subserving different physiological functions [Bean, 1989], it is highly probable that Ca^{2+} channel drugs for the T and N classes of channel will also emerge and may be involved in the control of pacemaker and neuronal functions respectively.

CALCIUM AND CELLULAR CONTROL

Major emphasis has been placed on the voltage-dependent Ca^{2+} channels as control points for therapeutic exploitation. However, there are other important loci of Ca^{2+} control including receptor operated Ca2+ channels and biochemically mediated Ca^{2+} mobilization processes, notably those involving the phosphatidylinositol pathway [Berridge, 1987]. Receptor-operated channels await the definition currently available for voltage-dependent Ca^{2+} channels, but in principle they may be viewed according to several models [**Figure 4**]. An important biochemical intermediate controlling intracellular Ca^{2+} release and possibly also plasmalemmal Ca^{2+} influx is inositol triphosphate [**Figure 5**]. This bifurcating pathway produces both cytosolic and membrane mediators of Ca^{2+} control and the linkages between these mediators and

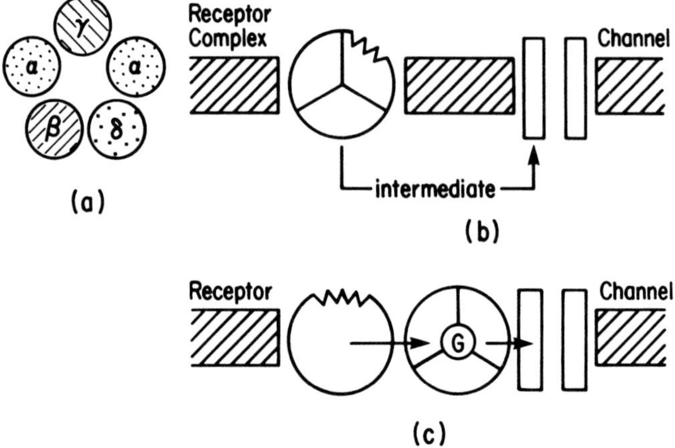

Figure 4. Representative organizations of receptor-operated Ca^{2+} channels indicating both direct and indirect association of receptor and channel and the involvement of guanine nucleotide binding proteins.

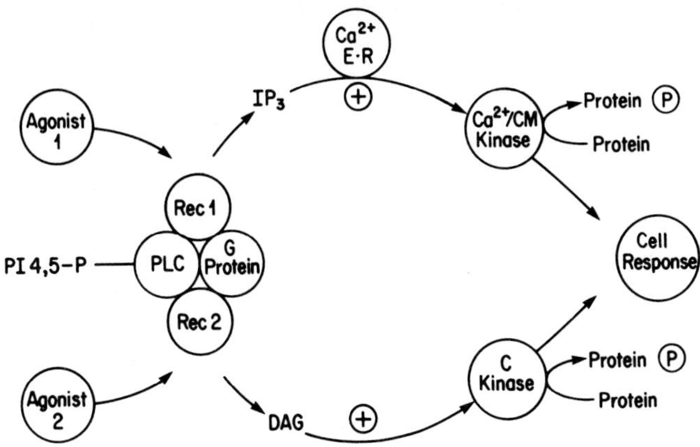

Figure 5. The pathway of receptor control of phosphatidyl-inositol metabolism with the production of IP_3 and 1,2-diacylglycerol.

receptor activation processes represent a major focus of cellular Ca^{2+} control at both physiologic and pathologic levels.

It is increasingly clear that the control of cellular Ca^{2+} is neither temporally nor spatially homogeneous [Berridge, 1989]. Thus, following stimulation the distribution of Ca^{2+} throughout the cell reveals local gradients and areas of high and low concentration. Additionally, the interaction of Ca^{2+} control processes makes possible Ca^{2+} oscillations following cell stimulation. Such oscillatory phenomena may reflect the ability of the cell to use Ca^{2+} as a cellular messenger whilst avoiding the detrimental effects of a persistent elevation of intracellular Ca^{2+}.

CALCIUM AND CELL PATHOLOGY

The defective control of cellular Ca^{2+} is increasingly realized to contribute to a variety of pathologic states. Thus, a central role for altered Ca^{2+} metabolism has long been assumed to underlie the hyperreactivity syndromes associated with hypertension

[**Figure 6**]. Additionally, it is argued that an abnormally elevated intracellular Ca^{2+} underliess the enhanced vasoconstriction of hypertension [Aopki and Frohlich, 1988]. Consistent with this, it is found that a correlation exists between blood pressure and the intracellular Ca^{2+} concentration of platelets from hypertensive patients [Erne _et al.,_ 1984]. However, several recent studies have suggested that hypertension, both clinical and experimental, may be associated with a reduced plasma concentration of ionized Ca^{2+} and even with a deficient dietary Ca^{2+} intake [McCarron, 1985]. The significance of all of these observations remains to be established. However, they are not necessarily inconsistent given the important role of extracellular Ca^{2+} in the maintenance of cellular integrity and excitability. The changes in plasma Ca^{2+} levels may also be associated with compensating changes in the serum levels of parathyroid hormone, vitamin D and calcitonin [Levine and Williams, 1983]. These finding indicate an association between cellular and organismic Ca^{2+} regulation that is of considerable importance to a more complete understanding of the global control of Ca^{2+} function.

Abnormally regulated Ca^{2+} may also underlie a variety of situations involving cell damage, loss of function and even cell death. The process of neuronal aging may be associated with elevated neuronal Ca^{2+} and loss of neuronal function. Thus, according to Gibson and Peterson [Resnick and Laragh, 1985]; "This suggests that age-related decreases in calcium movement across membranes are critical to normal cell function even if they are not the primary defect". It is thus of some interest that nimodipine, a 1,4-dihydropyridine Ca^{2+} channel antagonist with putative central selectivity, has beneficial effects against a variety of aging events [Scriabine _et al.,_ 1989] including enhancement of memory processes in aging rabbits [Deyo _et al.,_ 1989]. The role of Ca^{2+} modulation in the control of neuronal aging is of major potential significance.

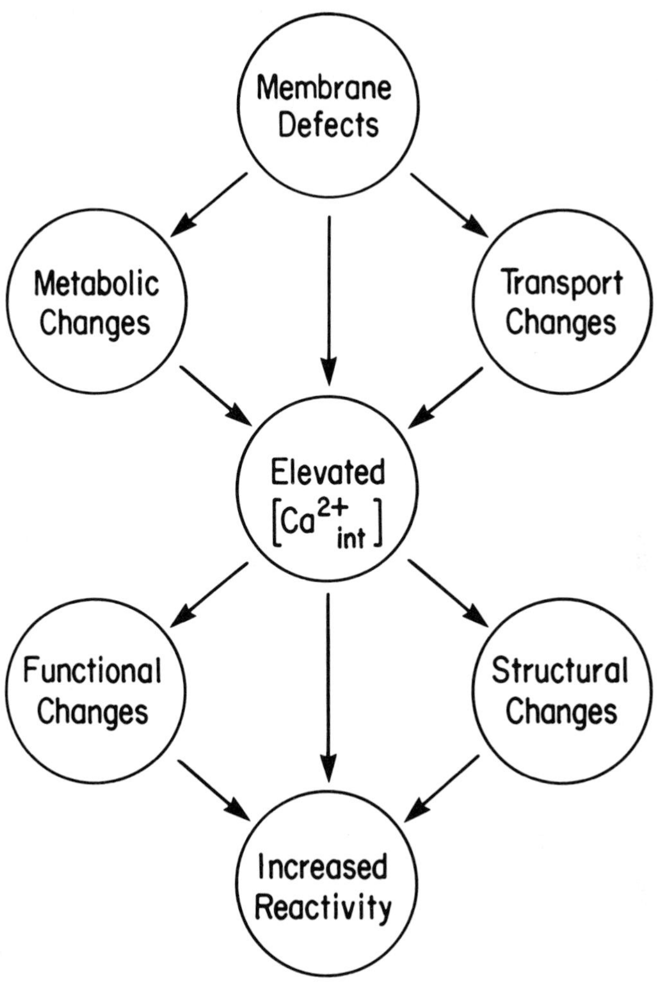

Figure 6. A potential sequence for the generation of vascular smooth muscle hyperreactivity with a central role assigned to the defective control of intracellular Ca^{2+}.

It is to be emphasised that a number of Ca^{2+} control processes operate in neuronal tissues in addition to the voltage-dependent Ca^{2+} channels. Of particular importance are the ion channels associated with excitatory amino acid receptors, notably the N-methyl-D-aspartate [NMDA] category. Ca^{2+} flux through

NMDA receptor-channel complexes has been associated with memory paradigms including long term potentiation [Brown et al., 1988; **Figure 7**]. However, the abnormal and persistent elevation subsequent to the elevation of excitatory amino acids during ischemia is associated with neuronal damage and cell death [Rothman and Olney, 1987].

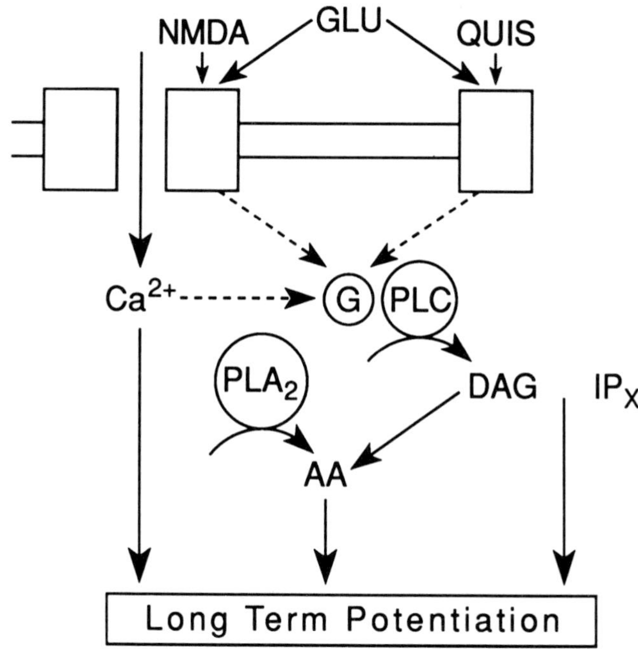

Figure 7. Organization of excitatory amino acid receptors [Glu, glutamate], [Quis, quisqualate] and [NMDA, N-methyl-D-aspartate] and the linkage to several biochemical events that may contribute to the process of long term potentiation. Over expression of these events contribute to neuronal damage and death.

Cell death following a variety of cellular insults is an inevitable consequence of a sufficiently persistent and prolonged elevation of intracellular Ca^{2+}. This dichotomy of the physiologic and pathologic roles of Ca^{2+} is clearly realized in the "Calcium Paradox" of cell death during reperfusion [Chapman and

Tinstall, 1987]. The necessary cellular insults can be many and diverse and include hypoxia, mechanical injury, energy depletion due to metabolic exhaustion or poisoning or heavy metals. Regardless, a general scheme may be established whereby the loss of energy control results in the loss of ionic control, the recruitment of both intracellular and extracellular Ca^{2+}, the activation of Ca^{2+}-dependent phospholipases and proteases, increased cell damage and the further loss of control, overloading of the mitochondrial with Ca^{2+} and cell death [**Figure 8**]. Considerable attention is being paid to possible drug control of these events and the resulting salvage of reversibly damaged tissue.

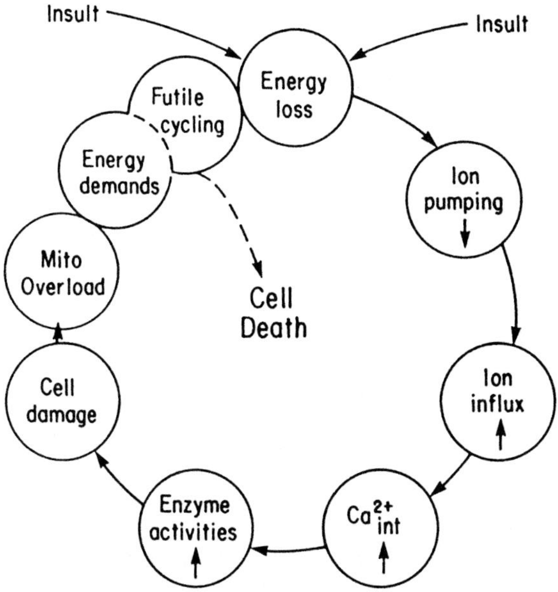

Figure 8. The cycle of events that contribute to Ca^{2+} overload and cell death.

ACKNOWLEDGEMENTS

This brief survey of the physiologic and pathologic control of cellular Ca^{2+} represents no more than the briefest of sketches. It is intended only to introduce the key players and to outline their characteristics. It is quite clear that Ca^{2+} is a cation of major cellular significance and the subsequent, and more authoritative, chapters yield ample testimony to this.

REFERENCES

Aoki K, Frohlich ED [Eds.] (1988). Calcium in Essential Hypertension, Academic Press Inc, Tokyo.

Bean BP (1989). Classes of calcium channels in vertebrate cells, Ann. Rev. Physiol., 51: 367-384.

Berridge MA (1987). Inositol trisphosphate and diacylglcerol: two interacting second messengers. Ann. Rev. Biochem., 56: 159-194.

Berridge MA (1989). Cytosolic calcium oscillators, FASEB J. 2: 3074-3082.

Brown TH, Chapman PF, Kairiss EW, Keenan CL (1988). Long term synaptic potentiation, Science 242: 724-728.

Campbell AK (1983). Intracellular Calcium: Its Universal Role as Regulator, J. Wiley & Sons, Inc, New York, NY.

Chapman RA, Tunstall J (1987). The calcium paradox of the heart, Prog. Biophys. Mol. Biol., 50: 67-96.

Deyo RA, Straube KT, Disterhoft JT (1989). Nimodipine facilitates associative learning in aging rabbits. Science 243: 809-811.

Erne P, Bolli P, Burgisser E, Buhler F (1985). Correlation of platelet calcium with blood pressure. Effect of antihypertensive therapy. New Eng. J. Med., 310: 1084-1088.

Gibson GE, Peterson C (1987). Calcium and the aging nervous system, Neurobiology of aging, 8: 329-343.

Hausman P (1982). The Calcium Bible, Rawson Associates, New York, NY.

Janis RA, Triggle DJ (1984). 1,4-Dihydropyridine Ca^{2+} channel antagonists and activators. Comparison of binding characteristics with pharmacology. Drug Dev. Res., 4: 254-274.

Kretsinger RH (1976). Calcium in biological systems. Coord. Chem. Revs., 18: 29-124.

Levine BA, Williams RJP (1983). Calcium binding to proteins and anionic centers, Calcium and Cell Function
[Ed. W. Y. Cheung] Vol II, pp. 2-38 Academic Press, London and New York.

McCarron DA (1985). Is calcium more essential than sodium in the pathogenesis of essential hypertension? Hypertension, 7: 607-627.

Resnick LM, Laragh JH (1985). Renin, calcium metabolism and the pathophysiologic basis of antihypertesnive therapy. Amer. J. Cardiol., 56: 68H-74H.

Ringer, S (1883). A further contribution regarding the influence of the different constituents of the blood on the contraction of the heart. J. Physiol. [Lond.] 4: 29-42.

Rothman SM, Olney JW (1987). Excitotoxicity and the NMDA receptor, Trends Neurosciences 10: 299-301.

Scriabine A, Schuurham T, Traber J (1989). Pharmacological basis of the use of nimodipine in central nervous system disorders. FASEB J, 3: 1799-1806.

Triggle DJ (1988). Endogenous ligands: myths and realities, In, The Calcium Channel: Structure, Function and Implications [Eds., M. Morad, W. Nayler, S. Kazda and M. Schramm], Springer-Verlag, Berlin.

Receptor-operated and voltage-dependent mechanisms

Calcium Channel Modulators in Heart and Smooth Muscle: Basic Mechanisms and Pharmacological Aspects
S. Abraham and G. Amitai, editors
© 1990, VCH, Weinheim/Deerfield Beach, FL and Balaban, Rehovot/Philadelphia

On the structure of calcium channels: The oligochannel concept

HANSGEORG SCHINDLER

Institute for Biophysics, University of Linz, 4040 Linz, Austria

INTRODUCTION

Recent advances in the biophysical analysis of ion transport processes have set the stage to trace down physiological observations on ion flux to molecular-mechanistic levels of understanding. First, I may introduce you to the kind of questions involved, second, how they can be approached and which kind of informations are obtained and, finally, which bearing these insights may have on physiology, pathology and pharmacology of functions and diseases related to ion channels.
I chose excitation-contraction coupling (ECC) in sceletal muscle or heart (cartoon in Fig.1) to indicate the kind of questions on the way towards a molecular mechanistic understanding (text-block in Fig.1). An electrical depolarisation pulse V(x,t) carried by Na^+-and K^+-channels, causes voltage gating of Ca^{++}-channels in the Transverse Tubule membrane. This signal induces opening of Ca^{++}-channels in the Sarcoplamic Reticulum membrane. Released Ca^{++}-ions bind to the contractile apparatus and trigger contraction. What is the structural and organisational context within which this coupling of the two Ca^{++}-channels can be understood in molecular-mechanistic terms? This central question includes the more specific questions, which primary and secondary constituents are involved, how they are organized to build the ECC-machinery, which system-and invironmental variables and processes couple to or regulate its function: intra-membrane variables like lipids and lateral membrane pressure, and regulatory processes occuring at the interfaces towards the three aqueous phases. The latter includes

second messengers, phosphokinases, phosphatases, G-proteins known to effect ECC from the myoplasmic side to which probably several more still unknown effectors have to be added in order to render a consistent description. This level of structure-function investigation, adressing the whole context for function, should be complemented by structure-

Figure 1. Exitation-Contraction-Coupling (ECC)

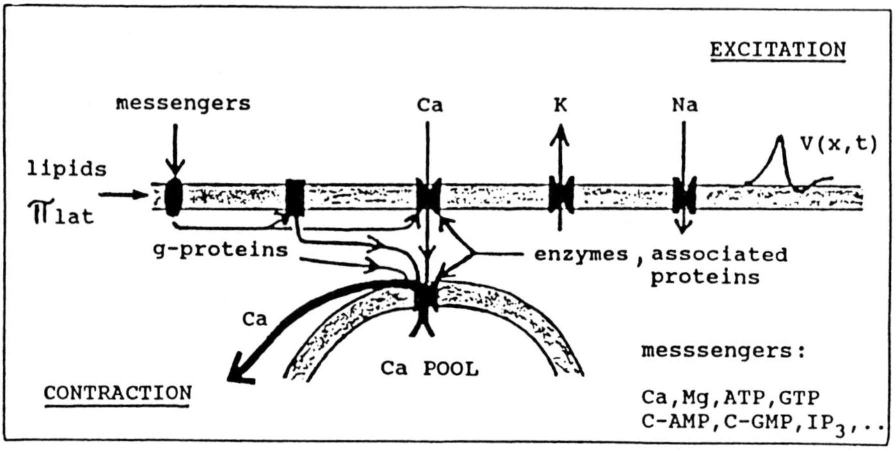

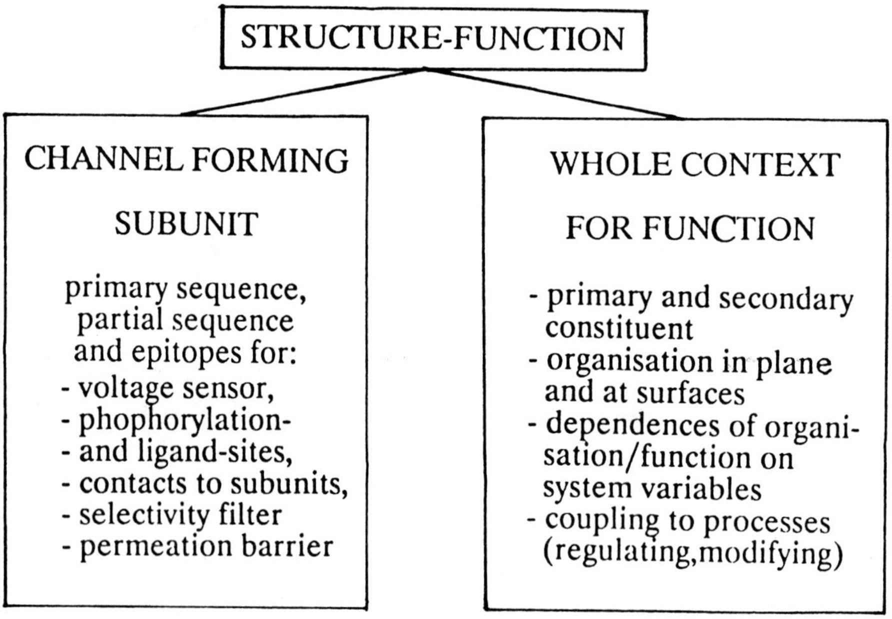

Figure 2. Molecular Model for ECC

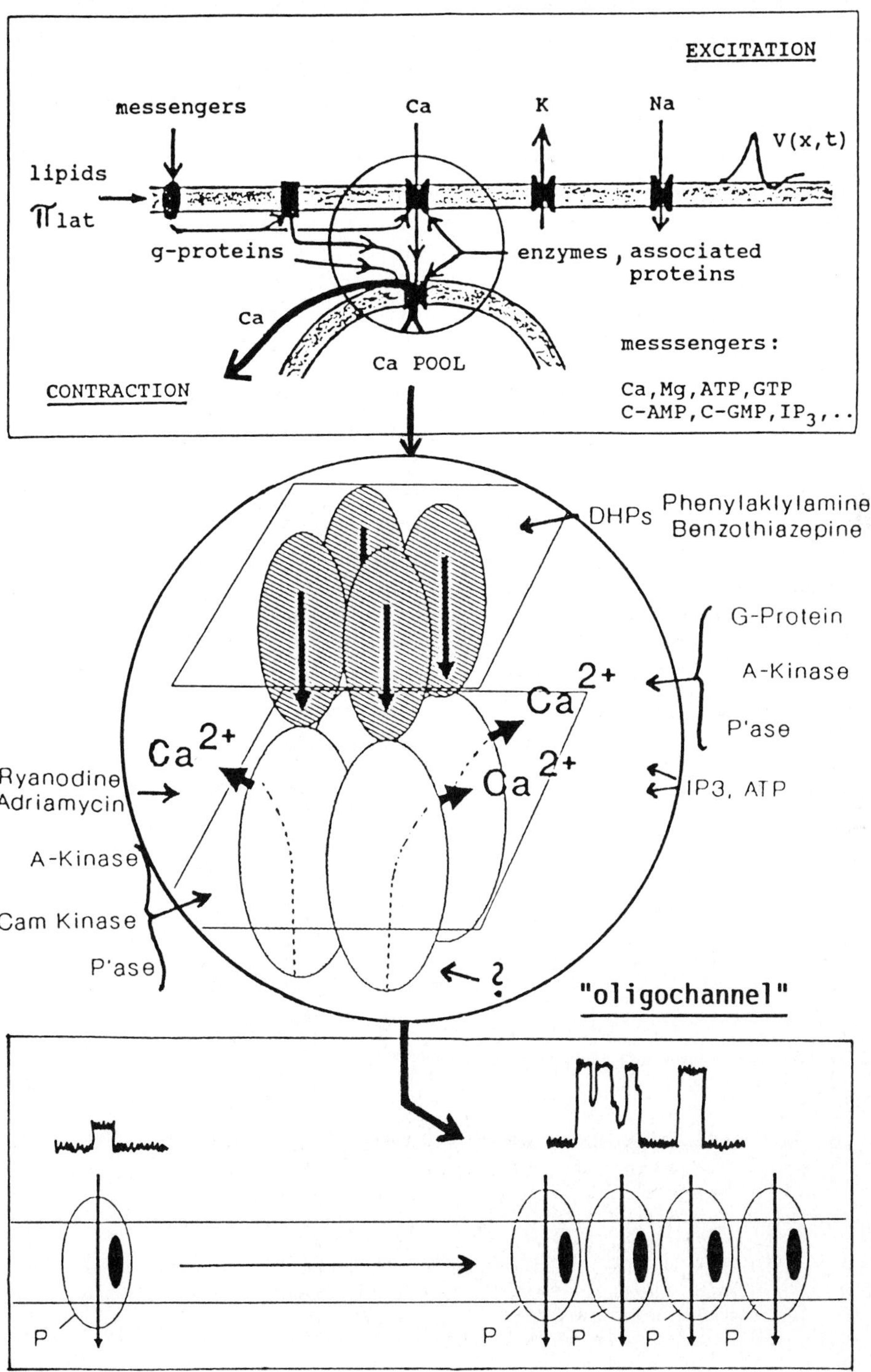

function knowledge about the ion pathway itself via primary sequences of channel forming subunits as basic information for identifying epitopes important for function and modulation (see list in Fig.1). - It is the combination of knowledge obtained at these two levels which is expected to yield qualitatively new insights in ECC at a molecular mechanistic level.
At both these levels the last year has brought a breakthrough by identifications of primary components via functional reconstitution of purified proteins, by determination of their primary sequences and by insights into the overall structural organisation of these channel proteins which has considerable bearing on viewing regulation/modulation. Our current view of the molecular machinery for ECC is illustrated in Fig.2. The main push came from the identification of the Ryanodine Receptor (Ry-Rec), being equivalent to the feet structures of the junctional terminal cisternae of SR (Inui et al., 1987), to constitute the Ca^{++}-release channel of SR. This was reported by three groups at about the same time for sceletal muscle: Lai et al,(1988), Smith et al. (1988), Hymel et al. (1988a) and for heart: Hymel et al. (1988b). The footstructure shows an approximately square crossection (27x27nm), with fourfold symmetry, it protudes by about 15nm into the myoplasm forming contacts to the T-Tubule membrane (see Saito et al. (1988) for ultrastructure and Wagenknecht et al. (1989) for three dimensional structure at 3.5nm resolution, for a review see Fleischer and Inui, 1989. In Fig.2 the foot is symbolized by the four larger ellipsoids. Its counterpart, the Ca^{++} channel in the T-Tubule membrane (L-type Ca^{++} channel), is constituted by the Dihydropyridine Receptor (DHP-Rec) as evidenced by several groups in the last three years: Flockerzi et al. (1986), Curtis and Catterall (1986) Talvenheimo et al. (1987), Smith et al. (1987), Hymel et al. (1988c). Meanwhile, primary sequences have been obtained for the channel forming α-subunit of the DHP-Rec (Tanabe et al., 1987) and for the Ry-Rec (Takeshima et al., 1989).
At present, efforts concentrate, besides on structural aspects of the channel subunits related to channel properties, on the overall organization of the two channels, their communication and their regulation. For this, we have applied a particularly designed reconstitution strategy (SVB-technique, for a review see Schindler, 1989) since the conventional strategy (fusion of lipid/protein vesicles to black lipid membranes) is a too limited approach. One

result, genuine to the potentials of the SVB-technique, was that Ca^{++}-channel function involves a particular organisation of channel proteins as indicated in Fig.2. Both channels, the L-type Ca^{++} channel, (Hymel et al.,1988c) and the SR Ca^{++}-release channel (Hymel et al., 1988a, b) are "oligochannels". The term "oligochannel" stands for an oligomer of channel proteins (tetramer in the figure) with several (4) parallel ion pathways which open and close in a synchronized or cooperative fashion, as illustrated in Fig.2. According to this, the ECC machinery is to be visualized as two physically linked "oligochannels", probably "tetrachannels" (see cartoon in Fig.2, taken from Hymel et al., 1989).

THE SVB-TECHNIQUE AS A TOOL FOR STRUCTURE-FUNCTION ASSIGNMENTS

Since there is no critical account available in the literature of using the conventional "Vesicle-Bilayer-Fusion" (VBF) technique in structure-function assignments I may try to outline this first using the cartoon in Fig.3. The assay starts by forming a black lipid membrane (basically a lipid bilayer with some solvent, normally decane, trapped in lenses). Vesicles, containing lipid and purified protein, are added to one side of the bilayer with an osmotic gradient across the vesicle membrane. Eventually ion channels are observed which is believed to result from fusion of vesicles to the black membrane. When applied to purified protein the extractable information is very limited; main restrictions are summarized in Fig.3b. Basic physical parameters of the system remain unassessible or uncertain: (i) number of proteins incorporated, (ii) their lateral distribution (whether monomers or oligomers relate to observed activities), (iii) the physical state of the membrane (lateral pressure) which has been shown to couple to channel functions, and (iv) lipid asymmetry. Moreover, recently it has been shown that vesicle fusion, under the conditions described, requires open channels to occur (Woodbury and Hall, 1988; Cohen et al., 1989; Niles et al., 1989). This function-dependent insertion implies selection among proteins of a real sample. This is a serious problem since the fraction of proteins observed as channels is normally only 10^{-12} to 10^{-10} of the proteins added. In consequence, impurities of another channel type or a tiny fraction of the channel proteins in particular states with

Figure 3a

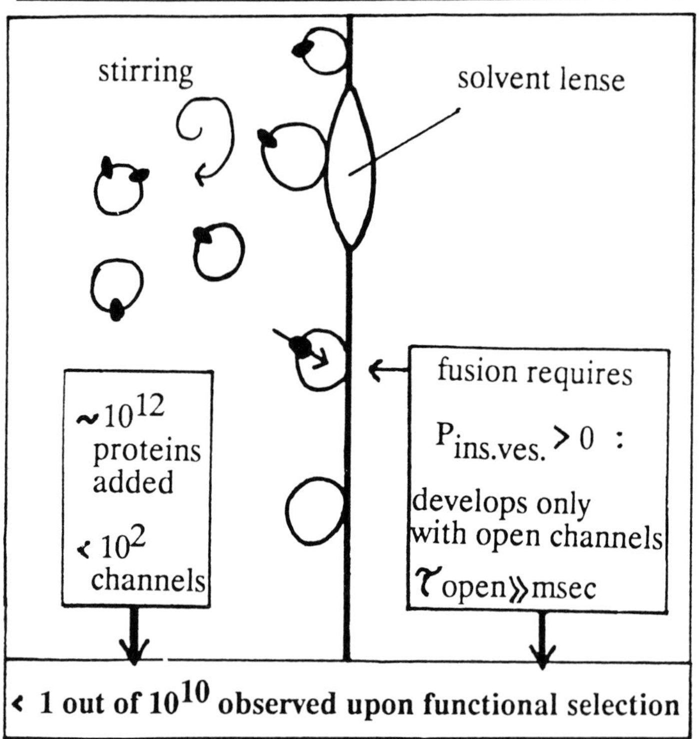

respect to structure, subunit composition, association, phosphorylation state, etc, may be responsible for the observed channel activity instead of the bulk part of the purified protein. Also, gating by ligands or by phosphorylation, etc. remains difficult to assess since open gates are required for insertion in the first place.

It is fair to add that vesicle-bilayer-fusion is a well suited technique for the screening for channels occuring in native membrane vesicles. The functional selection, a disadvantage when studying purified proteins, is here of advantage.

Figure 3b

VESICLE FUSION WITH BLACK MEMBRANES

When applied to purified protein

leaves

uncertain	- % recovery or viability of purified protein
difficult to assess	- components required for function - regulation of function (ligands,phosphoryl.,---)
unassesssible	- lipid asymmetry - lateral distribution - lateral pressure

but

is well suited to screen for the occurence of channels in native membrane vesicles and to characterize their properties, especially for cell membranes which are not assessible to patch clamp.

The SVB-technique (Septum supported, Vesicle derived Bilayer) uses selforganisation of surface films from vesicles followed by apposition of two such films to a bilayer of the same lipid/protein ratio as in the vesicle samples (see Fig.4, for details see Schindler, 1989). The physical parameters listed in Fig.4 are adjustable simply by choosing appropriate vesicles. For example, when vesicles are formed at a lipid/protein molar ratio of $10^8/1$ approximately 100 proeteins are present in the bilayer (of 100µm diameter containing 10^{10} lipids per monolayer). Moreover, these proteins are initially (just after bilayer formation) distributed at random over the bilayer plane, simply because they occured single in vesicles (100nm sized vesicles con-

Figure 4

VESICLE DERIVED BILAYER

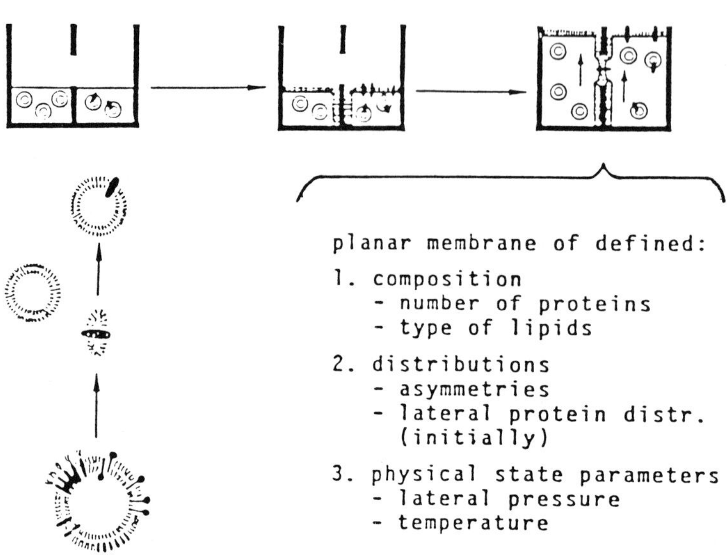

Figure 5

ASSOCIATION OF PORIN TRIMERS (+LPS)

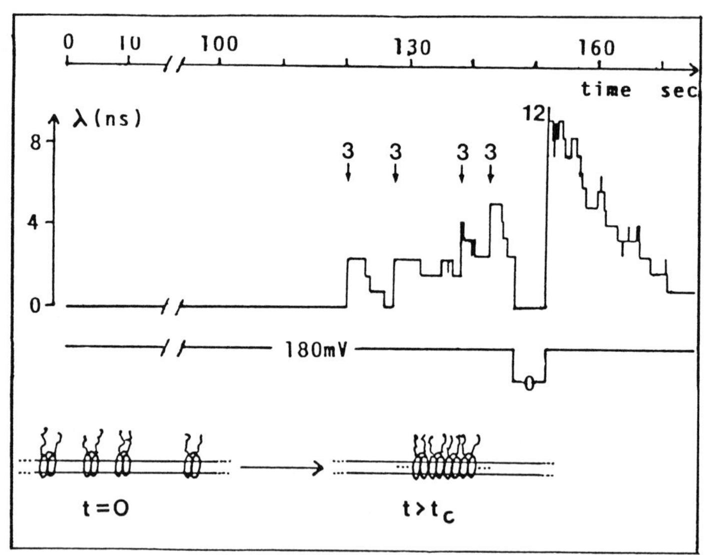

tain 10^5 lipids so that there is 1 protein per 10^3 vesicles). This allows to assay the role of channel protein association in channel function. Non-random initial distributions are generated by using vesicles containing more than one protein but correspondingly diluted with pure lipid vesicles to an overall lipid/protein ratio of, such as, again 10^8. This enables to crosscheck results from random initial distributions, such as, to test whether channel protein oligomerisation, found to influence channel function, is irreversible or reversible. The type of results obtained is shown in Fig.5. Starting from an initial random distribution of channel proteins (in this example 300 omp-F Porin trimers) the channel proteins need to associate via collisoins during lateral diffusion in order to form stable ion channels. Other conclusion were: voltage dependent closing after activation and 3 channels per trimer, meanwhile confirmed by X-ray crystallography and LPS (lipopolysaccharide) required for channel formation in coassociation with Porin (Schindler and Rosenbusch, 1978, 1981). One most interesting finding was, that channels in porin oligomers activate in a strictly cooperative fashion (see, such as, triple steps in Fig.5). This has been the first "oligochannel" which has been clearly identified.

ARE MOST ION CHANNELS "OLIGOCHANNELS"?

In the early days of "patch clamping" reports showed almost exclusively two state conductance transitions (rectangular events with unitary hights) in single channel traces. It was natural to assume that such elementary events (taken as definition for ion channels) represent ion flux through single ion pathways ("monochannel"). In the last years, however, increasing evidence appeared that channels of any type show conductance substates. These substates often are equally spaced or represent integer multiples of a smallest conductance unit. Whenever studied, all substates in such regular patterns showed the same ion selectivity and ion binding (for a first review on substates see Fox, 1987). These properties together can be regarded as a "fingerprint" of "oligochannels" and are extremely difficult to reconcile in terms of "monochannels", the only alternative. Certain Cl^- and K^+ channels of neurons show 16 exactly equally spaced levels in single conductance events (Kazachenko and Geletynk, 1984; Geletynk and Kazachenko, 1985). This leaves little doubt that such events represent activities of "oligochannels". Moreover, the same active conductance unit is seen to generate both 2 state events (full synchronization of the 16 pathways in the "oligochannel") and events showing all 16 steps (graded, uncomplete coupling of the 16 pathways). Besides these most pronounced examples of 16 substates there are other channels showing 6, 4 or 2 substates across the whole "zoo" of channels (see Fox, 1987). Not included in this review by Fox, updated to 1986, are Ca^{++}- and Na^{+} channels, which have to be added to the list of "oligochannel" candidates due to more recent studies. Na^{+}-channels from heart, for example, show 6 equally spaced sublevels with the same ion selectivity and binding (Schreibmayer et al., 1989). Fig.6 shows some data of this study, where substates have been induced by a veratridine/S-DPI mixture. Other modifiers induce different populations of substates, but always among the same 6 substates which are, therefore, genuine to the unmodified Na^{+} channel.

Figure 6. Substates of the Na-channel from heart plasmalemma

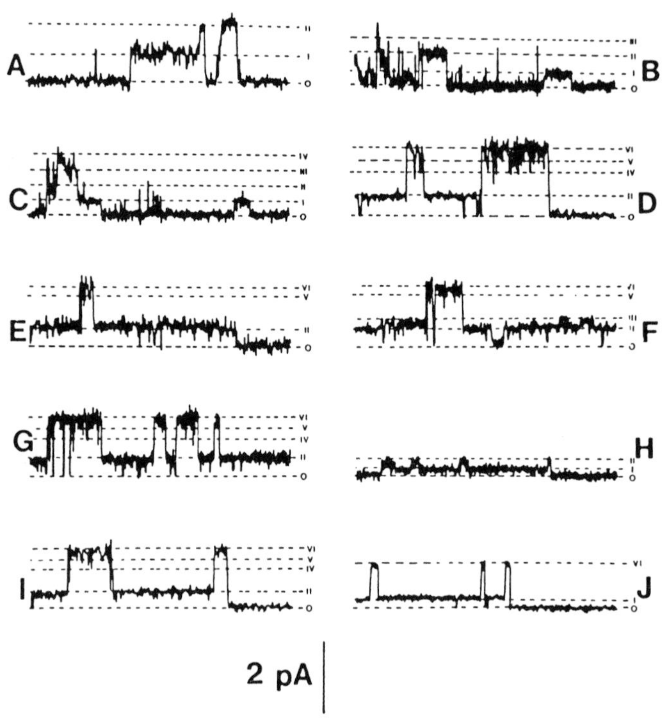

Inspite of all these indications from patch clamp studies, that "oligochannels" represent the appropriate structural basis of "single" channel activities, final decisions in this conceptional conflict "oligochannel" versus "monochannel" may not be expected from patch clamp studies. This obviously requires to directly relate electrical activity to protein organisation or, in other words, direct knowledge of the number of channel proteins and ion pathways in the molecular unit generating single electrical events. As outlined above, this is approached by the SVB-technique. Since its first application to Porin it has been applied to several channel systems (reviewed in Schindler, 1989). In all instances it was found that channel protein association influenced its function with one aspect in common: channels start to act synchronized. It appears that, whenever ion channels are in sufficiently close vicinity, they act cooperatively. The latest prominent examples are the two Ca^{++}-channels, the L-type Ca^{++} channel of sceletal muscle T-tubule membrane (Hymel et al., 1988c) and the Ca^{++} release channel of sceletal muscle and heart (Hymel et al., 1988a, b) bringing us back to ECC. From our data there is no doubt left that both channels are "oligochannels". L-type Ca^{++} channel events do, according to this, not represent the activity of one ion pathway but the cooperative gating of several pathways (probably 4) in DHP-receptor oligomers (probably tetramers, see Fig.2). Meanwhile, this view found support from electron-micrographs showing tetramer particles associated to foot structures to the T-tubule side, likely to be DHP-receptor tetramers (Block et al., 1988). Foot structures show two four fold occuring outlets (Wagenknecht et al., 1989) providing direct support for the "oligochannel" concept.

PHYSIOLOGICAL ROLES OF OLIGOCHANNELS

The voltage dependence or voltage gating of the L-type Ca^{++} channel, its essential characteristic, apparently results from interactions in DHP-receptor oligomers. The "oligochannel" shows the expected voltage dependence whereas the channel of a single DHP-receptor monomer exhibits a linear current-voltage relation (Hymel et al., 1988c). Oligomer induced voltage dependence is also evident for a voltage gated K^{+} channel in glial cells (Geletyuk and Kazachenko, 1989).

In this respect one may realize, that interactions in these oligomers are apparently rather weak and susceptable to changes of system variables like membrane fluidity (better descibed by lateral membrane pressure) or of interactions with surface proteins or of surface potentials and ionic conditions and may couple to, such as, phosphorylation reactions. Pathology related to voltage gating may thus be linked mechanistically to physical conditions of the membrane or aqueous phase, which is one line of our current efforts in ECC.
The second aspect to emphasize is related to the multitude of ligand and modification sites per channel as another consequence of "oligochannels". For example, for each L-type Ca^{++} channel one has now to consider 4 times the sites so far assumed with negative/positive cooperativity likely to occur. This may have offered many ways of regulation/adaptation to nature which, so far, have not been assigned to physiological observations. This, indeed, is found for the Ry-Rec or foot structure which binds only two ryanodine molecules although it represents a tetramer.
The natural tendency of protein association in membranes finds its reflection now in "oligochannels" which, in principle, offers many more ways to be exploited in regulation/ coupling/modulation/adaption/turnover, etc, than "monochannels" would offer. One may see here a paralism to enzyme cooperativity known for thirty years. The detailed study of "oligochannel" properties will probably play a key role in the molecular mechanistic understanding of ion transport functions and of their pathology leading to new conceptions in therapeutics and pharmacology of ion channel-related diseases.

ACKNOWLEDGEMENT

The data on ECC and the Na^{+} channel resulted from a joint research program supported by the Austrian Research Funds (S-45).

REFERENCES

Block BA, Imagava T, Campbell KP, Franzisi-Armstrong C (1988) J Cell Biol 107:2587-2600.
Cohen FS, Niles WD, Akabas MH (1988) J Gen Physiol 93:201-210.
Curtis BM, Catterall WA (1986) Biochemistry 25:3077-3083.

Fleischer S, Inui M (1989) Annu Rev Biophys Biophys-Chem 18:333-364.
Flockerzi V, Oeken H-J, Hofmann F, Pelzer D, Cavalie A, Trautwein W (1986) Nature 323:66-68.
Fox JA (1987) J Membr Biol 97:1-8
Geletynk VJ, Kazachenko VN (1985) J membr Biol 86:9-15.
Geletynk VI, Kazachenko VN (1989) Biochim Biophys Acta 981:343-350.
Hymel L, Inui M, Fleischer S, Schindler H (1988a) Proc Natl Acad Sci USA 85:441-445.
Hymel L, Schindler H, Inui M, Fleischer S (1988b) Biochem Biophys Res Comm 152:308-314.
Hymel L, Striessnig J, Glossmann H, Schindler H (1988c) Proc Natl Acad Sci USA 85:4290-4294.
Hymel L, Schindler H, Inui M, Fleischer S, Striessnig J, Glossmann H (1989) Ann New York Acad Sci 560:185-188.
Invi M, Saito A, Fleischer S (1987) J Biol Chem 262:1740-1747.
Kazachenko VN, Geletnynk VJ (1984) Biochim Biophys Acta 773:132-142.
Lai FA, Erickson HP, Rousseau E, Liu Q-Y, Meissner G (1988) Nature 331:315-319.
Niles WD, Cohen FS, Finkelstein A (1989) J Gen Physiol 93: 211-244.
Saito A, Inui M, Radermacher M, Frank J, Fleischer S (1988) J Cell Biol 107:211-219.
Schindler H, Rosenbusch JP (1978) Proc Natl Acad Sci USA 75:3751-3755.
Schindler H, Rosenbusch JP (1981) Proc Natl Acad Sci USA 78:2302-2306.
Schindler H (1989) Methods Enzym 171:225-253.
Smith JS, McKenna EJ,Ma J, Vilven J, Vaghy PL, Schwartz A, Coronado R (1987) Biochemistry 26:7182-7188.
Smith JS, Imagawa T, Ma J, Fill M, Campell KP, Coronado R (1988) J Gen Physiol 92:1-26.
Talvenheimo J, Worley JF, Nelson MT (1987) Biophys J 52:891.
Wagenknecht T, Grassucci R, Frank J, Saito A, Invi M, Fleischer S, (1989) Nature 338:167-170.
Woodbury DJ, Hall JE (1988) Biophys J 54:1053-1063.

Calcium Channel Modulators in Heart and Smooth Muscle:
Basic Mechanisms and Pharmacological Aspects
S. Abraham and G. Amitai, editors
© 1990, VCH, Weinheim/Deerfield Beach, FL and Balaban, Rehovot/Philadelphia

Possible role for phospholipase C in potentiation of neurotransmitter release - *in vitro* assay for synaptic faciliation

D. ATLAS[1], S. DIAMANT[2], I. LEV ARI[2], AND L. SCHWARTZ[2]

[1]The Otto Lowei Center for Neurobiology and [2]Department of Biological Chemistry, Hebrew University of Jerusalem, Jerusalem, Israel

INTRODUCTION

Recently neurotransmitter release was shown to be associated with the action of inostiol trisphosphate (IP_3) and diacylglycerol (DAG) producing agonists, and the involvement of PLC and PKC was demonstrated with their respective inhibitors (Diamant et al., 1988). Carbachol and acetylcholine induce [^{3}H]NE release in brain cortical slices in a calcium dependent fashion, and the muscarinic nature of the receptors responsible for the release was determined by the use of specific muscarinic antagonists such as quinuclidinyl benzylate, atropine and N-methyl-4-pyridyl benzylate (Diamant et al., 1988). Previously, it was demonstrated that the simultaneous presence of muscarinic agonists and elevated levels of K^+ induce larger amounts of inositol phosphates (Eva and Costa, 1986) with no potentiation by veratridine. A synergism of acetylcholine (ACh) and elevated K^+ was observed also in inositol tetrakisphosphate (IP_4) formation (Baird and Nahorski, 1986). In this report we show that (a) muscarinic agonists induce [^{3}H]noradrenaline ([^{3}H]NE), which is pertussis toxin senstive and cholera toxin insensitive, unlike the K^+ induced release which is not affected by either one of these toxins. (b) Simultaneous presence of receptor-mediated phosphatidyl inositol turnover and depolarizing agents, such as K^+, veratridine or oubain, induce a strong potentiation of inositol phosphate ([^{3}H]IP) formation. (c) Potentiation of neurotransmitter release is tightly correlated with potentiation of [^{3}H]inositol phosphate ([^{3}H]IP) accumulation. (d) Dissection of the stimulation period into two consecutive intervals, reveals desensitization of the muscarinic agonist carbachol (CCh)-induced release in the second period of stimulation. On the other hand, potentiation of release by CCh in the presence of K^+ persists in both stimulation periods, irrespective of the CCh-desensitization. These results point to the onset of activation of release initiated by CCh and K^+ together, which continues regardless of the lack of CCh-contribution to the release in the second period. It is thus proposed that membrane depolarization in combination with receptor-activated hydrolysis of phosphatidylinositol-4,5-bisphosphate (PIP_2) induces a change in synaptic transmission which persists in the absence of CCh, and hence can be regarded as a possible in vitro model for synaptic faciliation similar to long term potentiation (LTP) (Akers et al., 1986; Lynch et al., 1987; Malenka et al., 1988; Muller et al 1988).

METHODS

Preparation of cortical slices for [^{3}H]inositol phosphate formation

Cross-chopped slices (350 μm x 350 μm) of rat cerebral cortex were prepared using McIlwain tissue chopper, preincubated in 5 ml of oxygenated Krebs-Henseleit buffer (118 mM NaCl, 4.7 mM KCl, 1.2 mM $MgSO_4$, 1.3 mM $CaCl_2$, 25 mM $NaHCO_3$, 1.0 mM NaH_2PO_4 and 11.1 mM glucose) for 60 minutes at 37°C with buffer exchanging every 10 minutes. The slices were then incubate in the presence of [^{3}H]inositol (15 - 20 μCi/ml, 10 - 20 Ci/mmole) for 90 minutes at 37° C, washed by 2 x 5 ml of Krebs-Henseleit buffer containing 10 mM LiCl and distributed into assay tubes.

Measurement of [^{3}H]inositol phosphate formation

The assay was initiated by addition of 50 μl of packed slices into 0.5 ml buffer containing 10 mM LiCl and the appropriate ligands. The assay was terminated afer 30 minutes (if not otherwise indicated) by addition of 1.5 ml ice cold chloroform/methanol (1:2, by volume). Phases were separated by addition of 450 μl of chloroform and 450 μl of water followed by brief centrifugation. In the presence of LiCl, the major inositol metabolite in brain is IP, with minor production of IP_2 and IP_3. Therefore, a total water-soluble inositol phosphate fraction was eluted from AG 1 -x8 anion - exchange resin, formate form, at 1.2 M Ammonium Formate 0.1 M formic a as previously described (Fisher et al., 1981; Berridge et al., 1982). For assessment of total labeling of the phosphoinositides by [^{3}H]inositol, 100 μl aliquots of the lower organic phase were counted by liquid scintillation method.

Preparation of cortical slices and determination of [^{3}H]NE efflux

The assay of [^{3}H]NE release was carried out essentially as described (Schwartz and Atlas, 1989; Diamant et al., 1987). Male albino rats (150-200 g) were sacrificed and their brains rapidly removed. Slices of cerebral cortex were prepared using a McIlwain tissue chopper (350 μm) and were incubated for 15 min under oxygenated Krebs-Henseleit buffer (95% O_2 - 5% CO_2) containing the following (mM) 118 NaCl; 4.7 KCl; 1.2 $MgSO_4$; 25 $NaHCO_3$; 1.0 NaH_2PO_4; 0.004 Na_2EDTA and 11.1 Glucose with [^{3}H]NE (19.7 Ci/mmol, final concentration 0.25 μM) at 37°C. Termination of the loading period was followed by 3x3 ml washes for 10 min. The slices were then distributed into 24 incubation wells containing 1 ml of the oxygenated buffer with additional 1.3 mM $CaCl_2$.(For more details Diamant et al., 1988). The slices were washed for additional 3 x 10 min consecutive intervals in the incubation wells. At t=46 min the spontaneous release was monitored for 3 x 2 min intervals followed by agonist stimulation at t = 52 min for a 2 min duration, if not indicated otherwise. Control runs of slices under the same experimental conditions without stimulation show a steady basal efflux of 0.3±0.02% fractional release. The decline of the CCh induced release back to basal level was rapid and was observed already at the end of the second period after stimulation (t = 56 min). At t=60 min the remaining [^{3}H]NE was extracted from the tissue by 1ml of 0.1 N HCl for 16 h. Radioactivity was determined using liquid scintillation spectroscopy with 50% efficiency.

Analysis of the release data

Tritium efflux during 2 min period was expressed as percentage of tritium content of the slices at the beginning of the examined period. Tritium content in the slice was calculated by adding the amounts of tritium released during that period, the following periods and the tissue content at the end of the experiment.
Net evoked release was calculated from collected tritium during two intervals, the first stimulation period and the following one (2 x 2 min) after subtracting basal tritium efflux in the preceding period (one before stimulation) (x 2) assuming a constant spontaneous release. Each point represents an average of 4-12 determinations obtained in two or three separate experiments as indicated. Results are calculated as the average of all the experiments ±S.E.M.

RESULTS AND DISCUSSION

Pertussis Toxin (PTX) inhibition of carboachol stimulated [^{3}H]NE release

As we have previously reported (Diamant et al., 1988) CCh induced [^{3}H]NE release in a dose dependent manner in rat brain cortical slices. To examine whether ADP-ribosylation of a specific GTP binding protein by PTX, affects the atropine sensitive CCh mediated release, we measured [^{3}H]NE release from slices pretreated with PTX. As shown in Fig. 1, pretreatment of PTX (1.2 µg/ml) for 3 hours at 34°C caused a 70% inhibition of the [^{3}H]NE rlease induced by 1 mM CCh, as compared with the release induced by 1 mM CCh under similar conditions in the absence of PTX. Similarly, the K^+-induced [^{3}H]NE release was not affected (Fig. 1). In all the assays using PTX, there was no effect on the amount of [^{3}H]NE incorporated into the slices as compared to vehicle pretreatment. Cholera toxin at 0.5 µg/ml did not affect the net fractional release of [^{3}H]NE induced release by 1 mM CCh, (0.99 ± 0.06) as compared to treated cells (1.2 ± 0.09%) net fractional release (Lev-Ari et al., 1988).

Kinetics of synergistic-[^{3}H]IP-accumulation

Carbachol stimulates phosphatidylinositol turnover in [^{3}H]inositol labeled cortical slices via muscarinic receptors (review by Fisher and Agranoff, 1987) and K^+ at depolarizing concentrations induced IP-accumulation partly indirectly.

Figure 2A shows the time course of [^{3}H]IP accumulation in rat cortical slices in response to 0.8 mM CCh or 25 mM K^+ and in response to both of them added together. CCh-induced incorporation of [^{3}H]inositol into IP was significantly enchanced in the presence of depolarizing K^+ concentrations and the potentiation persisted almost linerarly up to 60 min (Fig. 2A). At 30 min the degree of potentiation was 5.8 fold above the additive value of [^{3}H] accumulated by K+ and CCh alone.

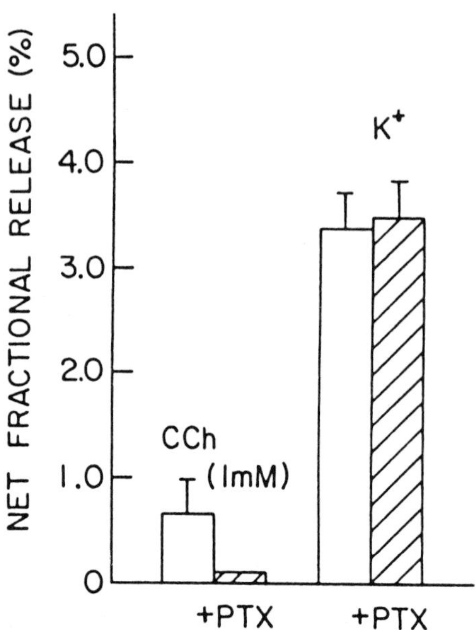

Figure 1: The effect of pertussis toxin CCh and K^+ induced [^{3}H]NE release.
Cortical slices after 15 min recovery period were randomly distributed to PTX-treated and non-treated incubation mixtures and were preincubated with 1.2 µg/ml PTX or vehicle respectively in 3 ml low Ca^{2+} Krebs Henseleit (250 µM $CaCl_2$) for 3 hrs at 34°C with stirring and constant oxygenation (O_2/CO_2 - 95%/5%). At the end of the incubation period the slices were washed by decantation x 3 and transferred to release-incubation assay loaded with [^{3}H]NE and assayed as described above! Each value represents the mean ± S.E.M. of 4-6 separate determinations.

Accumulation of [^{3}H]IP as a function of K^+ concentration

K^+ by itself at concentration (5-60 mM) induced a small gradual change in [^{3}H]IP accumulation (Fig. 2B) which could be attributed to a non-receptor mediated activation of PLC. Upon addition of 0.8 mM CCh, a significant synergism of [^{3}H]IP production was observed at K^+ concentrations above 5 mM KCl. The K^+ dose dependent curves in the presence and in the absence of CCh is different in the maximal rate of the reaction.

A possible but still remote possibility for explaining the increase in IP accumulation might be an increase in the number of PLC molecules coupled to muscarinic receptors under depolarizing conditions. Recent results have demonstrated a shift from synaptoneurosoms depolarized by 50 mM KCl, or by batrachotoxin (Cohen-Armon et al., 1987). Thus an increase in receptor population coupled to PLC might initiate further activation of the enzyme and as a consequence, an increased PIP_2-hydrolysis. A more active conformation of PLC induced by depolarization, and therefore a better rate of catalysis can also explain amplification of IP production.

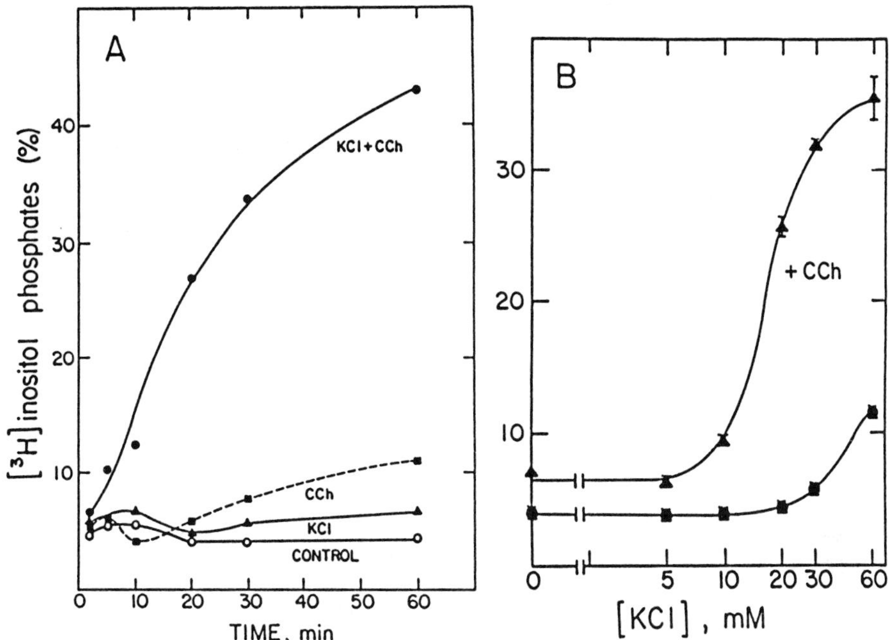

Figure 2A: <u>Time course for accumulation of [^{3}H]inostiol phosphate.</u> Cerebral cortical slices prelabeled with [^{3}H]inositol for 90 min, washed, then incubated in the presence of 10 mM LiCl with no agent added (O), 30 mM KCl (▲), 1 mM CCH (■) and in a combination of CCh and K^+ (●) for the time periods as indicated. [^{3}H]Inositol phosphate was separated and analysed. Values are means of three representative experiments.

Figure 2B: <u>Dose response curve for stimulation of phosphatidyl inositol hydrolysis by K^+.</u>
Cerebral cortical slices were preincubated with [^{3}H]inositol in standard incubation medium for 90 min at 37dC. KCl was added at the indicated concentrations in the absence (●) and in the presence of 1 mM CCh (▲). After 30 min, [^{3}H]IP was separated and analyzed. Values represent mean ± S.E.M. of three independent experiments.

<u>Synergistic accumulation of [^{3}H]IP as a function of CCh</u>
Total [^{3}H]IP production was measured as a function of CCh concentration (0.008 - 3 mM) (Fig. 3C). the CCh dose response curve displays an apparent half maximal effective concentration of 2 x 10^{-4} M which is not changed if receptor activation is carried out in the presence of 30 mM KCl (Fig. 3A) or 2 μM veratridine (Fig. 3B). On the other hand, a significant increase in V_{max} was observed in the presence of K^+ (~6-fo or veratridine (~2.5 fold).

Quinuclidinyl benzylate (QNB) a specific muscarinic antagonist inhibited the potentiation induced by CCh in the presence of depolarizing concentrations of K^+ (Fig. 3D), indicating the muscarinic receptor involvement in the potentiation process.

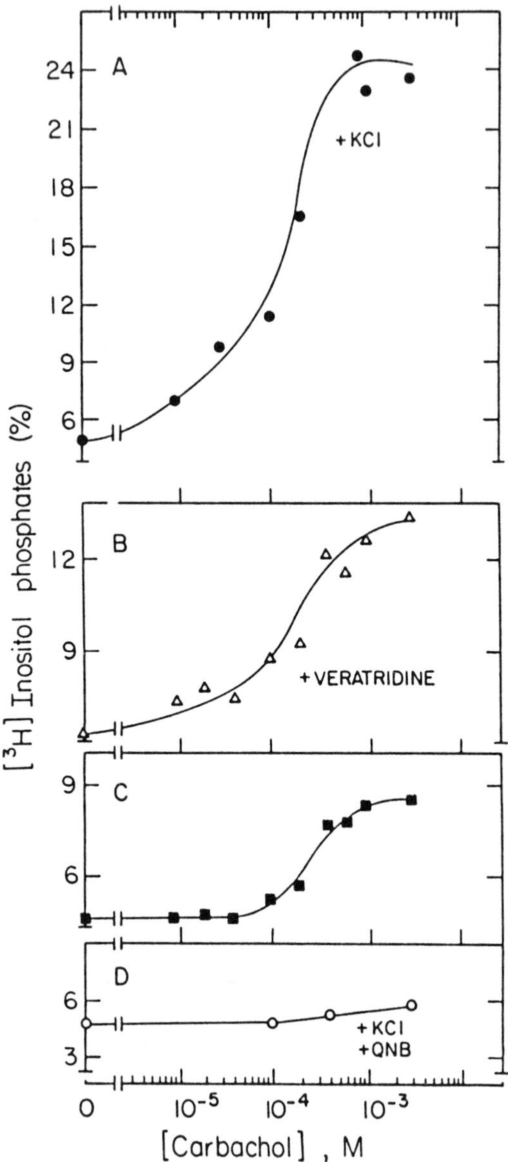

Figure 3: Induction of $[^3H]$IP accumulation as a function of carbachol in the presence and in the absence of KCL and veratridine.
Cerebral slices preloaded with $[^3H]$inositol were exposed to CCh at the indicated concentrations. (A) in the presence of 30 mM KCl, (B), in the presence of 2 μM veratridine (C) alone (D) in the presence of 30 mM KCl and 10 μM QNB. The reaction was terminated after 30 min and $[^3H]$inositol phosphate was separated and anlysed. values are means of representative experiments (n = 3).

Potentiation of neurotransmitter release

A recent report from our laboratory has shown that receptor mediated activation of PLC, mediate neurotransmitter release (Diamant et al., 1988). Hence, we attempted to examine whether potentiation of IP-production induced by muscarinic agonists in high K^+ or veratridine, would also be reflected in potentiation of neurotransmitter release.

Cortical slices prelabeled with [^{3}H]NE were stimulated at t=52 min for 2 min either by CCh (1 mM), K^+ (25 mM) or by 1 mM CCh and 25 mM KCl added together. Stimulation by K^+ and CCh combined induces an enhanced fractional release which is more than additive. CCh induced net fractional release of 0.74 ± 0.06%, K^+ alone induced 3.8 ± 0.3% and the two combined induced 7.2 ± 0.3% which is almost two fold above additivity (Fig. 4).

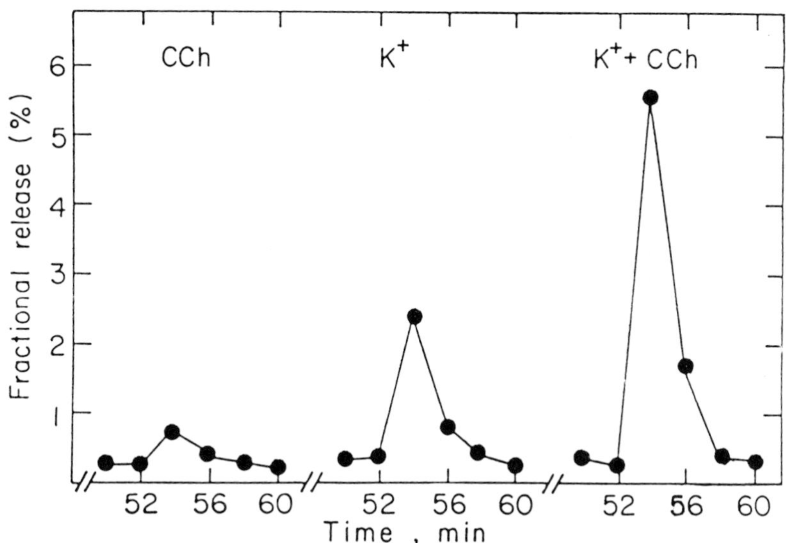

Figure 4: K^+, CCh- and K^+ + CCh induced [^{3}H]NE release.
Rat cortical slices preloaded with [^{3}H]NE in phyisological buffer, were stimulated for 2 min. by 1 mM CCh, 25 mM KCl, and 1 mM CCh in the presence of 25 mM KCl. The data is presented as a percent of total [^{3}H]NE incorporated into the vesicles. Each point represents the mean of duplicate determinations (n = 12). Net spontaneous release was 0.25 ± 0.05%.

The degree of potentiation produced by K^+ and CCh combined is 1.7 fold above the K^+-induced [^{3}H]NE release and 2.5-3 fold above the CCh induced release (Schwartz and Atlas, 1989). A longer stimulation period (4 min) shows no change in the CCh-induced fractional release as compared to 2 min. (Table 1). The potentiated release continues to be synergistic although no additonal release by CCh alone is observed.

This indicates that although CCh does not induce further [^{3}H]NE release in the additional 2 min stimulation interval, its presence is reflected in the continued synergy of release, alluding to a possible retention of its previous imprinting.

Table 1

CCh- K^+ and CCh + K^+ - induced [^{3}H]NE release in two consecutive stimulation intervals

	net fractional release (%)	
	stimulation time	
Ligand	2 min	4 min
CCh (1 mM)	0.74 ± 0.06	0.88 ± 0.08
KCl (25 mM)	3.8 ± 0.3	6.34 ± 0.3
KCl + CCh (25 mM)	7.2 ± 0.3	16.36 ± 0.6

Potentiation of CCh induced release as a function of KCl

As shown in Fig. 5, increasing K+ in the presence of 1 mM CCh induced potentiation of release already at 15 mM KCl. No potentiation of release was observed at 5 mM, which was used as the normal K^+ concentration of buffer system.

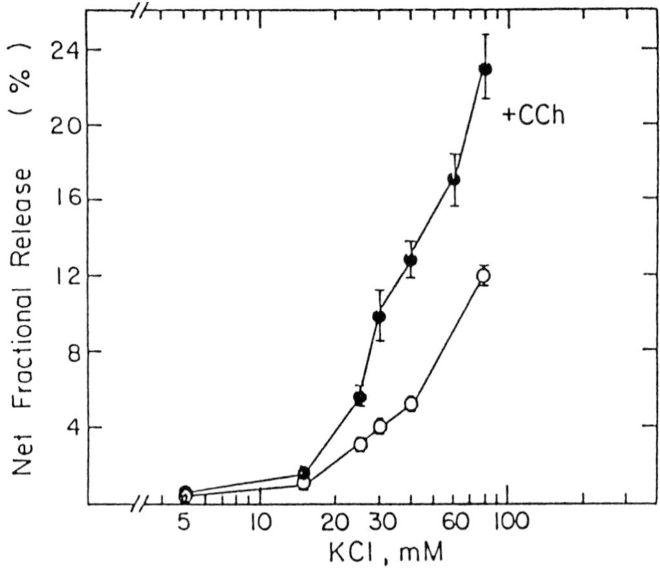

Figure 5: Dose response of K^+ -induced [^{3}H]NE release in the presence and in the absence of CCh.

Cortical slices preloaded with [^{3}H]NE in oxygenated buffer 95% O_2/5% CO_2 physiological buffer free Ca^{2+} at 37°C. At calcium (1.3 mM) was added at t=46 min and spontaneous release was monitored (2 min x 3). Stimulation was intiated at t = 52 min for a duration of 2 min in the absence of CCh (O) and in the presence of CCh at K^+ concentrations as indicated. Each point represents an average of 4-12 determinations obtained in two or three separate experiments as indicated. Results are calculated as the average ± S.E.M.

Induction of [^{3}H-IP accumulation by muscarinic agonists

Various cholinerigic muscarinic agonists were examined for their ability to induce [^{3}H]IP accumulation in the absence and in the presence of 30 mM KCl (Fig. 6A). The order potency of the various agonists as inducers of IP-accumulation was conserved under depolarizing conditions as compared to stimulaiton in normal buffer. At 30 mM KCl, potentiation was 5-7 fold and the order potency was ACh > CCh > oxo-M > arecoline >> pilocarpine. Similar results were observed by muscarinic agonists in the presence of 2 μM veratridine (Diamant and Atlas, 1989).

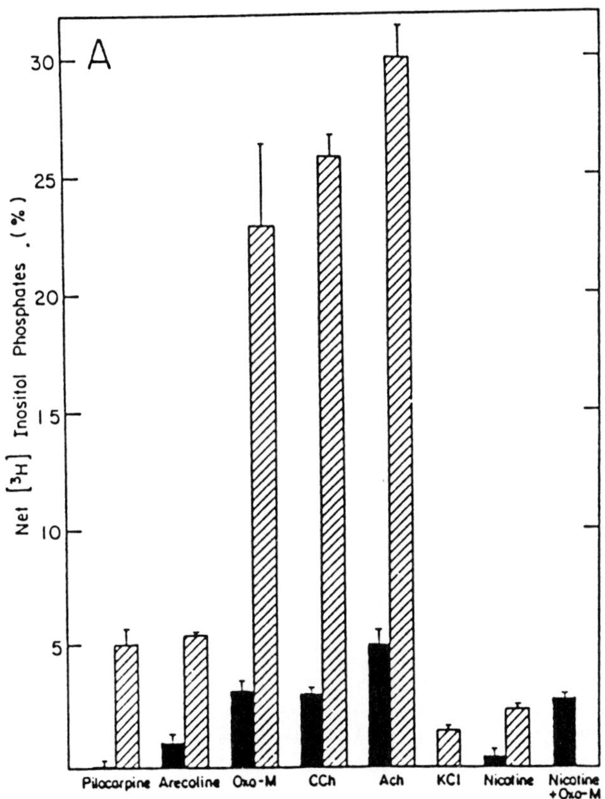

Figure 6: Potentiation of [^{3}H]IP accumulation and [^{3}H]NE release by various muscarinic agonists.

(A) Stimulation of phosphatidy inositol turnover by various muscarinic agoinists in the presence and inthe absence of KCl. Cerebral cortical slices preloaded with [^{3}H]inositol for 90 min, were washed and incubated in the presence of10 mM LiCl with 1 mM of various muscarinic agonists or 1 mM nicotine, in the absence (black column) and in the presence of 30 mM KCl (striated columns). The values represent an average of 3 experiments and the bars are the S.E.M.

Thus, incorporation of $[^3H]$inositol into IP by muscarinic agonists was significantly enhanced by KCl and veratridine. In view of these results we have examined in the presence of K^+ potentiation of IP-accumulation by the various agonists is reflected also as an enhancement of neurotransmitter release.

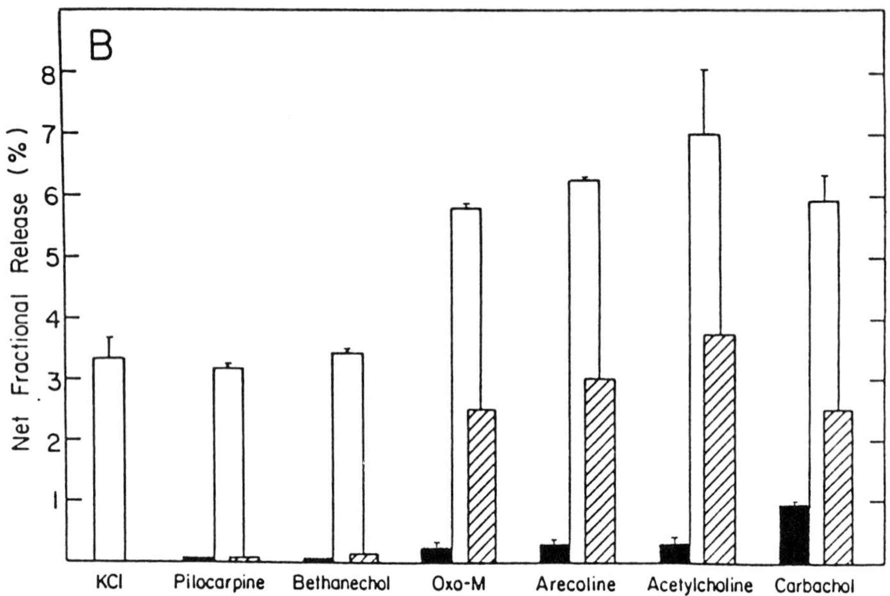

(B) Potentiation of $[^3H]$NE induced release by various muscarinic agonists in the presence and in the absence of KCl. Cortical slices preloaded with $[^3H]$NE (0.25 µM, 30 min) superfused with oxygenated (95% O_2/5% CO_2) physiological buffer were stimulated at t = 52 min for a two min interval by KCl (25 mM) alone, by muscarinic agonists (1 mM each, black columns) and by the various muscarinic agonists (1 mM) in the presence of KCl (25 mM, white columns). The shaded column represents the potentiated [3H]NE fractional release after substracting the net fractional release induced by K+ (25 mM) alone. Each column represents net fractional release (%) induced during 2 min stimulation period. The results are the mean S.E.M. of 6-8 independent determination.

As shown in Fig. 6B, combination of muscarinic agonists with 25 m KCl, similarly to their mutual effect on IP accumulation, induce potentiation of $[^3H]$NE release. Oxo-M, ACh, CCh and arecoline showed a 2 increase above the K^+ induced release. The extent of potentiation could evaluated by subtracting the release induced by K^+ only (shown in shaded area) as compared to agonist induced release (black area) (Fig. 6B). Pilocarpine and bethanechol display only a marginal effect on basal outflow of $[^3H]$NE which is not potentiated by K^+. Membrane depolarization by 2 µM veratridine which at this concentration induces a Ca^{2+} dependent $[^3H]$N release, showed potentiation by the muscarinic agonists (Scwartz and Atlas, 1989). The effects of nicotine and K^+ added together were additive in both induction of $[^3H]$IP accumulation (Fig. 6A) and $[^3H]$NE release (data not shown). IP-accumulation induced by oxo-M, a specific muscarinic agonist was not potentiated in the presence of nicotine (Fig. 6A). Thus cytoslic-calcium-level varies by different depolarizing agents and hence by the degree of potentiation. Thus, it seems that a change in cytosolic [Ca^{2+}] induced by nicotine, is not sufficient to enhance IP-accumulation initiated by muscarinic agonists. Further support for this result was observed by the use of calcium ionophore A-23187, which potentiates only slightly (~2 fold) the CCh-induced phosphatidyl inositol turnover (Diamant and Atlas, 1989).

Our proposed scheme for synergism observed is:
Scheme 1

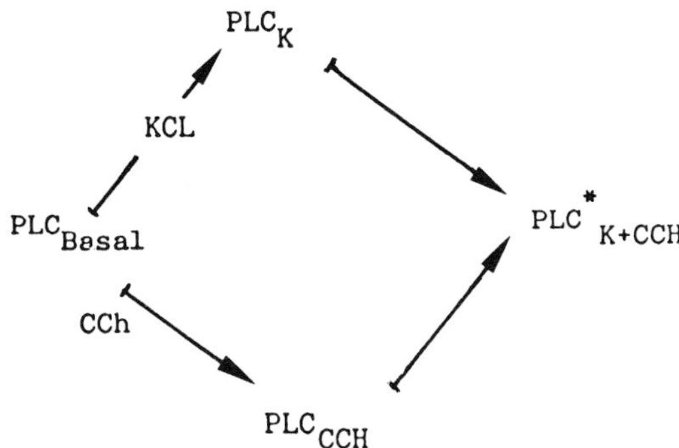

where KCL induces PLC activation (PLC_K) and CCh induced PLC activation (PLC_{CCH}), and both of them combined induce "super" activated enzyme (PLC^*_{K+CCH}).

Enhanced activation of PLC results in generation of elevated level of second messengers and in turn appears to lead to the enhancement of neurotransmitter release. It cannot be excluded that potentiation of release by K^+ and CCh might result from the closure of M-channels (Adams et al., 1982; Higashida and Brown, 1986) acting in concert with slowing o Ca^{2+} currents (Clapp et al., 1987). Both processes shown to be activated by agonists mediating PI hydrolysis (Adams et al., 1982; Higashida and Brown 1986). Recent studies have shown that follows hyperpolarization that follows action potential in CAI neurons in the hippocampus, under high frequency stimulation (LTP) is significantly reduced and PKC is persistently translocated from the cystosol to the membrane (LoTurco et al., 1988). These effects might result from a persistent potentiation of PLC activity as shown above by dual activation of muscarinic (or other IP producing agonists) and depolarization. More experiments should clarify the possible connection of enhanced IP-formation and potentiated neurotransmitter release within the context of long term potentiation in the cortex (Binder et al., 1987; Artola and Singer 1987).

Our results suggest three major conclusions: <u>First</u>, muscarinic agonists and depolarizing agents (of which KCL is the most effective) act in concert to induce potentiation of phosphatidyl inositol turnover and neurotransmitter release. Second, the simultaneous activation of both voltage dependent Ca^{2+} channels and receptor channels are obligatory for eliciting the potentiatory effect, which is blocked by receptor specific antagononist. Third, potentiation induced by K^+ + CCh persists in consecutive stimulation periods although the muscarinic receptor is already desensitized in the second stimulation interval.

Synergy of IP-accumulation in correlation with synergy of neurotransmitter release induced by muscarinic receptor activation and membrane depolarization suggests a possible role for PLC in the bifurcating control of neurotransmitter release and the possible involvement of PLC and voltage dependent channels in mediation of LTP.

REFERENCES

Adams P R, Brown D A and Constanti A. (1982) Pharmacological inhibition of the M-current. J. Physiol. 332: 223-262.

Akers R F, Lovinger D, Colley P, Linden D, Routtenberg A (1986) Translocation of protein kinase C activity may mediate hippocampal long term potentiation. Science 231: 587-589.

Artola A, Singer, W (1987) Long term potentiation and MDA receptors in rat visual cortex. Nature 333: 649-652.

Baird J G, Nahorski S R (1986) Potaasium depolarization markedly enhances muscarinic receptor stimulated inositol tetrakisphosphate accumulation in rat cerebral cortical slices. Biochem. Biophys. Res. Commun. 141: 1130-1137.

Berridge M J, Downs C P, Hanley M. R. (1982) Lithium amplifies agonist-dependent phosphatidylinositol responses in brain and salivary glands. Biochem. J. 206: 587-595.

Clapp L H, Vivaudou M B, Walsh J V Jr, Singer J. (1987) Acetylcholine increases voltage activated Ca2+ current in freshly dissociated smooth muscle cells. Proc. Natl. Acad. Sci. USA 84: 2092-2096.

Cohen-Armon M, Garty H, Sokolovsky M (1987) G-protein mediates voltage regulation of agonist binding to muscarinic receptors: Effects on receptor-Na^+-channel interaction. Biochemistry 27: 369-374.

Diamant S, Avraham B, Atlas D (1987)Neomycin inhibits K+ and veratridine-stimulated noradrenaline release in rat brain slices and rat brain synaptosomes. FEBS Lett. 219: 445-450.

Diamant S, Lev-Ari I, Uzielli I, Atlas D (1988) Muscarinic agonists evoke neurotransmitter release: Possible roles for phosphatidyl inositol bisphosphate breakdown products. J. Neurochem. 51: 795-802.

Diamant S, Atlas D (1989) Synergy of [^{3}H]inositol phosphate formation by receptor activation and membrane depolarization in brain cortical slices. Brain Res (in press).

Eva C, Costa E (1986) Potassium ion facilitation of phosphoinositide turnover activation by muscarinic receptor agonists in rat brain. J. Neurochem. 46: 1429-1435.

Fisher S K, Agranoff, B W (1981) Enhancement of the muscarinic synaptosomal phospholipid labeling effect by the inophores A-23187. J.Neurochem. 37: 968-977.

Fisher S K, Agranoff B W (1987) Receptor activation and inositol lipid hydrolysis in neuronal tissues. J. Neurochem. 48: 999-1017.

Higashida H, Brown D A (1986) Two phsophatidylinositide metabolites control two K^+ currents in a neuronal cell. Nature 323: 333-335.

Lev-Ari I, Schwartz L, Atlas D [^{3}H]norepinephrine induced release in rat brain cortical slices is mediated via a pertussis toxin sensitive GTP-binding protein and involves activation of protein kinase C. Cellular Signalling (in press).

LoTurco J J, Coulter D A, Alkon D L (1988) Enhancement of synaptic potential in rabbit CA1 pyramidal neurons following classical conditioning. Proc. Natl. Acad. Sci. USA 85: 1672-1676.

Lynch M A, Clements M P, Errington M L, Bliss T V P (1988) Increased hydrolysis of phosphatidylinositol-4,5-biphosphate in long term potentiation. Neuroscience Lett. 84: 291-296.

Malenka R C, Madison D V, Nicoll R A (1986) Potentiation of synaptic transmission in the hippocampus by phorbol esters. Nature 321: 175-177.

Muller D, Turnbull J, Baudry M, Lynch G (1988) Phorobol ester induced synaptic facilitation is different than long term potentiation. Proc. Natl. Acad. Sci. USA. 85: 6997-7000.

Schwartz L, Atlas D (1989) Membrane depolarization and muscarinic receptor activation lead to synergism of neurotransmitter release. Brain Res. (in press).

Calcium Channel Modulators in Heart and Smooth Muscle: Basic Mechanisms and Pharmacological Aspects
S. Abraham and G. Amitai, editors
© 1990, VCH, Weinheim/Deerfield Beach, FL and Balaban, Rehovot/Philadelphia

Fingerprinting of ligand mediated cellular activation: Relationships between inositol phosphates, phosphatidylinositols, intracellular calcium and cell function

PALMER TAYLOR[1], ALEXIS TRAYNOR-KAPLAN[2], S. KELLY AMBLER[1], BARBARA THOMPSON[1], MARTIN POENIE[3], ROGER Y. TSIEN[3], AND LARRY SKLAR[2]

[1]Department of Pharmacology, University of California, San Diego, CA, 92093; [2]Department of Immunology, Scripps Clinic and Research Foundation, La Jolla, CA, 92037; and [3]Department of Physiology-Anatomy, University of California, Berkeley, Berkeley, CA, USA

SUMMARY

The diversity in inositol-containing phospholipids, the coupling of cellular activation with formation of increased quantities of phosphatidylinositol mono- and polyphosphates and the plethora of inositol phosphates that can be generated upon activation of phospholipase C suggest that each cell, upon activation, is capable of generating a unique fingerprint of mediators. Coupled to such fingerprints is the control of transient or persistent changes in intracellular free Ca^{2+} which is essential to both the rapid differentiated functions of the cell and slower cellular events such as mitogenesis. Some unique aspects of inositol phospholipid hydrolysis, intracellular Ca^{2+} cycling and oscillations are examined in the BC3H-1 cell, a model for smooth muscle. Finally, we demonstrate formylpeptide elicited formation and breakdown of unique phosphatidylinositol polyphosphates which appear associated with activation in the primed neutrophil. Thus, signal transduction is linked to both stimulation of phosphatidylinositol-3 kinase and phospholipase C activities.

INTRODUCTION

It is widely recognized that the breakdown of phosphatidylinositol polyphosphates to form various inositol phosphate isomers is linked to signal transduction for a wide variety of cellular receptors (Berridge and Irvine, 1984). One of the inositol phosphates, inositol 1,4,5 trisphosphate (Ins1,4,5P_3), and possibly several of the inositol tetrakisphosphates, are capable either by themselves or in conjunction with other mediators of releasing intracellular Ca^{2+} from bound stores. The released Ca^{2+} is the critical link not only in cellular activation processes such as secretion, contraction and exocytosis, but also when elevated Ca^{2+} is sustained it may be important in committing the cell to mitogenesis. In addition, receptor activation appears to be linked to phosphorylation of phosphatidylinositol yielding phosphatidylinositol polyphosphates. Hence, both the breakdown and formation of the phosphatidylinositol polyphosphates appear coupled to receptor mediated cellular activation. In fact, it is this very diversity of mediator generation that enables disparate cellular activation events to share portions of common signaling pathways. Hence, each cell upon activation produces a unique fingerprint characterized by the complement of mediators generated.

In this study we examine the relationship between inositol phosphate formation, the mobilization and cycling of intracellular Ca^{2+} and the production of novel phosphatidylinositol polyphosphates by activation of a phosphatidylinositol-3′ kinase. Activation processes in two cell types, BC3H-1 cells (a smooth muscle derived line) and the human neutrophil, are examined. The temporal relationships between inositol phosphate formation, the mobilization of intracellular Ca^{2+}, and the oscillatory behavior of free intracellular Ca^{2+} are examined in BC3H-1 cells. In the neutrophil, we investigate the relationship between formyl peptide activation, inositol phosphate formation, intracellular Ca^{2+} mobilization, and the generation of phosphatidylinositol-3-phosphates.

METHODS

BC3H-1 Cells

Procedures for cell growth on 35 mm plates and coverslips have been described previously (Ambler et al., 1986; 1987;1988). All of the activation events involve measurements on cells adhered to surfaces for we found in early studies that suspension of the cells by either trypsin or by EDTA results in variable responses. Inositol phosphate isomers were measured following labeling of the cells with [^{3}H]-inositol for 24-48 hours, extraction from the activated or control cells and then separation and quantitation of the inositol phosphates by high pressure liquid chromatography (HPLC) on partisil SAX columns. Measurements of Ca^{2+} mobilization for cells <u>en</u> <u>mass</u> (~10^5 cells) were conducted for cells grown to confluence on coverslips. To measure Ca^{2+} oscillations, measurements were conducted in individual preconfluent cells to prevent membrane fusion and intercellular coupling of responses. Details on the optical detection systems have been presented (Ambler et al., 1988).

Human Neutrophils

Conditions for neutrophil preparation and activation have been presented previously (Traynor-Kaplan et al., 1988; 1989). Cells were labeled with ^{32}P prior to the activation interval. Phospholipids and inositol phosphates were analyzed as described (Traynor-Kaplan et al., 1988; 1989). The phospholipids were converted to their corresponding glycerophosphates by methylamine deacylation while the water soluble glycerophosphates could be converted to the respective inositol phosphates by mild periodate treatment. Identification of the inositol containing phospholipids relied on this sequence and co-elution of glycerophosphates and inositol phosphates on HPLC with the appropriate chromatographic standards.

RESULTS AND DISCUSSION

Calcium Mobilization in BC3H-1 Cells

BC3H-1 cells grown on coverslips were loaded with either quin or fura-2 dyes, placed in a cuvette holder to which the agonist can be injected into the surrounding solution. Addition of phenylephrine results in a rapid increase in intracellular Ca^{2+} ($t_{1/2}$-2-3 sec) which is sustained for a 30 second interval (fig. 1). The free Ca^{2+} declines slowly over the next 3-5 minutes even though the agonist is maintained. The slow decline reflects in previous observations that the mobilized Ca^{2+} is pumped out of the cell over this interval (Brown et al., 1984). Addition of antagonist results in a rapid return of Ca^{2+} to basal levels.

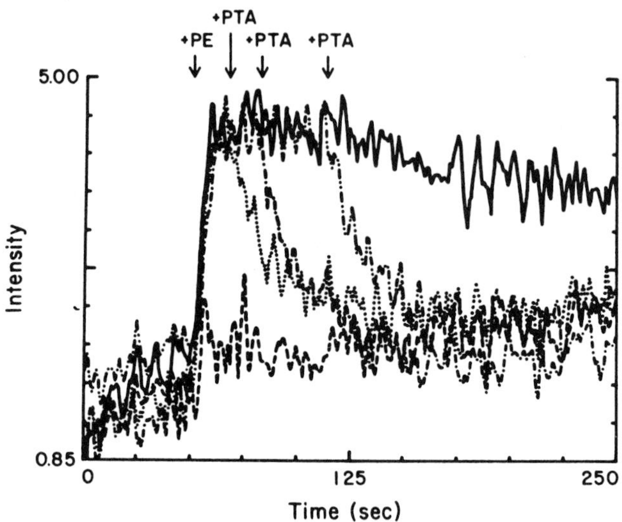

Figure 1: Phenylephrine (PE) stimulation and phentolamine (PTA) antagonism of Ca^{2+} mobilization in BC3H-1 cells. Quin 2 fluorescence was monitored with 330 nm excitation and 500 nm emission wave lengths. The ligands were added to adherant monolayers of cells on coverslips employing a cuvette in which ligand rapidly perifused the cells and fluorescence was monitored continuously. PE was applied to the cells at the arrow. At 0 (---), 15 (...), 30 (-.-.) and 60 (-..-) seconds following addition, 10 μM PTA was added (from Ambler et al., 1986, reproduced with permission).

If similar measurements are made in fura 2 loaded cells in a preconfluent state where the responses of individual cells can be analyzed, several unusual observations become manifest. First, the cells were highly variable in their temporal responses. Some responded immediately to agonist and others responded after a time delay. The response was not a continuous elevation in each cell but rather distinct oscillatory events (fig. 2). No oscillations were observed in the absence of agonist whereas 91 and 57 percent of the cells showed oscillations in response to phenylephrine and histamine, respectively (Table 1). Although the duration of oscillations was variable, with some cells oscillating for as long as 15 minute intervals, the frequency and amplitude of the oscillations were confined to defined windows (Table I). Moreover, consistent with earlier observations that the elevation and subsequent efflux of intracellular free Ca^{2+} elicited by agonist was largely due to its mobilization from internal bound stores, removal of external Ca^{2+} by short intervals of EGTA treatment did not abolish the oscillations nor did it change their frequencies or initial amplitudes. Rather Ca^{2+} removal only served to cause gradual diminutions in oscillation amplitude and to the average rise in intracellular Ca^{2+} (Ambler et al., 1988).

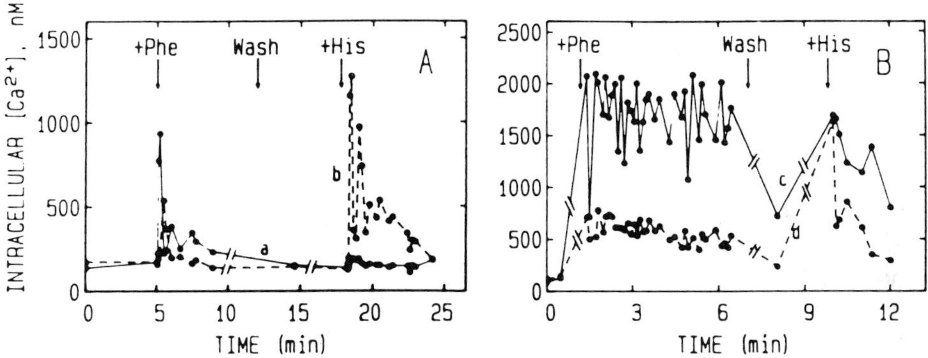

Figure 2: Kinetics of agonist stimulated alterations in Ca^{2+} in single subconfluent BC3H-1 cells. $[Ca^{2+}]_i$ was determined in Ambler et al., 1988. A, phenylephrine (10 μM) was added and responses recorded for two cells (a and b) in the viewing frame. After washing 100μM histamine was added and the responses of cells a and b recorded. B, phenylephrine was added to a series of cells and the Ca^{2+} oscillations on a highly responsive (c) and non-responsive cells (d) were monitored. After 6.5 min buffer was added. Histamine was added at 10 min and the responses measured. The cell that was least responsive to phenylephrine was found to be more responsive to histamine.

Table I

Average parameters of agonist-induced alterations in $[Ca^{2+}]_i$ in individual subconfluent BC3H-1 cells.

The peak $[Ca^{2+}]_i$, oscillation frequency, and the amplitude of the oscillations elicited by 10 μM phenylephrine or 100 μM histamine were determined for each individual cell and the average values ± standard deviations were subsequently determined. A cell was counted as oscillatory if it produced at least two separate spikes of elevated $[Ca^{2+}]_i$ of minimum amplitudes of 100 nM during a single period of agonist exposure. The amplitude is defined as the change in $[Ca^{2+}]_i$ from the trough to the peak of the oscillation. The frequency is defined as the number of full oscillations (half the number of transitions) per minute. Basal $[Ca^{2+}]_i$ showed no oscillations and was found to be 107 ± 16 nM.

Treatment	Cells treated	Responsive cells	Peak $[Ca^{2+}]_i$	Oscillating Cells	Oscillating amplitude	Oscillation frequency
	number	%	nM	%	$nMCa^{2+}$	min^{-1}
$[Ca^{2+}]_{out}$ = 1.8 mM						
+ 10 μM phenylephrine	54	100	858 ± 332	91	243 ± 109	1.91 ± 0.65
+ 100 μM histamine	42	88	1159 ± 459	57	220 ± 100	1.73 ± 0.45
$[Ca^{2+}]_{out}$ ≤1 nM						
+ 10 μM phenylephrine	76	100	679 ± 180	87	217 ± 72	1.95 ± 0.47
+ 100 μM histamine	50	74	515 ± 170	22	252 ± 94	2.04 ± 0.50

In addition, observations of the responsiveness of the individual cells revealed that each cell sets its own pattern of oscillations. For example, following cell division the daughter cells, despite being perfect clones and possessing identity of genetic material, show independent responses to the various agonists. Immediately following the establishment of the separating membrane the two halves of the previously dividing cell not only show asynchrony in their oscillations but a different sensitivity to the two test agonists, histamine and norepinephrine (Ambler et al., 1988).

It remains to be determined what are the molecular bases for driving the oscillations. For example, a primary candidate for causing Ca^{2+} release is Ins $1,4,5P_3$. However, we lack the capacity to monitor cellular inositol polyphosphates by non-destructive methods in single cells and are forced to analyze them by extraction and chemical assay. The BC3H-1 cell shows agonist elicited Ins $1,4,5P_3$ production but the accumulation is slow and significant changes in putative mediator production cannot be seen until after 30 seconds of stimulation (cf Table 2) (Ambler et al., 1987).

Table II

Effect of agonist stimulation on cellular Ins-P_n formation.

BC3H-1 cells were loaded with [^{3}H]inositol, as outlined in Methods. Cells were treated with buffer ± 10 μM phenylephrine (Phe) for 5 or 30 sec, as indicated. The individual [^{3}H]ins-P_n were extracted and separated by HPLC as described in Methods.

Ins-P_n	5 sec		30 sec	
	-Phe	+Phe	-Phe	+Phe
Ins-1,4,5-P_3	1083 ± 109	1123 ± 164	1077 ± 81	1125 ± 124
Ins-1,3,4-P_3	ND[a]	ND	ND	ND
Ins-P_4	ND	ND	ND	ND
Ins-P_2	675 ± 80	682 ± 88	661 ± 108	802 ± 73
Total	1758 ± 189	1805 ± 252	1738 ± 189	1927 ± 197
% Increase above basal		3		11

[a]ND = not detected above background cpm.

The slow kinetics of Ins 1,4,5P_3 accumulation contrast with the rapid generation of intracellular Ca^{2+} (fig. 1) and suggest several possibilities. First, Ins 1,4,5P_3 may not be the sole mediator of Ca^{2+} release. This hypothesis in the absence of other candidate mediators or the demonstration of depolarization induced release of Ca^{2+} can only be speculative. We know that saponin permeabilized BC3H-1 cells show a sensitivity to Ins 1,4,5P_3 equivalent to other cells for the release of Ca^{2+}, but again evidence for other mediators associated with activation could not be adduced. Second, the Ins 1,4,5P_3 could be released in restricted environments in sufficient local concentrations to release Ca^{2+} yet not be detectible when total cytoplasmic Ins-1,4,5P_3 is measured. Subsequent accumulation of Ins 1,4,5P_3 occurs in the cytoplasm via cumulative release events. Third, the rapid release of Ins 1,4,5P_3 and its equally rapid metabolism could result in sufficiently short pulses of the mediator in individual cells which also might preclude its detection as an early event. In two other cell lines the increase in Ins 1,4,5P_3 elicited by agonists was shown to follow the increase in intracellular Ca^{2+} and, here again, it was established that the Ca^{2+} comes from internal stores rather than from the outside through Ca^{2+} channels (Tashjian et al., 1987; Nanberg and Rosengurt, 1988).

Our studies in the neutrophil have revealed a rapid mobilization of intracellular Ca^{2+} and a far slower production of an essential product of neutrophil activation, superoxide (fig. 3). Superoxide production is blocked by using an intracellular chelator MAPTAM to buffer the internal free Ca^{2+} (Traynor-Kaplan et al., 1988; 1989). Breakdown of phosphatidylinositol bis 4,5 phosphate (PtIns4,5P_2) is seen also in response to FMLP. Again, its breakdown appears slower than the increase in Ca^{2+}.

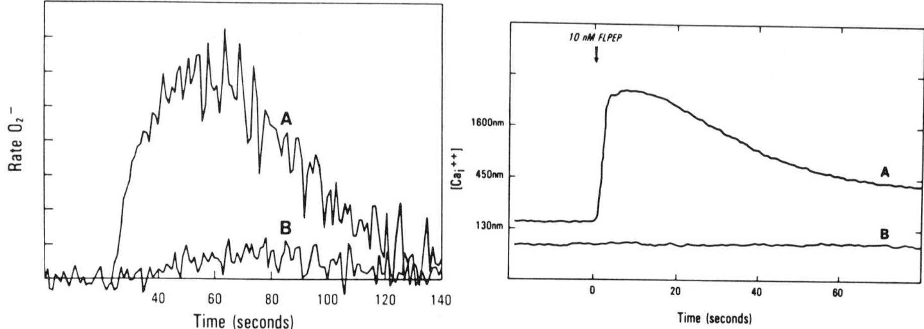

Figure 3: Response of primed neutrophils to addition of a formyl peptide analogue N-formylnorLeu-Leu-Phe-norLeu-Try-Lys-fluorescein. Right: increase in intracellular Ca^{2+}. The time course was measured by the dye Indo I in a suspension of neutrophils. Left: Measurements of superoxide production after addition of the formyl peptide analogue. A. control cells; B. Ca^{2+} cheloter, MAPTAM treated cells (details are given in Traynor-Kaplan et al., 1988,1989 reproduced with permission).

Analysis of phospholipid composition associated with neutrophil activation reveals the production of two new bands separable on thin layer chromatography. These lipids can be separated and characterized by deacylation with methylamine to form water soluble compounds (Traynor-Kaplan et al., 1988,1989). The two compounds migrate in the expected positions of glycerophosphoinositol-P_2 and glycerophosphoinositol-P_3. The glycerophosphoinositol P_2 elutes on HPLC before the glycerophosphoinositol 4,5P_2 and the identification of the isomer is achieved by subsequent mild periodic acid treatment to form Ins #1,3,4P_3. This can be confirmed with standard [^{3}H]Ins 1,3,4P_3.

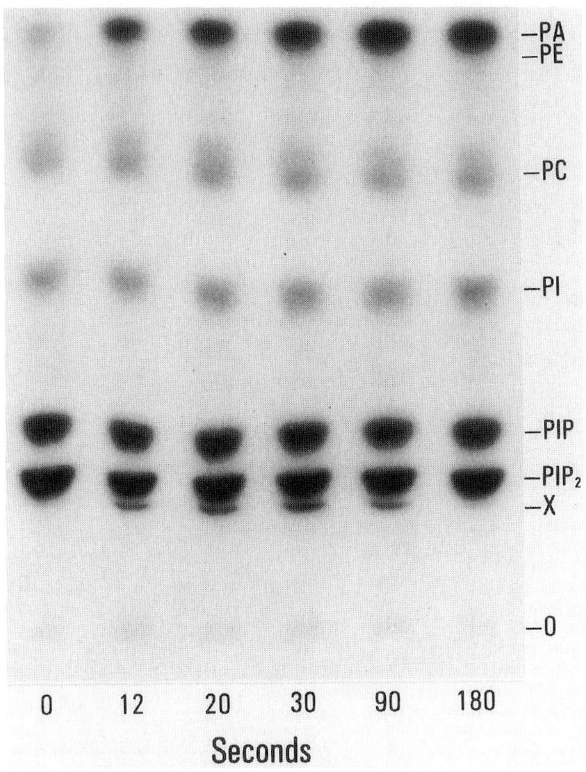

Figure 4. Time course for the generation of the phospholipid, phosphatidylinositol trisphosphate, $PtdInsP_3$. The formation of this phospholipid (x) becomes maximal after 30-45 seconds. Its characterization is based on removal of x from the thin layer plate, deacylation by methylamine to form glycerophosphoinositol trisphosphate ($GroPInsP_3$). $GroPInsP_3$ is then treated with dilute periodate to form inositoltetrakisphosphate. The IP_4 chromatographs with authentic $[^3H]IP_4$ (cf Traynor-Kaplan et al., 1988, 1989 for details).

The same treatment of glycerolphosphoinositol-P_3 yields inositol tetrakisphosphate (IP_4). Although the individual isomers of IP_4 have not been identified, the present phospholipid is most likely PtdIns3,4,5P_3. The increases in these two novel phosphoinositides through a phosphatidylinositol-3′ kinase follows the increase in Ca^{2+} becoming maximal in 45-60 seconds and then declines thereafter. Hence the time course parallels that of superoxide production. Although buffering internal Ca^{2+} with MAMTAM blocks the formation of superoxide (fig. 3), it does not block formation of the phosphatidylinositol bis and tris phosphates (PtdIns3,4P_2 and PtdInsP_3). This suggests that the full activation response requires both a rapid calcium spike and other slower events. Previous studies have shown that Ca^{++} increases alone do not cause neutrophil activation and that activation of protein kinase C and Ca^{2+} mobilization are also insufficient (Grinstein and Furuya, 1988; Korchak et al, 1988; Della Bianca, et al., 1988). Therefore, the identification of potential mediators with the appropriate temporal responses becomes critical to delineating the signal transduction pathway.

In this connection, we also observe that neutrophil activation by the formyl peptide and the formation of PtdIns3,4P_2 and PtdInsP_3 are blocked by prior treatment of the cells with pertussis toxin. Hence, the pathway involves a G protein (G_i) linkage. Finally, phorbol ester treatment of the neutrophil does not result in the formation of these phospholipids.

In separate studies, Cantley and his colleagues have observed that platelet derived growth factor stimulation of smooth muscle cells results in mitogenic sequelae. Associated with this process is the activation of a phosphatidylinositol-3′ kinase resulting in formation of PtdIns3P, PtdIns3,4P_2 and PtdIns3,4,5P_3 in vivo. Moreover, tissue fractions enriched in phosphotyrosine contain this kinase activity (Auger et al., 1989; Whitman et al., 1988). In the case of the smooth muscle cell, phosphatidylinositol-3′kinase activity appears to be associated with mitogenic and cell proliferation. By contrast, the neutrophil is an end stage cell and appears to be employing a similar pathway in its normal activation events. Hence, signal transduction not only relies on activation of specific phospholipases to form soluble inositol phosphates, but also activation of kinases which yield higher phosphorylated products of phosphatidylinositol. These phospholipids are either absent or in very low abundance in resting cells.

REFERENCES

Ambler SK, and Taylor P (1986). Mobilization of intracellular calcium by α_1-adrenergic activation in muscle cell monolayers. J Biol Chem 261:5866-5871.

Ambler SK, Thompson B, Solski PA, Brown JH and Taylor P (1987). Receptor-mediated inositol phosphate formation in relation to calcium mobilization: a comparison in two cell lines. Mol Pharmacol 32:376-383.

Ambler SK, Poenie M, Tsien RY and Taylor P (1988). Agonist-stimulated oscillations and cycling of intracellular free calcium in individual cultured muscle cells. J Biol Chem 263:1952-1959.

Ambler SK, Taylor P (1989). Release of non-mitochondrial sequestered Ca^{2+} from permeabilized muscle cells in culture. Mol Pharmacol 35:369-374.

Auger KR, Seranian LA, Soltoff SP, Libby P and Cantley LC (1989). PDGF dependent tyrosine phosphorylation stimulates production of novel polyphosphoinositides in intact cells. 57, 167-175.

Brown RD, Berger KD and Taylor P (1984). $Alpha_1$-adrenergic receptor activation mobilizes intracellular Ca^{2+} in a muscle cell line. J Biol Chem 259:7554-7562.

Berridge MJ and Irvine RF (1984). Inositol trisphosphate, a novel second messenger in cellular signal transduction. Nature (Lond) 312:315-321.

Della Bianca VM, Gizeskowait M, Duri S, Rossi F (1988). Fluoride can activate the respiratory burst, independent of Ca^{2+}, stimulation of phosphoinositide turnover and protein kinase C translocation in primed neutrophils. Biochem Biophys Res Comm 150:955-964.

Grinstein S, Furuya W (1988). Receptor-mediated activation of electropermeabilized neutrophils. J Biol Chem 263:1779-1783.

Korchak HM, Vosshall MLB, Zagon, G., Ljubich, P., Rich, A.M. and Weissman G. (1988). Activation of the neutrophil by calcium mobilizing ligands. J Biol Chem 263:11080-11097.

Nanberg E, Rozengurt E (1988). Temporal relationship between inositol polyphosphate formation and increases in cytosolic Ca^{2+} in quiescent 3T3 cells stimulated by platelet derived growth factors, bombesin and vasopressin. EMBO J 7:2741-2747.

Tashjian AM Jr, Heslop JP, Berridge MJ (1987). Subsecond and second changes in inositol polyphosphates. GH_4C_1

cells induced by thyrotropin releasing hormone. Biochem J 243:305-308.

Traynor-Kaplan AE, Harris AL, Thompson BL, Taylor P, Sklar LA (1988). An inositol tetrakis phosphate containing phospholipid in activated neutrophils. Nature 334:353-356.

Traynor-Kaplan AE, Thompson BL, Harris AL, Omann GM, Taylor P and Sklar LA (1989). Transient increases in $PI(3,4)P_2$ and PIP_3 during activation of human neutrophils. J Biol Chem, in press.

Whitman M, Downes CP, Keeler M, Keller T, Cantley L (1988). Type I phosphatidylinositol kinase makes a novel inositol phospholipid, phosphatidylinositol-3-phosphate. Nature 322:644-646.

Calcium Channel Modulators in Heart and Smooth Muscle: Basic Mechanisms and Pharmacological Aspects
S. Abraham and G. Amitai, editors
© 1990, VCH, Weinheim/Deerfield Beach, FL and Balaban, Rehovot/Philadelphia

Receptor-operated calcium channels in vascular smooth muscle: from physiology to pharmacology

URS T. RUEGG, ANDREAS WALLNOEFER, CYNTHIA CAUVIN, AND PAUL W. MANLEY
Preclinical Research, Sandoz Ltd., 4002 Basel, Switzerland

Receptor-operated channels and underlying mechanisms

The existence of receptor-operated Ca^{2+} permeable channels (ROCs), which are distinct from potential-sensitive Ca^{2+} channels (PSCs) was postulated in vascular smooth muscle 10 years ago (Bolton, 1979; van Breemen et al., 1979). However, while PSCs have been studied and subclassified extensively by the use of organic Ca^{2+} entry blockers, lack of specific pharmacological tools hindered the characterization of ROCs.

ROCs are activated by the interaction of vasoconstrictor agonists with their receptors thus allowing, by definition, significant amounts of Ca^{2+} (and probably other ions, e.g. Na^{+}) to enter the cytosol (Hallam and Rink, 1989). With respect to the mechanism of opening of ROCs, it has been suggested that inositol-1,4,5-trisphosphate, ($Ins(1,4,5)P_3$ (Penner et al. 1988, Kuno and Gardner 1987), which is formed by phosphoinositide hydrolysis, or one of its metabolites, e.g. $Ins(1,3,4,5)P_4$ (Irvine and Moor, 1986) could act as second messengers. Alternatively, it has been proposed that the opening of ROCs is triggered by a rise of intracellular Ca^{2+} concentration $[Ca^{2+}]_i$ (von Tscharner et al., 1986), which is caused by the $Ins(1,4,5)P_3$ induced release of internal Ca^{2+}. Similarly, Putney (1986) suggested that Ca^{2+} entry after agonist stimulation occurs via a type of gap junction and is caused by the refilling of previously emptied Ca^{2+} stores.

On the other hand, however, Benham and Tsien (1987), who so far have provided the only direct electrophysiological measurements of ROCs in vascular smooth muscle, proposed a direct activation of the channel, which may possibly be linked to the receptor via a G-protein.

For our study of ROCs, rat aortic smooth muscle cells (SMCs) and mesenteric resistance vessels (MRVs) were used as models. Ca^{2+} fluxes, cytosolic Ca^{2+} concentration ($[Ca^{2+}]_i$) and second messenger systems were studied in cultured rat aortic SMCs. These cells contain dihydropyridine-sensitive ("L-type") PSCs (Rüegg et al., 1985). Stimulation of the cells with an agonist, e.g. $[Arg^8]$vasopressin (AVP), angiotensin II (A II), ATP or endothelin led to the formation of Ins(1,4,5)P_3 (Wallnöfer et al., 1989) and to $^{45}Ca^{2+}$ release from internal stores (Doyle and Rüegg, 1985, Wallnöfer et al., 1989). Additionally, after stimulation with agonists enhanced dihydropyridine-insensitive Ca^{2+} entry is observed in these cells (Wallnöfer et al., 1987). This effect can be attributed to the activation of ROCs. The results obtained in the cellular model were correlated to rat MRVs, where a corresponding Ca^{2+}-antagonist insensitive part of the contraction was observed (Cauvin et al., 1985). These vessels can be stimulated by noradrenaline (Cauvin and Malik, 1984), by AVP (Cauvin et al., 1988) or by endothelin (Wallnöfer et al., 1989).

Binding of the agonist to its receptor leads to activation of phospholipase C and to formation of Ins(1,4,5)P_3 (Berridge, 1986). Ca^{2+} release from internal pools causes a transient rise of cytosolic $[Ca^{2+}]$ as we have measured with the fluorescent Ca^{2+} indicator fura-2 (Grynkiewicz et al. 1985) in the aortic SMCs (Rüegg et al., 1989, Wallnöfer et al. 1989). Similarly, Ca^{2+} release from internal stores contributes to the phasic contraction of the MRVs (Cauvin et al. 1988). However, the sustained rise in $[Ca^{2+}]_i$, which is observed after agonist-activation (Wallnöfer et al. 1989) and the tonic part of MRV contraction (Cauvin et al., 1988) is dependent on Ca^{2+} influx from the extracellular medium. Whereas in the aortic SMCs Ca^{2+} entry appears to occur mainly via ROCs, in the MRVs PSCs and ROCs both contribute to the tonic contraction (Rüegg et al., 1989). This observation can be explained by the fact that the aorta depolarizes less than MRVs when challenged with an agonist (Cauvin et al. 1985).

ROC blockers

Enhanced Ca^{2+} entry stimulated by receptor-activation could be inhibited by the inorganic ions La^{3+}, Cd^{2+}, Mn^{2+}, Co^{2+}, Ni^{2+}, and Mg^{2+} (in this rank order of potency) (Wallnöfer et al., 1989). Searching for organic inhibitors of ROCs, we found that agonist-sensitive elevation of $^{45}Ca^{2+}$ influx and the sustained elevation of $[Ca^{2+}]_i$ in the SMCs is reduced in the presence of 2-nitro-4-carboxyphenyl-N,N-diphenylcarbamate (NCDC), a proposed inhibitor of phospholipase C. NCDC was about 8-fold more potent in inhibiting ROCs (pIC_{50} 4.78 ± 0.21) than PSCs (pIC_{50} = 3.88 ± 0.20) (Rüegg at al. 1989). NCDC (10-100μM) also abolished the contraction and the depolarization induced by agonists in the MRVs (Rüegg et al. 1989).

It appears therefore that ROC-activation is a fundamental mechanism by which vascular smooth muscle contraction can occur. Specific inhibitors of ROCs would not only provide

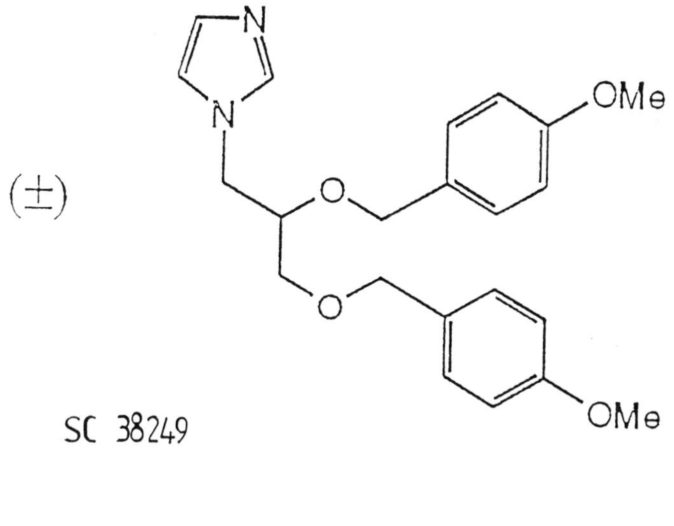

(±)

N N OMe O O OMe

SC 38249

Fig. 1 SC 38249 {(±)-1-(2,3-bis[(4-methoxy-phenyl)-methoxy]propyl]-1H-imidazole)}

useful tools for the further characterization of these channels and their mechanism of activation but may also have some therapeutic potential, e.g as possible antihypertensive, anti-atherosclerotic, antiasthmatic or antiinflammatory agents.

SC 38249 {(±)-1-(2,3-bis[(4-methoxyphenyl)-methoxy]propyl]-1H-imidazole)} (Fig. 1), originally prepared at Searle, has been characterized as a broad spectrum platelet aggregation inhibitor (Lad et al.. 1988). While some of its activities were attributable to the inhibition of thromboxane synthetase others were not.

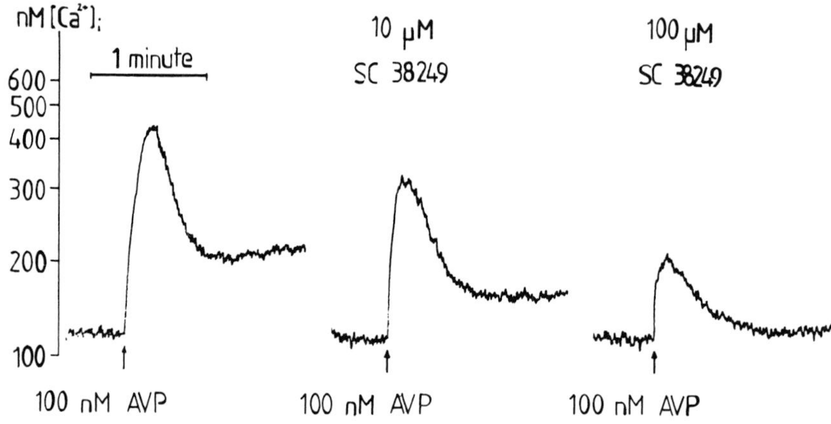

Fig. 2 Effect of SC 38249 on the cytosolic [Ca^{2+}] rise after stimulation of rat cultured aortic smooth muscle cells in suspension with 100 nM [Arg^8]vasopressin. Cells were loaded in suspension with 10 μM fura-2 AM and $[Ca^{2+}]_i$ was recorded fluorimetrically as described in Rüegg et al. 1989 or Wallnöfer et al. 1989. SC 38249 was added 2 min before stimulation with 100 nM AVP. The sustained phase is dependent on Ca^{2+} influx from the extracellular medium (Rüegg et al. 1989).

Since its mechanism was dependent on the presence of extracellular Ca^{2+}, we have investigated its effects on ROCs.

The stimulation of $^{45}Ca^{2+}$ entry into SMCs seen with agonists such as AVP, A II, ATP and endothelin were blocked by 10^{-4} M SC 38249. The IC_{50} for this inhibition corresponded well to the inhibition of platelet aggregation (i.e. ca. 10^{-5} M) (Lad et al., 1988). Measuring cytosolic $[Ca^{2+}]_i$ revealed that SC 38249 blocked the sustained phase which is dependent on influx of extracellular Ca^{2+} (Fig. 2).

In MRVs, the corresponding Ca^{2+} antagonist insensitive, agonist-induced (noradrenaline and AVP) tonic tension was was also inhibited by SC 38249. In the presence of diltiazem (10^{-6} M; to block potential-sensitive channels) SC 38249 reduced the contractions induced by such agonists in a concentration dependent manner (Fig. 3).

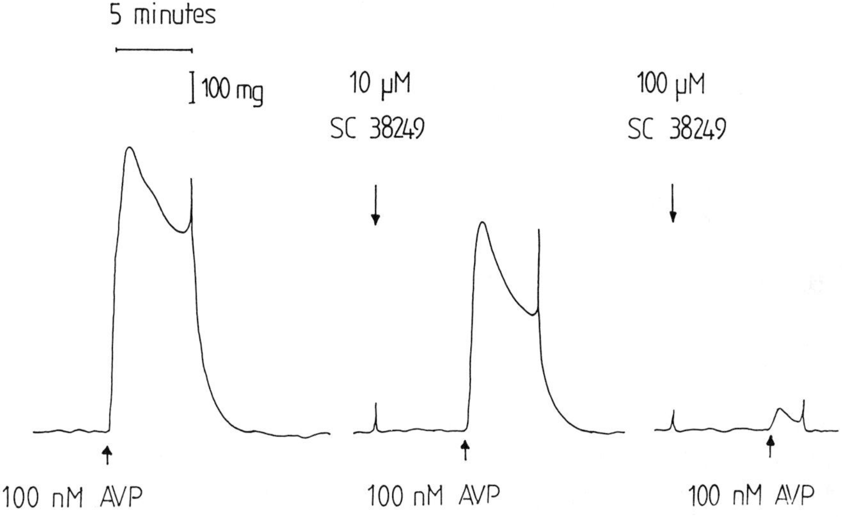

Fig. 3 Effect of SC 38249 on [Arg^{8}]vasopressin induced contraction of a rat mesenteric resistance vessel (MRVs). MRVs were prepared as described earlier (Cauvin et al. 1988). SC 38249 was added 5 minutes before stimulation with 100 nM AVP.

As seen when the intracellular $[Ca^{2+}]$ was measured the late tonic phase was sensitive to the action of this compound (IC_{50} for AVP 10^{-5} M) while the initial peak phase was less affected (IC_{50} $4 \cdot 10^{-5}$ M). Similarly, agonist-stimulated $^{45}Ca^{2+}$ efflux from cells which have been previously loaded with the tracer (Doyle and Rüegg, 1985) was less affected than agonist-stimulated $^{45}Ca^{2+}$ influx. Table 1 summarizes the results.

Table 1: Summary of the effects of SC 38249 on ROCs in MRVs and SMCs

TEST		pIC_{50} [-log M]
RAT MESENTERIC RESISTANCE VESSELS		
Contraction		
10 μM noradrenaline	phasic	4.5
	tonic	5.3
100 nM vasopressin	phasic	4.7
	tonic	5.2
RAT AORTIC SMOOTH MUSCLE CELLS		
Stimulated $^{45}Ca^{2+}$ influx		
10 nM endothelin		4.3
10 μM ATP		4.5
Stimulated $^{45}Ca^{2+}$ efflux		
1 μM angiotensin II		3.0
Cytosolic Ca^{2+} measurements with fura-2		
100 nM vasopressin	phasic	4.5
	tonic	4.8

In conclusion, this study suggests that SC 38249 is a blocker of ROCs in vascular smooth muscle. The fact that it inhibits different types of agonists at very similar concentrations in platelets (Lad et al. 1988) as well as in smooth muscle indicates that either the ROCs in the preparations are very similar or that a common denominator of the mechanism triggering ROC opening is affected by this compound.

References

Benham C, Tsien RW (1987). ATP-receptor-operated channels permeable to calcium in arterial smooth muscle. Nature 328:275-278.

Berridge M (1986). Intracellular signalling through inositol trisphosphate and diacylglycerol. Biol Chem Hoppe-Seyler 367:447-456.

Bolton T. (1979). Mechanisms of action of transmitters and other substances on smooth muscle. Physiol Rev 59:607-718.

Cauvin C, Malik S (1984). Norepinephrine activates α_1-adrenoceptors to induce Ca^{2+} influx and intracellular Ca^{2+} release in isolated rat aorta and mesenteric resistance vessels. J Pharmacol Exp Ther 230:413-418.

Cauvin C, Lukeman S, Cameron J, Hwang OK, van Breemen C (1985). Differences in norepinephrine activation and diltiazem inhibition of Ca^{2+} channels in isolated rabbit aorta and mesenteric resistance vessels. Circ Res 56: 822-828.

Cauvin C, Weir S, Wallnöfer A, Rüegg U (1988). Agonist-induced activation of rat mesenteric resistance vessels: comparison between noradrenaline and vasopressin. J Cardiovasc Pharmacol 12 (Suppl.5): 128-133.

Doyle V, Rüegg UT (1985). Vasopressin induced production of inositol trisphosphate and calcium efflux in a smooth muscle cell line. Biochem Biophys Res Comm 131:469-476.

Grynkiewicz G, Poenie M, Tsien RY (1985). A new generation of Ca^{2+} indicators with greatly improved fluorescence properties. J Biol Chem 260:3440-3450.

Hallam TJ, Rink TJ (1989). Receptor-mediated Ca^{2+} entry: diversity of function and mechanism. Trends in Pharmacol. Sci. 10: 8-10.

Irvine RF, Moor RM, (1986). Micro-injection of inositol 1,3,4,5-tetrakisphosphate activates sea urchine eggs by a mechanism dependent on external Ca^{2+}. Biochem J 240: 917-920.

Kuno M, Gardner P, Ion channels activated by inositol 1,4,5-trisphosphate in plasma membrane of human T-lymphocytes. Nature 326: 301-304.

Lad N, Honey AC, Lunt DO, Booth J, Westwick J, Manley PW & Tuffin DP (1988). Effect of SC 38249, a novel substituted imidazole, on platelet aggregation in vitro and in vivo. Thrombosis & Haemostasis 59:164-170.

Penner R, Matthews G, Neher E, (1988) Regulation of calcium influx by second messengers in rat mast cells. Nature 334: 499-504.

Putney JW (1986). A model for receptor-operated calcium entry. Cell Calcium 7: 1-12.

Rüegg UT, Doyle V, Hof R (1985). Effects of calcium entry blockers on calcium fluxes in a vascular smooth muscle cell line. J Hyperten 3(Suppl 3):S57-S59.

Rüegg UT, Doyle V, Zuber J-F, Hof R (1985). A smooth muscle cell line suitable for the study of voltage sensitive calcium channels. Biochem Biophys Res Comm 130:447-453.

Rüegg UT, Wallnöfer A, Weir S, Cauvin C (1989). Receptor-operated Ca^{2+} channels in vascular smooth muscle. J Cardiovasc. Pharmacol. 14 (Suppl. 6) S49-S58.

von Tscharner V., Prod'hom B, Baggiolini M, Reuter H (1987). Ion channels in human neutrophils activated by a rise in free cytosolic Ca^{2+} concentration. Nature 324: 369-372.

Wallnöfer A, Cauvin C, Rüegg U (1987). Vasopressin increases $^{45}Ca^{2+}$ influx in rat aortic smooth muscle cells. Biochem Biophys Res Comm 148:273-278.

Wallnöfer A, Weir S., Rüegg U, Cauvin C., (1989). The mechanism of action of endothelin-1 as compared with other agonists in vascular smooth muscle. J. Cardiovasc. Pharmacol. 13 (Suppl 5) S23-S31.

Wallnöfer A, Cauvin C, Lategan TW, Rüegg UT (1989). Differential blockade of agonist and depolarization induced $^{45}Ca^{2+}$ influx in smooth muscle cells. Am J Physiol. 257 (in press).

van Breemen C, Aaronson P, Loutzenhiser R (1979). Sodium-calcium interactions in mammalian smooth muscle. Pharmacol Rev 30:167-208.

Calcium Channel Modulators in Heart and Smooth Muscle:
Basic Mechanisms and Pharmacological Aspects
S. Abraham and G. Amitai, editors
© 1990, VCH, Weinheim/Deerfield Beach, FL and Balaban, Rehovot/Philadelphia

Temporal integration of receptor responses in BC3H-1 muscle cells

R. DALE BROWN[1], HELEN RISTIC[1], SHAHROKH TORANJI[2], AND DORIS A. TAYLOR[3]

[1]Department of Pharmacology, University of Illinois at Chicago, Chicago, IL, 60612; [2]Department of Pharmacology, University of California at San Diego, La Jolla, CA; and [3]Department of Microbiology and Immunology, Albert Einstein College of Medicine, Bronx, NY, USA

INTRODUCTION

The importance of intracellular ionized calcium concentration, $[Ca^{2+}]_i$, as a key determinant of vascular contractile state has long been recognized (Bolton, 1979). Many vasoactive drugs and hormones act through this fundamental pathway. Despite these common modes of action, these agents may exhibit greatly divergent patterns of temporal responses. Smooth muscle is notable in its ability to sustain contractions over prolonged intervals of agonist exposure, yet examples of tachyphylaxis also occur. Elevation of cytosolic Ca^{2+} upon vascular stimulation may occur by transmembrane influx of Ca^{2+} down its inwardly directed concentration gradient, or by mobilization from intracellular organelles, particularly sarcoplasmic reticulum. Experimental evidence suggests that inositol trisphosphate (IP_3), as well as other metabolites derived from hydrolysis of membrane phosphoinositides (PI), act as second messengers to couple plasma membrane receptors with mobilization of intracellular Ca^{2+} and with regulation of Ca^{2+} entry. (reviewed by Berridge, 1987) Thus an understanding of receptor regulation of phosphoinositide metabolism should provide insight into the biochemical basis of the integrated cellular response.

To approach this problem, we have utilized cultured BC3H-1 muscle cells as a model system (Schubert et al., 1974). The cells grow as adherent monolayers, allowing uniform exposure of genetically homogeneous populations of cells to experimental solutions and avoiding the complexities of intact smooth muscle preparations. The cells express alpha$_1$-adrenergic and H$_1$-histaminergic receptors which regulate intracellular Ca^{2+}. Alpha$_1$ receptor activation of cellular Ca^{2+} translocation is sustained during prolonged intervals of adrenergic agonist exposure, whereas H$_1$ receptor function becomes refractory following brief exposure to histamine (Brown et al.,1986). In the present study we examine regulation of phosphoinositide metabolism by these two receptor types in order to determine the basis of their disparate kinetic behaviors.

METHODS

BC3H-1 cell culture and experimental agonist exposure protocols were performed as described by Brown et al. (1986). Equilibration of cultures with (^{3}H)inositol and measurement of (^{3}H)inositol incorporation into cellular lipids were performed as described by Ambler et al. (1984). Production of (^{3}H)inositol phosphates was measured by extraction into 10% $HClO_4$ (Ambler et al. 1988) and separation of inositol mono-, bis-, and trisphosphate on Dowex anion exchange columns (Downes and Michell, 1981). Glycogen phosphorylase activity was measured in BC3H-1 cell samples as described by Toranji and Brown (1989), as follows. Experimental treatment was terminated by rapidly immersing the culture dish in liquid nitrogen. Frozen cells were scraped from the dish on ice in 300 ul of homogenization buffer (content in mM: NaF, 10; beta-glycerophosphate, 40; beta-mercaptoethanol, 10; EDTA, 2; bovine serum albumin, 0.1% w/v; pH 6.8). The sample was homogenized, clarified by centrifugation, and 10 ul aliquots of supernatant assayed in the direction of glucose-1-phosphate production by a fluorimetric coupled enzyme assay (Hardman et al. 1964). Data are expressed as the ratio of phosphorylase in the active a form relative to total enzyme (a + b forms).

RESULTS

Occupation of alpha$_1$-adrenergic or H$_1$-histaminergic receptors on BC3H-1 cells elevates intracellular Ca^{2+}, which can be monitored functionally through conversion of the metabolic enzyme glycogen phosphorylase to its active a form (Toranji and Brown, 1989). Figure 1 shows the kinetics of phosphorylase activation by receptor stimulation or by direct elevation of intracellular Ca^{2+} with the ionophore A23187. Phosphorylase activation occurs rapidly; maximum response is seen within 5-10 sec. Phosphorylase activity remains elevated during 20-30 min exposure to alpha-adrenergic agonist or Ca^{2+} ionophore. By contrast, phosphorylase is only transiently activated upon addition of histamine, and returns to basal

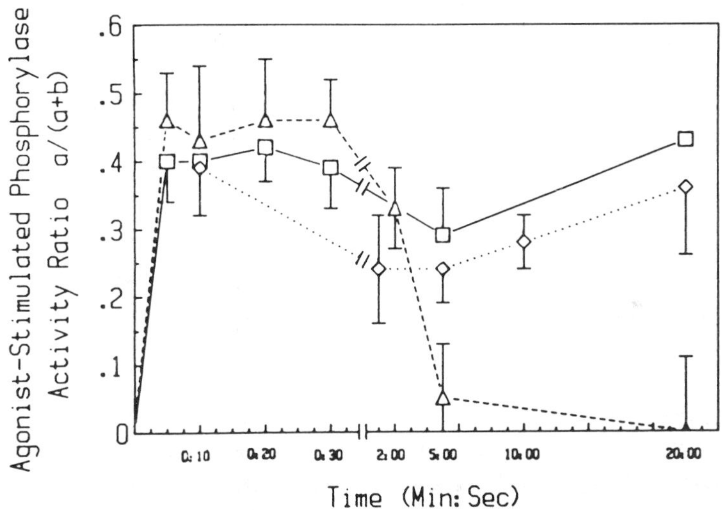

Figure 1. Kinetics of glycogen phosphorylase activation. BC3H-1 cultures were exposed to physiological buffer containing specified ligands over the indicated time intervals followed by rapid freezing, homogenization, and assay of phosphorylase activity as described in Methods. Agonist-stimulated activity ratios were corrected for basal activity in paired sets of control cultures. Data are taken from Toranji and Brown, (1989). (□), Norepinephrine, 3 μM. Propranolol, 1 μM, was added with the adrenergic agonist to block beta-adrenergic receptor activation. (△) Histamine, 100 μM (◇), A23187, 10 μM.

activity within 5 min. Cultures which have become refractory following extended histamine exposure are insensitive to further histamine addition yet remain fully responsive to subsequent alpha adrenergic stimulation (Data not shown.)

Phosphoinositide turnover stimulated by $alpha_1$ or H_1 receptors was assessed by following the uptake of (^{3}H) inositol into cellular lipids as shown in Figure 2. Stimulation of either receptor type initially elicits comparable phosphoinositide synthesis. $Alpha_1$ receptor-mediated PI synthesis is sustained over at least 30 min of agonist exposure, whereas accumulation of (^{3}H)inositol into lipid ceases within 10 min of histamine exposure. Combined addition of maximally effective concentrations of the two agonists does not

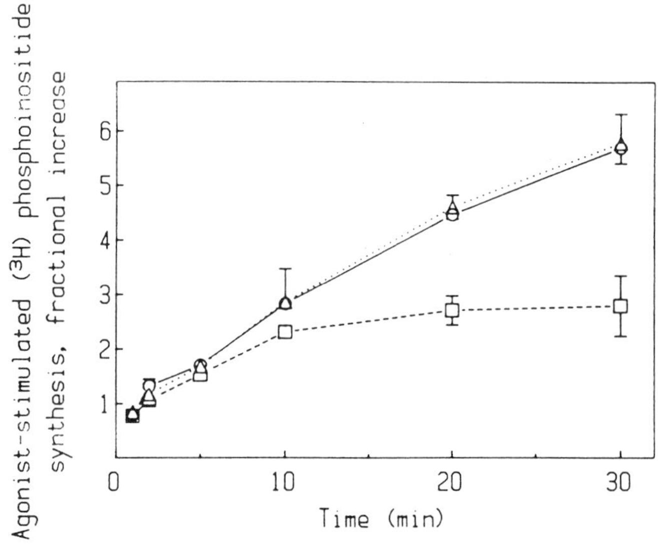

Figure 2. Kinetics of (^{3}H)phosphoinositide synthesis. Incorporation of (^{3}H)inositol into organic-extractable cellular lipids following indicated agonist exposures was measured as described in Methods. Data are expressed as fold-stimulation relative to unstimulated controls at zero time. (○) Phenylephrine, 10 μM; (□) Histamine, 100 μM; (△) phenylephrine plus histamine.

further increase PI synthesis over short or prolonged intervals of agonist exposure.

The measurement of phosphoinositide synthesis provides an indirect index of metabolic activity through this pathway. As shown by Berridge (1981), the hydrolysis of phosphatidylinositol-bis-phosphate to generate inositol trisphosphate and the subsequent production of inositol bis- and monophosphates are more proximal responses to receptor occupancy. We therefore turned to examine directly the receptor-elicited production of inositol phosphate metabolites with anion exchange chromatography. Results are shown in Figure 3. Both receptor types produce small but consistent elevations of inositol trisphosphate over an initial 5 min interval. The elevation of IP_3 in response to alpha-adrenergic agonist is sustained over 30 min, whereas IP_3 returns to basal values during this interval in response to histamine. Combined addition of the two agonists fails to produce additive increases in IP_3 formation. A consistent pattern of responses is obtained from measurement of inositol bis-phosphate production, also shown in Figure 3. We have recently used anion exchange hplc methodology to confirm these results and to separate the isomeric inositol 1,4,5-

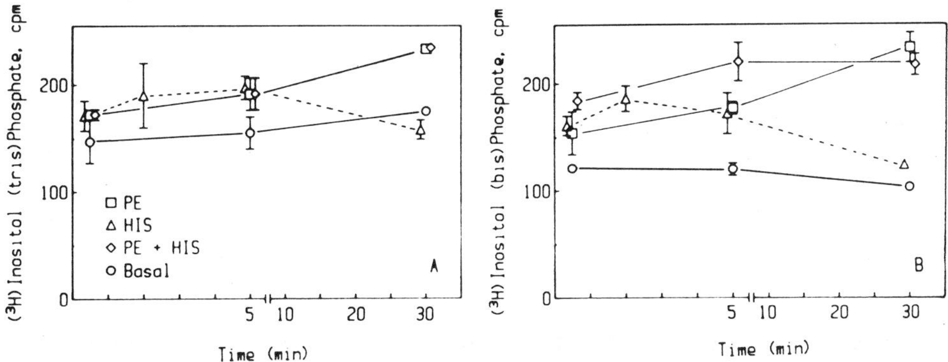

Figure 3. Kinetics of inositol phosphate production. Cultures equilibrated with (^{3}H)inositol were treated with the specified agonists for indicated times. Aqueous-soluble inositol phosphates were extracted and analyzed as described in Methods. (□), Phenylephrine, 10 μM; (△), histamine, 100 μM; (◇), phenylephrine plus histamine; (○), no added agonist. A. Inositol triphosphate B. Inositol bisphosphate.

and 1,3,4-tris-phosphates. We find that both receptor types predominantly produce the 1,4,5-IP_3 isomer, which has been shown to possess greater biological activity in releasing intracellular Ca^{2+} (Irvine et al., 1986). Production inositol of 1,3,4,5-tetrakisphosphate is not detected in these cells (Brown, 1989).

The agonist exposure and challenge paradigms described in Figure 4 were employed to more clearly define the refractoriness which occurs with prolonged histamine exposure. The data show that 20 min exposure of BC3H-1 cultures to histamine abolishes histamine receptor responsiveness yet the cells remain responsive to a subsequent alpha-adrenergic challenge. Moreover, $alpha_1$

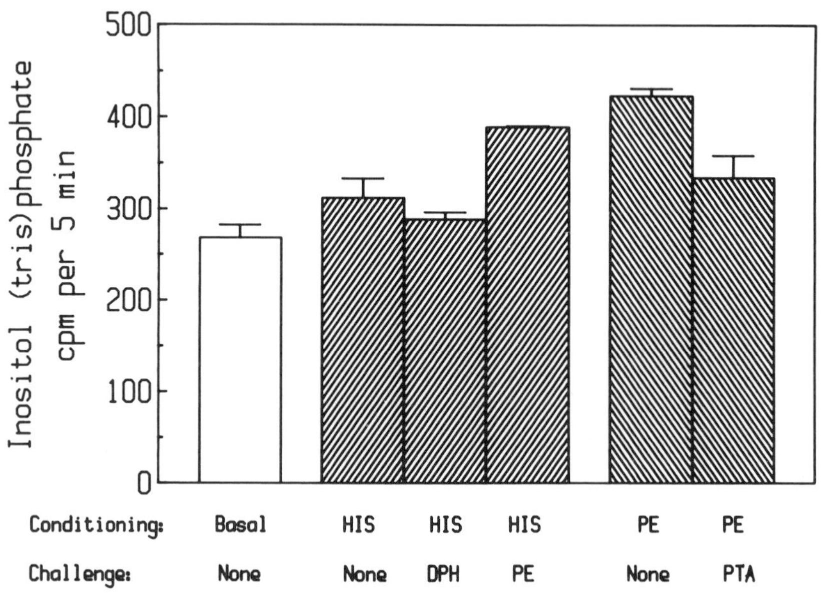

Figure 4. Influence of prior agonist exposure on receptor responsiveness. BC3H-1 cultures equilibrated with (^{3}H)inositol were exposed to physiological buffer containing the specified conditioning agonist for 20 min. The conditioning solution was removed and inositol trisphosphate production was measured in response to 5 min exposure with the specified challenging agonist. Agonist additions are as follows. HIS, 100 μM histamine; PE, 10 μM phenylephrine; PTA, 10 μM phentolamine; DPH, 10 μM diphenhydramine; Basal, physiological buffer containing no added agonist; NONE, reaction stopped immediately after conditioning interval with no agonist challenge.

receptors support elevated production of IP_3 over a 30 min interval and remain sensitive to reversal by subsequent addition of the alpha receptor antagonist phentolamine. Thus the refractoriness of histamine responsiveness is selective for the H_1 receptor and does not reflect a generalized uncoupling of the PI signalling pathway.

The kinetics of H_1-receptor desensitization were measured as shown in Figure 5. Cultures were exposed to either the adrenergic or histaminergic agonist for specified time intervals and then the rate of (^{3}H)inositol incorporation into phosphoinositides was measured over a 10 min interval in response to either the same agonist or the respective receptor antagonist. Phosphoinositide synthesis

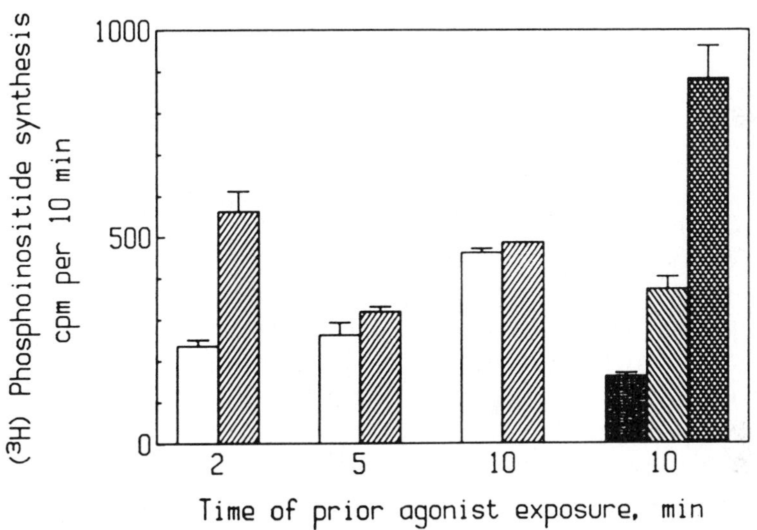

Figure 5. Kinetics of desensitization of receptor-mediated phosphoinositide turnover. Cultures equilibrated with (^{3}H)inositol were exposed to 100 μM histamine (▨□), 10 μM phenylephrine (▧▩), or buffer containing no agonist addition (▤) for the specified time intervals. The solution was removed and the net increase in (^{3}H)phosphoinositide synthesis was measured over the succeeding 10 min interval in response to a fresh solution containing 100 μM histamine (▨), 10 μM diphenhydramine (□), 10 μM phenylephrine (▩), 10 μM phentolamine (▧), or no added agonist (▤). The data were corrected for radiolabel incorporation occurring during the prior exposure interval.

mediated by the H_1 receptor is greatly attenuated within 5 min histamine exposure; the residual enhancement of PI synthesis relative to naive cultures is insensitive to reversal by the H_1 antagonist, diphenhydramine. Thus the H_1 receptor becomes refractory to stimulation by agonist or reversal by antagonist within 5 min exposure to histamine. By contrast, phenylephrine stimulation of the alpha$_1$ receptor to elevate PI synthesis is linear over this interval and the alpha$_1$ receptor remains sensitive to reversal by the antagonist phentolamine.

DISCUSSION

BC3H-1 cells provide a simplified system in which to quantitate the intracellular events linking receptor occupancy with response. The cells express both alpha$_1$-adrenergic and H_1-histaminergic receptors which regulate intracellular Ca^{2+}, offering the possibility to explore the mechanisms by which vascular cells integrate multiple inputs in the continuous regulation of contractile state. In the present study we have examined the coupling of these two receptors to phosphoinositide metabolism.

Alpha$_1$- and H_1-receptor occupancy activate the cycle of phosphoinositide synthesis and breakdown. Within the precision of the radiolabelling method, maximally effective concentrations of the respective agonists elicit comparable magnitudes of initial responses. Combined stimulation by both agonists elicits no further increase in phosphoinositide metabolism. These results suggest that the two receptors operate on phosphoinositide metabolism through a common mode, and that the maximum velocity of radiolabelled PI turnover is limited by other constraints in addition to total receptor occupancy. We previously quantitated a unitary relationship between PI turnover and intracellular Ca^{2+} mobilization in response to alpha$_1$ receptor occupancy (Ambler et al., 1984). The present data suggest that this relationship also holds for H_1 receptors. In the general situation where multiple receptor populations couple to a common functional response, one would predict an invariant relation between production of the limiting activation intermediate (IP_3, for example) and functional response, independent of the particular combination of occupied receptors. The observed

response would represent a weighted sum of the occupancies of all receptors coupled through the common pathway, modulated by their respective efficacies. This scheme would provide a simple means for the cell to integrate regulatory inputs.

Elevation of intracellular Ca^{2+} and activation of Ca^{2+}-dependent responses occur within seconds upon $alpha_1$ or H_1 receptor stimulation. Termination of $alpha_1$ receptor response also occurs rapidly upon displacement of agonist by addition of an excess concentration of antagonist (Ambler and Taylor, 1986, Toranji and Brown, 1989). By contrast, agonist stimulation of radiolabelled phosphoinositide turnover is paradoxically slow and of small magnitude relative to unstimulated cultures. Inositol phosphate accumulation is barely detectable after 30 sec and requires 2-5 min of receptor stimulation to reach a maximum. These observations are consistent with previous studies of $alpha_1$ receptor-mediated phosphoinositide metabolism in BC3H-1 cells and may relate to nonuniform radiolabelling of cellular phosphoinositides or to agonist-elicited production of inositol phosphates within a limited intracellular compartment (Ambler et al., 1984, 1987) . Ongoing research in the laboratory is aimed at quantitation of inositol phosphate production by chemical mass determinations in order to eliminate the uncertainties of radiolabelling methods. A further possibility would be that additional mechanisms besides IP_3 production mediate intracellular Ca^{2+} mobilization. In this regard, current studies are aimed at characterizing and quantitating lipid metabolites produced upon receptor stimulation.

Intervals of histamine exposure uncouple the H_1 receptor from activation of PI metabolism. The kinetics of H_1 receptor desensitization correlate respectably with the reversal of phosphorylase activation, although radiolabelled phosphoinositide turnover remains slightly elevated at a time when phosphorylase activity is largely inhibited. This discrepancy may again arise from nonuniform radioisotope distribution or may reflect compensatory resynthesis of substrates. These data stand in clear contrast to the sustained phosphoinositide turnover and Ca^{2+} signalling mediated by $alpha_1$ receptors (Brown et al., 1986, 1989). We cannot rigorously exclude the possibility that the loss of histamine responsiveness arises from selective depletion of histamine-sensitive phosphoinositide substrates. However, the body of data strongly

suggest that the refractoriness of H_1 response occurs at a step proximal to the H_1 receptor and its coupling to phospholipase C. It will be of interest to utilize radioligand binding methods in order to test directly whether prior histamine exposure alters the affinity of the H_1 receptor for agonists or antagonists.

In sum, $alpha_1$ and H_1 receptors in BC3H-1 cells regulate cellular responses through common intracellular pathways. $Alpha_1$ receptor activation is sustained whereas H_1 receptor response is transient. These specific temporal responses appear to originate in the properties of the individual receptors or their coupling to phosphoinositide metabolism, a proximal step in vascular signal transduction.

REFERENCES

Ambler SK, Brown RD, Taylor P (1984). The relationship between phosphatidylinositol metabolism and mobilization of intracellular calcium elicited by $alpha_1$-adrenergic receptor stimulation in BC3H-1 muscle cells. Mol Pharmacol 26:405-413.

Ambler SK, Taylor P (1986) Mobilization of intracellular calcium by $alpha_1$-adrenergic receptor activation in muscle cell monolayers. J Biol Chem 261:5866-5871.

Ambler SK, Thompson B, Solski PA, Brown JH, Taylor P (1987). Receptor-mediated inositol phosphate formation in relation to calcium mobilization: A comparison of two cell lines. Mol Pharmacol 32:376-383.

Berridge MJ (1981). Rapid accumulation of inositol trisphosphate reveals that agonists hydrolyse polyphosphoinositides instead of phosphatidylinositol. Biochem J 212:849-858.

Berridge MJ (1987). Inositol trisphosphate and diacylglycerol: Two interacting second messengers. Ann Rev Biochem 56:159-193.

Bolton TB (1979). Mechanisms of action of transmitters and other substances on smooth muscle. Physiol Rev 59:606-718.

Brown RD, Prendiville P, Cain C (1986). $Alpha_1$-adrenergic and H_1-histamine receptor control of intracellular Ca^{2+} in a muscle cell line: The influence of prior agonist exposure on receptor responsiveness. Mol Pharmacol 29:531-539.

Brown RD (1989) Activation and desensitization of receptor-mediated phosphatidylinositol metabolism in BC3H-1 muscle cells. J Cell Biol 107:66a (abstract).

Downes CP, Michell R (1981). The polyphosphoinositide phosphodiesterase of erythrocyte membranes. Biochem J 198:133-141.

Hardman JG, Mayer SE, Clark B (1965). Cocaine potentiation of the cardiac inotropic and phosphorylase responses to catecholamines as related to uptake of ^{3}H-catecholamines. J Pharmacol Exp Ther 150:341-348.

Irvine RF, Letcher AJ, Lander DJ, Berridge MJ (1986). Specificity of inositol phosphate-stimulated Ca^{2+} mobilization from Swiss mouse 3T3 cells. Biochem. J. 240: 301-304.

Schubert DA, Harris AJ, Devine CF, Heinemann S (1974). Characterization of a unique muscle cell line. J Cell Biol 61:398-413.

Toranji S, Brown RD (1989) Temporal integration of alpha$_1$-adrenergic responses in BC3H-1 muscle cells: Regulation of glycogen phosphorylase activity. J Biol Chem, in press.

ACKNOWLEDGEMENTS

The authors are grateful to Kelly Ambler for helpful discussions on measurement of phosphoinositide turnover. This work was supported by NIH (GM 41470) and American Heart Association, California Affiliate (Grant-in-Aid 87-S108).

Calcium Channel Modulators in Heart and Smooth Muscle:
Basic Mechanisms and Pharmacological Aspects
S. Abraham and G. Amitai, editors
© 1990, VCH, Weinheim/Deerfield Beach, FL and Balaban, Rehovot/Philadelphia

Calcium utilization of α-adrenoceptors

PIETER B.M.W.M. TIMMERMANS

Division of Cardiovascular Science, Medical Products Department, E.I. du Pont de Nemours & Company, Experimental Station, Wilmington, DE, USA 19880-0400

INTRODUCTION

The pharmacological, functional and anatomical differences of pre- and postsynaptically located α-adrenoceptors in synapses of (peripheral) noradrenergic nerve endings originally led to the suggestion to denominate the postsynaptic receptors, which mediate the excitatory responses, α_1-adrenoceptors and classify the presynaptic receptors, which induce the inhibitory effects, α_2-adrenoceptors (Langer, 1974). Presently, however, the terms pre- and postsynaptic refer to the location of the particular α-adrenoceptor with respect to the synapse. The designation α_1- or α_2- is solely based upon the drug selectivities of the α-adrenoceptor in question, irrespective of its anatomical position or function (Timmermans and van Zwieten, 1982).

In the CNS and the periphery, the prejunctional α-adrenoceptor population consists mainly of the α_2 subtype. Presynaptic α_2-adrenoceptors on a particular neuron which are activated by its own neurotransmitter, i.e. norepinephrine, are called autoreceptors. These have to be distinguished from presynaptic α_2-heteroreceptors which are activated by transmitters released from neighboring nerve terminals or blood-born substances (Göthert, 1985). In the CNS, somadendritic α_2-autoreceptors modulate impulse generation. Postsynaptically, both α_1- and α_2-adrenoceptors mediate a variety of responses. For example, α_1- and α_2-adrenoceptors have been demonstrated to be located postjunctionally on vascular smooth muscle, both subtypes mediating vasoconstriction upon stimulation (McGrath, 1982).

SIGNAL TRANSFORMATION OF α_1-ADRENOCEPTORS

As has already been alluded to, α_1-adrenoceptors are mainly found at postsynaptic sites (smooth muscle, liver, heart, salivary glands, adipose tissue, sweat glands, kidney, brain) mediating a host of (mainly stimulatory) physiological and biochemical actions (contraction/relaxation, glucogenolysis, positive inotropy/chronotropy, secretion, Na^+ reabsorption, neurotransmission). There is now general agreement that α_1-adrenoceptor mediated responses involve a rise in cytosolic free Ca^{2+} and that Ca^{2+} mainly exerts its intracellular effects by binding to the ubiquitous Ca^{2+} dependent regulatory protein calmodulin. The Ca^{2+}-calmodulin complex, in turn, interacts with a variety of enzymes and other cellular proteins altering their activities and leading to the various physiological responses observed (Exton, 1981). In the case of smooth muscle in many tissues, such as blood vessels, iris, uterus, vas deferens, α_1-adrenoceptor agonists cause contraction which results from activation of myosin light-chain kinase by the Ca^{2+}-calmodulin complex, which in turn phosphorylates the 20 kDa light chain myosin. Phosphorylation of myosin is required for actin activation of myosin ATPase activity and the formation of cross bridges between actin and myosin that results in contraction. The α_1-adrenoceptor-induced rise in cytosolic Ca^{2+} is initially due to the release of Ca^{2+} from internal stores (sarcoplasmic reticulum, mitochondria, plasma membrane) which is responsible for the initial phasic (fast) component of contraction. This is followed by a slower tonic component which is carried by a transmembraneous influx of Ca^{2+} through receptor-operated Ca^{2+} channels (Bülbring and Tomita, 1987). The α_1-adrenoceptor-induced relaxation of most gastrointestinal smooth muscle is associated with hyperpolarization caused by an increase in K^+ and Cl^- conductances. These conductance changes are dependent on Ca^{2+} which presumably opens a K^+ channel. α_1-Adrenoceptor-induced stimulation of liver glycogen breakdown is the result of a rise in cytosolic Ca^{2+} which allosterically activates phosphorylase _b_ kinase converting phosphorylase _b_ into the more active form phosphorylase _a_, which increases glycogen breakdown. In salivary and lacrimal glands, α_1-adrenoceptor agonists cause efflux of K^+ and H_2O due primarily to increased plasma membrane permeability. This is attributable to a rise in cytosolic Ca^{2+} which presumably opens K^+ channels (Putney, 1987).

Positive inotropic and chronotropic responses in the heart are predominantly mediated by β-adrenoceptors. However, it has been established that stimulation of cardiac postsynaptic α_1-adrenoceptors can also give rise to an increase of force development and to an increase in heart rate although to a much lesser degree (Wilffert, 1986). The α_1-adrenoceptor-induced positive inotropic effect develops much more slowly than that induced by β-adrenoceptor stimulation and is particularly pronounced at low frequencies and is accompanied by slower relaxation. The details of the signal transduction to α_1-adrenoceptor activation leading to the positive inotropic effect are largely unknown. Whereas β-adrenoceptor agonists elicit this response by an increase in cAMP, the positive inotropic effects mediated by α_1-adrenoceptors is not accompanied by an enhanced intracellular cAMP level. There are indications that the positive inotropic effect caused by α_1-adrenoceptor stimulation partly results from an influx of extracellular Ca^{2+}, since this effect is attenuated by Ca^{2+} channel blockers (Tung et al., 1985) and to an increase in phosphatidylinositol turnover (see below) which does not require the presence of extracellular Ca^{2+} (Scholz et al., 1988).

It has been realized than enhanced phosphatidylinositol (PI) breakdown has an intrinsic function in the operation of many cell surface receptor systems, such as α_1-adrenoceptors, which mobilize Ca^{2+} (Michell, 1975). Data are accumulating suggesting that (α_1-adrenoceptor) agonist stimulated Ca^{2+} entry may be intimately related to mobilization of Ca^{2+} from intracellular pools and that both are regulated by the water-soluble product of phosphatidylinositol 4,5-biphosphate (PIP_2) hydrolysis. The α_1-adrenoceptor coupled event is presumably activation of a plasma membrane phospholipase C in which a guanine nucleotide regulatory protein is involved. The activated phospholipase C hydrolyzes PIP_2 generating two second messenger molecules, i.e. inositol 1,4,5-triphosphate (1,4,5IP_3) and 1,2-diacylglycerol (DG) (fig. 1). 1,4,5IP_3 has been shown to release Ca^{2+} from intracellular pools. A specific receptor exists for 1,4,5 IP_3 on the endoplasmic reticulum (ER) which mediates the opening of a Ca^{2+} channel in the ER membrane (Berridge, 1987; Exton, 1988).

The action of 1,4,5IP_3 is terminated by the combined action of two separate pathways. It can be sequentially

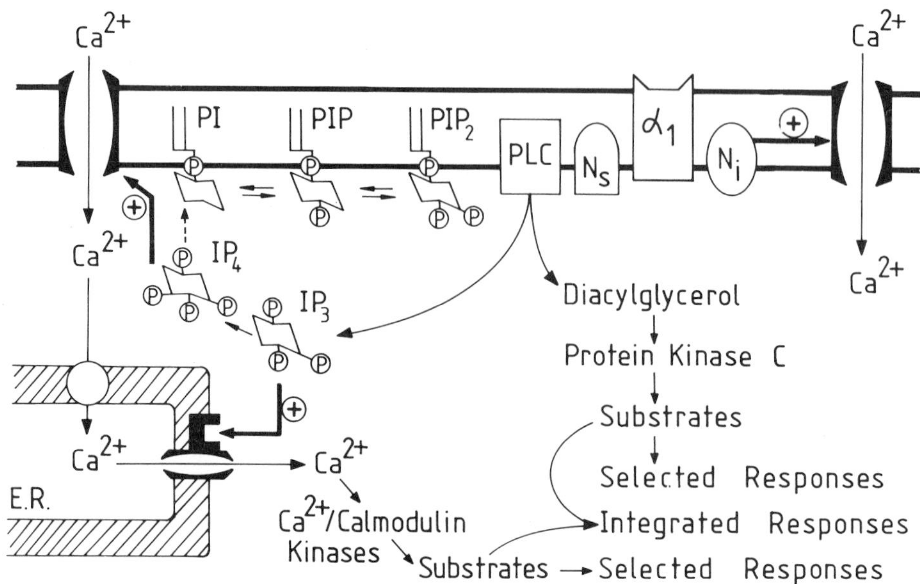

Fig. 1. Schematic overview of the signal transduction mechanisms to stimulation of α_1-adrenoceptors. For details see text.

dephosphorylated to free inositol, the first step is mediated by an inositol triphosphatase which removes the 5-phosphate to give $1,4IP_2$ which has no Ca^{2+} mobilizing activity. The alternative pathway begins with an inositol triphosphate kinase which phosphorylates $1,4,5IP_3$ to inositol 1,3,4,5-tetrakisphosphate ($1,3,4,5IP_4$). The significance of the latter is that it may function as a second messenger to regulate the entry of Ca^{2+} ions across the plasma membrane. Since the $1,4,5IP_3$ kinase is activated by Ca^{2+} it would seem that the entry of external Ca^{2+} might depend upon a prior elevation of intracellular Ca^{2+}. Inositol phosphates thus play a central role in Ca^{2+} signalling by controlling not only the release of internal Ca^{2+} but may also regulate the entry of external Ca^{2+} (Putney, 1987; Exton, 1988).

The DG which remains within the plane of the membrane constitutes the other limb of this bifurcating signal pathway which acts by stimulating protein kinase C to phosphorylate a variety of specific proteins. Many of the actions of this DG/protein kinase C pathway seem to be designed to modulate various aspects of the Ca^{2+} signalling pathway to give an integrated and highly versatile receptor mechanism which is employed to control many cellular processes including secretion, metabolism, contraction, neuronal excitability and cell growth. The physiological effects of activation of protein kinase C by DG are far from being fully clarified, and the identity and function of many of the target proteins phosphorylated by protein kinase C under physiological conditions of (α_1-adrenoceptor) agonist stimulation are not known. Because many effects of agonists are mimicked by phorbol esters or synthetic DG's, it is likely that there are many physiologically important substrates for protein kinase C, but the majority of these remain to be defined (Kikkawa and Nishizuka, 1986).

Although the role of inositol lipids in α_1-adrenoceptor mediated Ca^{2+} mobilization has mainly been tested in nonexcitable mammalian tissues, such as liver, pancreas and lacrimal glands, it may also apply to excitable (vascular) smooth muscles. The mobilization of Ca^{2+} from intracellular stores by $1,4,5IP_3$ may correspond to the rapid phase of smooth muscle contraction whereas the opening of the slow Ca^{2+} channel by $1,3,4,5IP_4$ may apply to the tonic phase. However, a remarkable feature of the response of vascular smooth muscle to α_1-adrenoceptor stimulation, which is distinctly different from the response to α_2-adrenoceptor activation (see below), is the heterogeneity among blood vessels from the same and different species with respect to the dependence of the contractile response on entry of extracellular Ca^{2+} or release of intracellular Ca^{2+} (Timmermans et al., 1987). At one extreme, α_1-adrenoceptor-induced vasoconstriction can be virtually insensitive to Ca^{2+} entry blockade (e.g. rabbit and guinea pig aortas) whereas at the other extreme this susceptibility can be as pronounced as that found for α_2-adrenoceptor-mediated vascular contraction (e.g. dog large coronary artery) (see below). A further complicating factor is that α_1-adrenoceptor agonists differentially employ the available intra- and extracellular Ca^{2+} sources for their contraction. Some of them primarily or exclusively activate Ca^{2+} influx, whereas others are also able to promote an intracellular mobilization of Ca^{2+} which corresponds with their ability to cause breakdown of PI (Chiu et al., 1986, 1987). It has been found recently that the α_1-adrenoceptor induced Ca^{2+} influx in vascular smooth muscle can be achieved without a detectable PI turnover suggesting that α_1-adrenoceptor activation can directly open a slow Ca^{2+} channel without the involvement of inositol lipids (Chiu et al., 1987). Furthermore, although α_1-adrenoceptor-mediated vasoconstriction has been found insensitive to impairment by "pertussis toxin" (Boyer et al., 1983), stimulation of α_1-adrenoceptors (rabbit pulmonary artery) by (higher concentrations of) α_2-adrenoceptor agonists is attenuated by this agent providing support for the hypothesis that these latter agonists open receptor-operated Ca^{2+} channels by the mediation of a guanine nucleotide binding inhibitory protein, N_i.

The data summarized above suggest that stimulation of a common α_1-adrenoceptor initiates different responses in the subsequent effector systems which depends on the agonist employed and the tissue studied. Fig. 1 summarizes the major features of the mechanisms by which α_1-adrenoceptor agonists are thought to exert their effects in their target cells.

SIGNAL TRANSFORMATION OF α_2-ADRENOCEPTORS

α_2-Adrenoceptors in general are believed to be coupled to adenylate cyclase in an inhibitory manner in that stimulation results in inhibition of cAMP production (Limbird, 1988). cAMP is thought to produce its effects by activation of kinases that catalyze the phosphorylation of intracellular proteins. Tissues in which α_2-adrenoceptor agonist-induced inhibition of cAMP accumulation has been detected include platelets, pancreatic islets, adipocytes and neuroblastoma x glioma cells. Binding of ligands to α_2-adrenoceptor sites is influenced by guanine nucleotides. Occupancy of the α_2-adrenoceptor with agonists is thought to promote and stabilize interaction of the receptor with a guanine nucleotide binding inhibitory protein, N_i, and to promote the release or exchange of GDP for GTP at the guanine nucleotide binding site. The binding of GTP to N_i leads to destabilization of its interaction with the receptor and as a result the affinity of the receptor for the α_2-adrenoceptor agonist decreases. Another consequence of GTP binding is that there is dissociation of the α- and β, γ-subunits of N_i. Inhibition of adenylate cyclase is thought to result from an inhibitory action of the α-subunit of N_i on the catalytic moiety. Agonist occupancy of the α_2-adrenoceptor also stimulates the GTPase activity of N_i. This leads to reassociation of the N_i subunits and termination of the inhibition of the cyclase. Islet-activating protein ("pertussis toxin") abolishes the inhibitory effect of α_2-adrenoceptor agonists on adenylate cyclase apparently because it inhibits the release or exchange of GDP for GTP and dissociation of N_i through ADP-ribosylation of the α-subunit (Gilman, 1984).

Although α_2-adrenoceptors are negatively involved in the adenylate cyclase activity in several tissues and the inhibition of this enzyme apparently accounts for the functions of α_2-adrenoceptors, certainly not all α_2-adrenoceptor-induced responses can be explained by inhibition of the enzyme. For example, the inhibition of insulin release from pancreatic islets by α_2-adrenoceptor agonists occurs at a step distal to the generation of cAMP, as is the inhibitory effect of intestinal α_2-adrenoceptors on secretory diarrhea (Limbird, 1988).

Although α_2-adrenoceptor-induced stimulation of (human) platelets causes both inhibition of adenylate cyclase and induction of aggregation, evidence is now available that this aggregatory response can be accomplished without affecting the level of cAMP. Compounds have been found which behave as agonists for the aggregatory response, but as apparent antagonists for inhibition of adenylate cyclase mediated by the α_2-adrenoceptor; or as antagonists for the aggregatory response, but as agonists for the inhibition of adenylate cyclase (Clare et al., 1984). It has been suggested that unique α_2-adrenoceptors mediate these two responses. The nature of the second messenger system is unknown but recent data seems to exclude Ca^{2+} for this role.

Although there is as yet no direct evidence to show that inhibition of adenylate cyclase decreases α_2-adrenoceptor-mediated transmitter release from nerve terminals and the experimental evidence supporting an involvement of cAMP (or cGMP) is limited and rather conflicting (Timmermans and van Zwieten, 1982), it seems that central presynaptic α_2-adrenoceptors are coupled to a guanine nucleotide binding inhibitory protein, N_i (Schoffelmeer et al., 1986), whereas a "pertussis toxin" sensitive binding protein appears not to be involved in the signal transduction of presynaptic α_2-adrenoceptors at peripheral sites (Musgrave et al., 1987). On the other hand, regulation of norepinephrine release mediated by presynaptic α_2-adrenoceptor stimulation can only be demonstrated for the Ca^{2+}-dependent processes of transmitter release (e.g. nerve stimulation or depolarization with either K^+ or veratridine), while the Ca^{2+}-independent release of norepinephrine induced by tyramine cannot be affected. Thus, presynaptic α_2-adrenoceptor activation would inhibit the transmembrane inward current of Ca^{2+} ions via potential-sensitive permeability channels.

Patch clamp studies have revealed three distinct types of Ca^{2+} channel in a variety of neurons. The channels have been labelled L (large Ba conductance, long-lasting whole cell current), T (tiny Ba conductance, rapid transient whole cell current), and N (novel type of channel, found only in neurons so far, intermediate Ba conductance and rate of inactivation). The L-type Ca^{2+} channel only is selectively modulated by the dihydropyridines. Sympathetic neurons display only N and L-type Ca^{2+} channels. The dihydropyridine insensitive N-type channels are the dominant pathway for Ca^{2+} entry triggering release of norepinephrine from (rat) sympathetic neurons. Norepinephrine inhibits N-type Ca^{2+} current, consistent with its well-known negative feedback effect on its own release (see Miller, 1987).

Promotion of K^+ conductance and stimulation of (Na^+,K^+) ATPase activity are additional mechanisms which have been suggested to link presynaptic α_2-adrenoceptor stimulation with modulation of transmitter release. More recently, a role for protein kinase C has been implemented in the exocytotic release of norepinephrine from sympathetic nerve endings (Fredholm and Lindgren, 1988).

Taken together, the data favor the view that stimulation of presynaptic α_2-adrenoceptors leads to an inhibition of neurotransmitter release by decreasing the cAMP dependent-enhancement (phosphorylation?) of the opening of N-type slow Ca^{2+} channels (fig. 2).

Whereas stimulation of presynaptic α_2-adrenoceptors is likely to limit the availability of extracellular Ca^{2+}, activation of postsynaptic α_2-adrenoceptors in vascular smooth muscle promotes the influx of Ca^{2+} ions. In contrast to α_1-adrenoceptor-mediated pressor effects (see above) those to stimulation of α_2-adrenoceptors have been found invariably and equally sensitive to complete inhibition by Ca^{2+} entry blockers, such as nifedipine, verapamil and diltiazem (Timmermans et al., 1987). In addition, stimulation of α_2-adrenoceptors has been shown <u>in vitro</u> to cause uptake of $^{45}Ca^{2+}$ in vascular smooth muscle sensitive to blockade by nifedipine. As a consequence thereof, it has

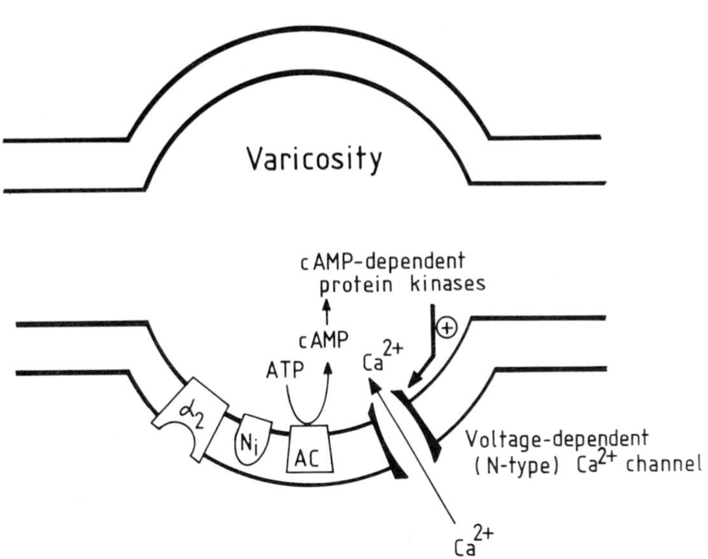

Fig. 2. Schematic diagram illustrating a hypothetical model linking activation of presynaptic α_2-adrenoceptors to inhibition of transmitter release through attenuation of Ca^{2+} influx. For details see text.

been submitted that all vascular postjunctional α_2-adrenoceptors subserve a single vasoconstrictor mechanism, primarily, if not exclusively, governed by an influx of extracellular Ca^{2+} ions. Activation of vascular α_2-adrenoceptors most likely does no mobilize intracellular Ca^{2+} through breakdown of PI. α_2-Adrenoceptor stimulation in vascular tissue gives rise to the opening of specific receptor-operated Ca^{2+} channels through which Ca^{2+} enters the cell and this process is not accompanied by depolarization of the cell membrane (Timmermans et al., 1987). On the other hand, it has been demonstrated that the pressor effects to α_2-adrenoceptor agonists are particularly sensitive to attenuation by "pertussis toxin" (Boyer et al., 1983), suggesting participation of a guanine nucleotide binding inhibitory protein, N_i, in this vasoconstrictor process. Until now no data are available showing that α_2-adrenoceptor stimulation affects cAMP levels in vascular smooth muscle. Fig. 3 summarizes the main intracellular events by which postsynaptic α_2-adrenoceptor agonists are presently believed to exert their signal transduction.

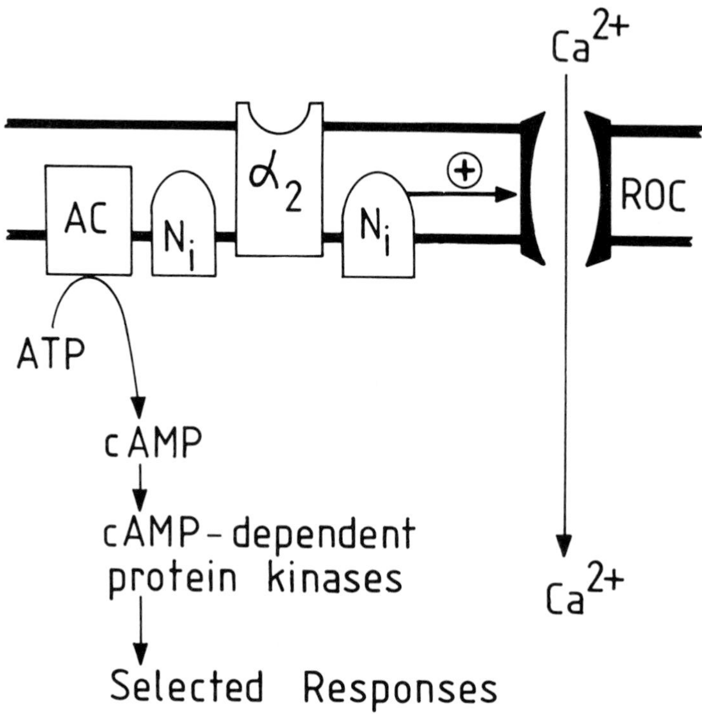

Fig. 3. Schematic overview of the signal transduction mechanisms to stimulation of postsynaptic α_2-adrenoceptors. For details see text.

CONCLUSIONS

Elevation of cytosolic free Ca^{2+} and decrease of cAMP are the best characterized signal transducing responses to activation of (stimulatory) α_1- and (inhibitory) α_2-adrenoceptors, respectively. The increase in cytosolic Ca^{2+} due to release of Ca^{2+} from extracellular stores and influx of Ca^{2+} results in the activation of Ca^{2+}/calmodulin-dependent protein kinases and the phosphorylation of key regulatory proteins. The role of protein kinase C is still ill-defined and many of the target proteins phosphorylated by this enzyme are not known. Similarly, the levels of cAMP regulate the activity of cAMP-dependent kinases which mediate the phosphorylation of target proteins. One of the challenges for future research is to evaluate how different receptor-generated second-messenger pathways that activate a variety of protein kinases and phosphatases interact with each other.

REFERENCES

Berridge MJ (1987). Inositol triphosphate and diacylgycerol: two interacting second messengers. Ann Rev Biochem 56: 169-193.

Boyer JL, Cardenas C, Posadas C, Garcia-Sainz JA (1983). Pertussis toxin induces tachycardia and impairs the increase in blood pressure produced by α_2-adrenergic agents. Life Sci 33: 2627-2633.

Bülbring E, Tomita T (1987). Catecholamine action on smooth muscle. Pharmacol Rev 39: 49-96.

Chiu AT, McCall DE, Thoolen MJMC, Timmermans PBMWM (1986). Ca^{++} Utilization in the constriction of rat aorta to full and partial alpha-1 adrenoceptor agonists. J Pharmacol Exp Ther 238: 224-231.

Chiu AT, Bozarth JM, Timmermans PBMWM (1987). Relationship between phosphatidylinositol turnover and Ca^{++} mobilization induced by alpha-1 adrenoceptor stimulation in the rat aorta. J Pharmacol Exp Ther 240: 123-127.

Clare KA, Scrutton MC, Thompson NT (1984). Effects of α_2-adrenoceptor agonists and of related compounds on aggregation of, and on adenylate cyclase activity in human platelets. Brit J Pharmacol 82: 467-476.

Exton JH (1981). Molecular Mechanisms involved in α-adrenergic responses. Mol Cell Endocrinol 23: 233-264.

Exton JH (1988). Mechanisms of action of calcium-mobilizing agonists: some variations on a young theme. FASEB J 2: 2670-2676.

Fredholm BB, Lindgren E (1988). Protein kinase C activation increases noradrenaline release from rat hippocampus and modifies the inhibitory effect of α_2-adrenoceptor and adenosine A_1-receptor agonists. Naunyn-Schmiedeberg's Arch Pharmacol 337: 477-483.

Gilman AG (1984). Guanine nucleotide-binding regulatory proteins and anal control of adenylate cyclase. J Clin Invest 73: 1-4.

Göthert M (1985). Role of autoreceptors in the function of the peripheral and central nervous system. Arzneim-Forsch (Drug Res) 35: 1909-1916.

Kikkawa U, Nishizuka Y (1986). The role of protein kinase in transmembrane signalling. Ann Rev Cell Biol 2: 149-178.

Langer SZ (1974). Commentary: Presynaptic regulation of catecholamine release. Biochem Pharmacol 23: 1793-1800.

Limbird LE (1988). Receptors linked to inhibition of adenylate cyclase: additional signaling mechanisms. FASEB J 2: 2686-2695.

McGrath JC (1982). Evidence for more than one type of postjunctional alpha-adrenoceptor. Biochem Pharmacol 31: 467-484.

Michell R (1975). Inositol phospholipids and cell surface receptor function. Biochim. Biophys. Acta 415: 81-147.

Miller RJ (1987). Multiple calcium channels and neuronal function. Science 235: 46-52.

Musgrave I, Marley P, Majewski H (1987). Pertussis toxin does not attenuate α_2-adrenoceptor mediated inhibition of noradrenaline release in mouse atria. Naunyn-Schmiedeberg's Arch Pharmacol 336: 280-286.

Putney JW (1987). Calcium-mobilizing receptors. Trends Pharmacol Sci 8: 481-486.

Schoffelmeer ANM, Wierenga EA, Mulder AH (1986). Role of adenylate cyclase in presynaptic α_2-adrenoceptor and μ-opioid mediated inhibition of ^{3}H-noradrenaline release from rat brain cortex slices. J Neurochem 46: 1711-1717.

Scholz J. Schaefer B, Schmitz W, Scholz H, Steinfath M, Lohse M, Schwabe U, Puurunen J (1988). Alpha-1 Adrenoceptor-mediated positive inotropic effect and inositol triphosphate increase in mammalian heart. J Pharmacol Exp Ther 245: 327-335.

Timmermans PBMWM, van Zwieten PA (1982). α_2-Adrenoceptors: Classification, localization, mechanisms and targets for drugs. J Med Chem 25: 1389-1401.

Timmermans PBMWM, Chiu AT, Thoolen MJMC (1987). Calcium handling in vasoconstriction to stimulation of alpha-1 and alpha-2 adrenoceptors. Can J Physiol Pharmacol 65: 1649-1657.

Tung LH, Rand MJ, Louis WJ (1985). Cardiac α-adrenoceptors involving positive chronotropic responses. J Cardiovasc Pharmacol 6: S121-S126.

Wilffert B (1986). Adrenoceptors in the heart. Progress Pharmacol 6: 47-64.

Calcium Channel Modulators in Heart and Smooth Muscle:
Basic Mechanisms and Pharmacological Aspects
S. Abraham and G. Amitai, editors
© 1990, VCH, Weinheim/Deerfield Beach, FL and Balaban, Rehovot/Philadelphia

Relationships between catecholamine secretion and distinct calcium fluxes in cultured medullary chromaffin cells

E. HELDMAN[1], R. ZIMLICHMAN[2], M. LEVINE[3], L. RAVEH[1], AND H.B. POLLARD[3]

[1]Israel Institute for Biological Research, Ness-Ziona, Israel; [2]Soroka Medical Center, Ben-Gurion University, Beer-Sheva, Israel; [3]Laboratory of Cell Biology and Genetics, NIDDK, NIH, Bethesda, MD, USA

INTRODUCTION

The role of calcium as a mediator between stimulus and secretion of hormones or neurotransmitters is one of the most pursued topics in cell biology today. Chromaffin cells from the adrenal medulla have been widely used to study some aspects concerning the role of calcium ions in stimulus-secretion coupling (for review see Pollard et al., 1985). Chromaffin cells in the adrenal medulla receive innervation from cholinergic endings of the splanchnic nerve and secrete catecholamines physiologically into the blood in response to receptor stimulation by acetylcholine (Coupland, 1965). It is now accepted that consequently to the stimulus, calcium ions enter into the cells and initiate a cascade of biochemical reactions which lead to the release of the vesicular content into the extracellular milieu (Baker and Knight, 1984). However, questions such as how calcium enter chromaffin cells, is extracellular calcium the sole source for its elevated intracellular level, how much of it is free, and over what time course it remains free, are as yet unanswered with certainty.

In order to answer some of these questions, we measured calcium influx, intracellular calcium levels, and catecholamine secretion in chromaffin cells which were stimulated by various secretagogues. Correlations among rate of calcium influx, intracellular free calcium, and catecholamine secretion suggest that calcium ions act at or near their site of entry to initiate the secretory process. In addition, our results suggest that calcium ions enter chromaffin cells via two separate routes (voltage-dependent and receptor-associated), each of which independently initiates a secretory response. However, a unidirectional communication, from the voltage-dependent to the receptor associated route, restricts the amounts of the catecholamines which are secreted during a physiological stimulation. Calcium ions are the basis of such biochemical communication between the otherwise two separate routes of calcium entry.

METHODS

Cell culture: Cultured chromaffin cells were prepared from bovine adrenal glands as previously described (Greenberg and Zinder, 1982). Dissociated bovine chromaffin cells were seeded for monolayer cultures in plates containing 24 wells (each well 1.8 cm in diameter) at a density of 10^6 cells per well, or for cell suspension in bacteriological flasks (250 ml) at a density of 5×10^7 cells per flask, and grown under 5% CO_2 atmosphere in Minimal Eagle's medium containing 5% fetal calf serum, 2.9 mM glutamine, 100 µg/ml streptomycin, 5 µg/ml gentamicin and 10 µg/ml cytosine arabinoside for 3-4 days before the experiment.

Calcium uptake: Plated cells were incubated in balanced salt solution containing 125 mM NaCl, 4.75 mM KCl, 1.4 mM $MgCl_2$, 2 mM $CaCl_2$, 25 mM 4-(2-hydroxyethyl)-1-piperazine-ethanesulfonic acid (HEPES) pH 7.35 and 10 mM glucose at 37°C for 15 min. The cells were then stimulated by replacing the incubation medium with fresh medium containing secretagogue and $^{45}CaCl_2$ (1-2 µCi). The stimulation was stopped by rapid suction of the labeled medium and addition of cold balanced salt solution containing 2 mM $LaCl_3$. The cells were then rapidly washed four times with lanthanum-containing medium and lysed in 2% acetic acid. Radioactivity was determined in the lysate.

Secretion of catecholamines: Cultured cells were washed and preincubated for 15 min at 37°C in a standard medium with or without inhibitors. After 15 min, the medium was replaced with a secretagogue-containing medium, and the cells were further incubated for 15 min. Media were then collected for the determination of released catecholamines.

The intact cells were lysed in 10% acetic acid and a freeze-thaw cycle for the determination of cellular content of catecholamines. Catecholamines were determined by the modified trihydroxyindole method (Kelner et al., 1985), using epinephrine as a standard.

Determination of intracellular calcium: Cells in suspension were collected from the tissue culture flask and washed with balanced salt solution. The cells were resuspended in balanced salt solution containing 20 µM quin 2/AM (Calbiochem), at a concentration of 4 x 10^6 cells/ml, and incubated in a light-protected water bath for 1h at 37°. At the end of the incubation, the cells were washed two times with balanced salt solution and resuspended at a concentration of 4 x 10^6 cells/ml. The cells were then transferred to a quartz cuvette, placed in a fluorometer for a measurement of fluorescence at 335/495 nm under constant stirring at 37°C. Secretagogues or inhibitors were added to the cuvette while measuring the fluorescence. Portions of the cells were incubated in parallel without quin 2 for the correction of light scattering and self fluorescence. At the end of the incubation the cells were lysed with 3% triton X-100 to determine the maximum fluorescence and then 25 mM Mg-EGTA was added to obtain minimum fluorescence. Intracellular calcium levels were calculated as previously described by Tsien et al. (1982).

RESULTS AND DISCUSSION

Kinetics of catecholamine secretion and calcium influx

Chromaffin cells from bovine adrenal medulla can be stimulated to secrete catecholamines either by activation of their nicotinic receptors with an agonist, or by inducing membrane depolarization. Figure 1 shows the kinetics of catecholamine secretion induced by three different secretagogues. As can be seen, activation of the nicotinic receptors by 62 µM nicotine induced prompt secretory response with a faster kinetic than that observed after depolarization by either 50 mM KCl or 20 µM veratridine. However, the amounts of the catecholamines which were secreted during stimulation with veratridine were much larger than those which were secreted during stimulation with nicotine or high K^+ (insert to Figure 1).

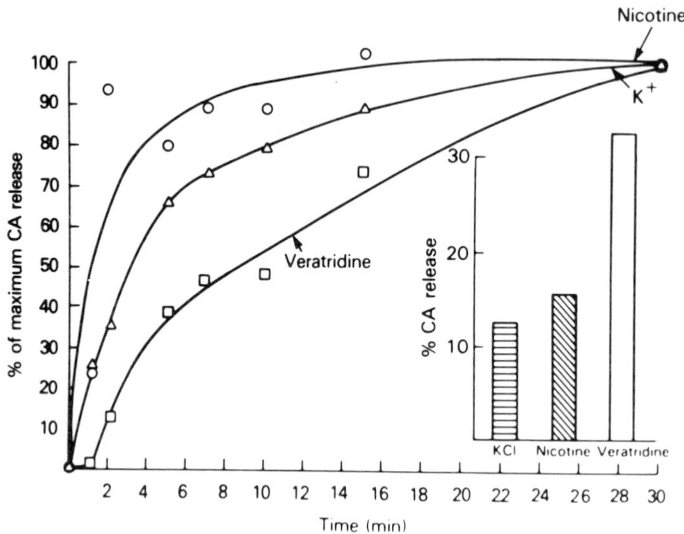

Figure 1. Kinetics of catecholamine secretion induced by various secretagogues. Percent release for each time point was normalized according to the 30 min time point. The insert represents percents of total catecholamines released during 15 min stimulation with each of the designated secretagogue.

All these secretagogues also induced calcium influx. As can be seen from Figure 2, the kinetics of calcium influx induced by the above mentioned secretagogues correlated with the corresponding kinetics of the catecholamine secretion, but not with the total amounts of the catecholamines which were secreted in response to each of the tested secretagogue.

These results could be interpreted either by assuming that each secretagogue induced a different rate of calcium influx via a different calcium channel, or by assuming that each secretagogue induced a different degree of membrane depolarization resulting in a different intracellular calcium level. To distinguish between these two possibilities, we measured calcium influx under conditions which may differentiate between different routes of calcium entry. In parallel, we measured the resulting intracellular calcium levels under the same conditions.

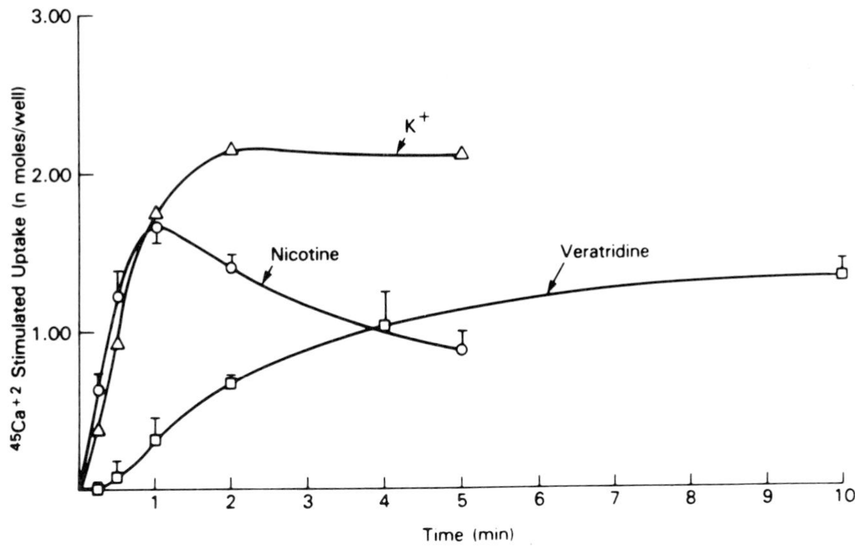

Figure 2. Kinetics of calcium influx induced by various secretagogues.

Evidence for two different calcium entry pathways in chromaffin cells

As could be seen from Figure 1, the rate of catecholamine secretion decreases with the time of stimulation with either high K^+ or nicotine. This decrease could be a result of a diminishing releasable pool of catecholamines, or alternatively a result of an inactivation of one or more processes within the cascade of events leading to exocytosis. The fact that veratridine (which depolarizes the cell membrane by allowing Na^+ to enter the cell) induced release of significantly larger amounts of catecholamines than high K^+ (which also depolarizes the cell membrane but via a different mechanism), suggests that the releasable pool is not diminishing with time, but rather an inactivation mechanism is responsible for the decreased rate of catecholamine secretion. Since the first event in the cascade of reactions that lead to exocytosis is calcium entry, we examined if this process is inactivated with time during cell stimulation. Figure 3 shows that calcium influx induced either by high K^+ or nicotine was inactivated during the second of two repetitive homologous stimuli.

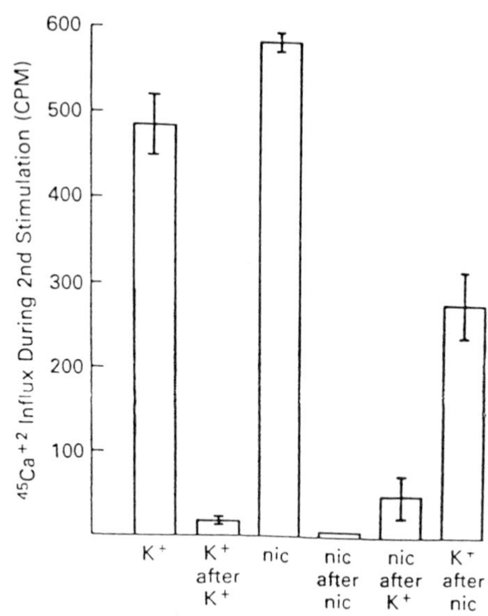

Figure 3. Calcium influx during two repetitive challenges by either 50 mM KCl or nicotine.

However, if nicotine stimulation preceded the stimulation with high K^+, the second pulse of the high K^+-induced calcium influx was only partially inactivated. These results suggested that calcium ions may enter chromaffin cells through two different pathways, and that the voltage-dependent pathway may be inactivated independently of the nicotinic receptor-associated pathway. Consistent with these interpretation, secretion of catecholamines was decreased dramatically during the second of two repetitive homologous stimulations, but remained similar to control level during a second stimulation with high K^+ which followed a first stimulation by nicotine (Figure 4). These results suggested that inactivation of the calcium channels is responsible for the decreasing rate of catecholamine secretion, and that each pathway of calcium entry may be activated and initiate secretion independently.

In search for a tool which may differentiate between the two pathways of calcium entry, we found that high osmotic strength, adjusted by either NaCl or sucrose inhibited the high K^+-induced calcium influx but had no effect on nicotine-induced calcium influx (Figure 5). These results strongly supported our hypothesis that in chromaffin cells there are at least two distinct pathways for calcium entry.

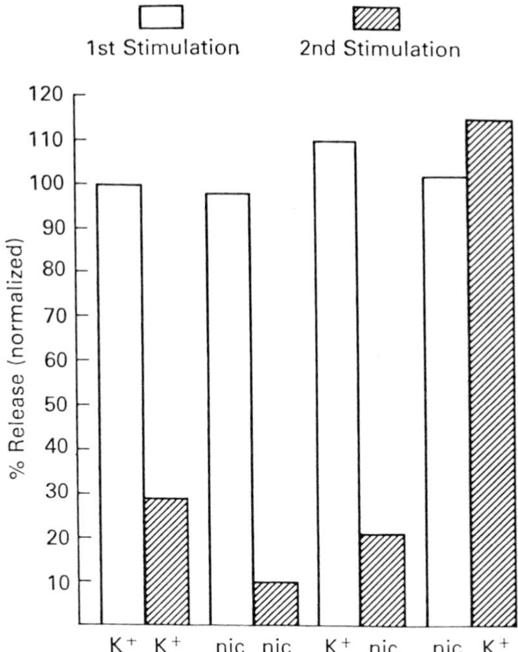

Figure 4. Catecholamine secretion during two repetitive challenges by either 50 mM KCl or nicotine.

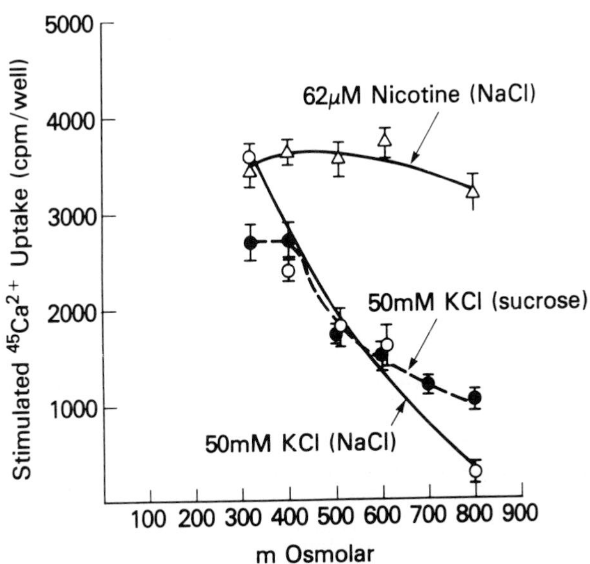

Figure 5. Effect of different osmolarities, adjusted with either NaCl or sucrose, on calcium influx induced by either 50 mM KCl or 62 μM nicotine.

Another evidence for the concept that calcium ions enter chromaffin cells via two different pathways came from our experiments with barium. Barium is a potent secretagogue which may induce secretion of approximately 70% of the total catecholamines in the absence of calcium (Heldman et al., 1989). We found previously that barium and calcium acts intracellularly at different sites. Therefore, calcium-dependent and barium-dependent catecholamine secretion should be additive, unless if calcium and barium compete at a common site for entering chromaffin cells. Here we demonstrate that nicotine-induced catecholamine secretion is not affected by barium ions (Figure 6), whereas high K^+-induced catecholamine secretion is significantly inhibited by 1 and 3 mM Ba^{2+} (Figure 7).

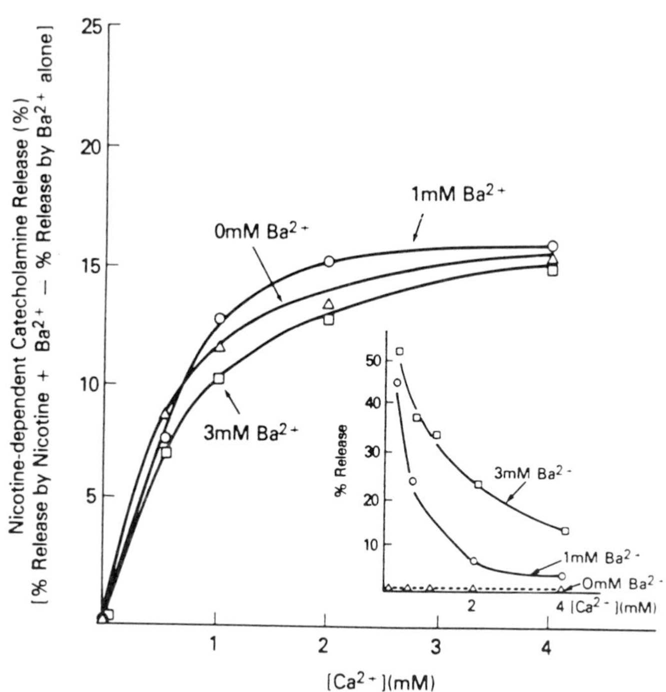

Figure 6. Nicotine-induced secretion in the presence of various concentrations of barium and calcium. The insert demonstrates secretion by barium alone in the presence of various calcium concentrations.

As can also be seen from Figure 7, calcium ions attenuated the inhibition of the high K^+-induced secretion which was induced by barium. These results suggest that barium and calcium compete at the voltage-dependent calcium channel but not at the nicotinic receptor-associated calcium channel, adding to the wealth of evidence that calcium ions use two different pathways to enter through the membrane of chromaffin cells.

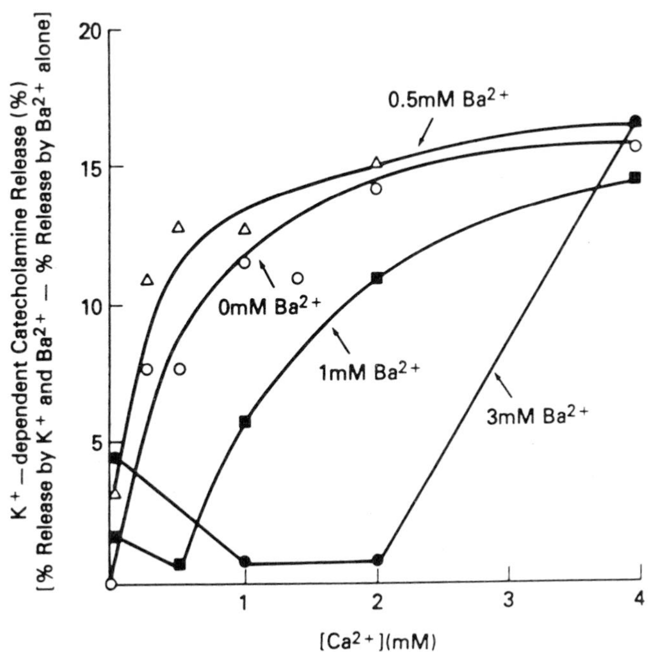

Figure 7. Catecholamine secretion induced by 50 mM KCl in the presence of various concentrations of barium and calcium.

Intracellular calcium levels after stimulation with various secretagogues and their relevance to catecholamine secretion

Nicotine induced prompt elevation in intracellular free calcium followed by a quick decline back to the control level (Figure 8). High K^+ also induced a quick rise in intracellular free calcium. However, the decline period was longer than that observed after stimulation with nicotine (Figure 8). Veratridine caused slow elevation in the level of intracellular free calcium with a lower peak value than those obtained after stimulation with nicotine or high K^+. The decline period resembled to that observed after stimulation with high K^+ (Figure 8).

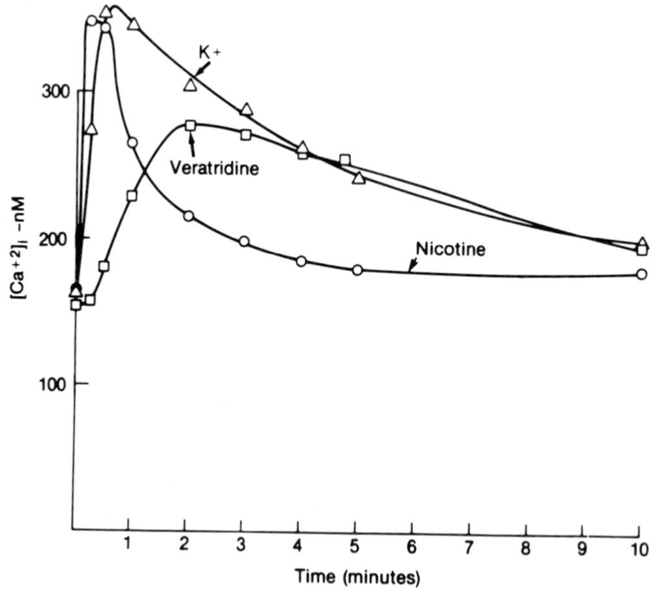

Figure 8. Kinetics of changes in intracellular calcium levels ($[Ca^{2+}]i$), measured with quin 2, during stimulation with various secretagogues.

Since veratridine, in comparison to nicotine or high K^+, induces secretion of larger amounts of catecholamines but lower levels of intracellular free calcium, it did not seem to us that the level of intracellular free calcium is directly related to the amount of catecholamine secretion. Indeed, the calcium ionophore A23187 (0.5 µM) caused marked elevation in intracellular calcium but it did not induce significant catecholamine secretion (data not shown).

As extracellular calcium and calcium influx are mandatory for obtaining catecholamine secretion from cultured chromaffin cells (Kirshner et al., 1982; Holz et al., 1982), it should be assumed that calcium ions which enter the cell during stimulation affect some intracellular processes associated with the secretory response. To initiate calcium-dependent processes, threshold levels of calcium ions are required. Therefore, accumulation of the entering calcium ions at some intracellular site should play some role in the initiation of the secretory process.

Indeed, we found that if calcium influx is abruptly blocked by adding calcium channel blockers such as verapamil, secretion is also blocked abruptly. However, when we measured intracellular calcium levels after nicotine or high K^+ stimulations and consecutive addition of verapamil (500 μM), we found that although catecholamine secretion stopped abruptly upon addition of the calcium channel blocker, intracellular calcium levels were not dropped so rapidly (Figure 9). Moreover, in the absence of extracellular calcium, high K^+, but not nicotine, evoked rise in intracellular free calcium (data not shown).

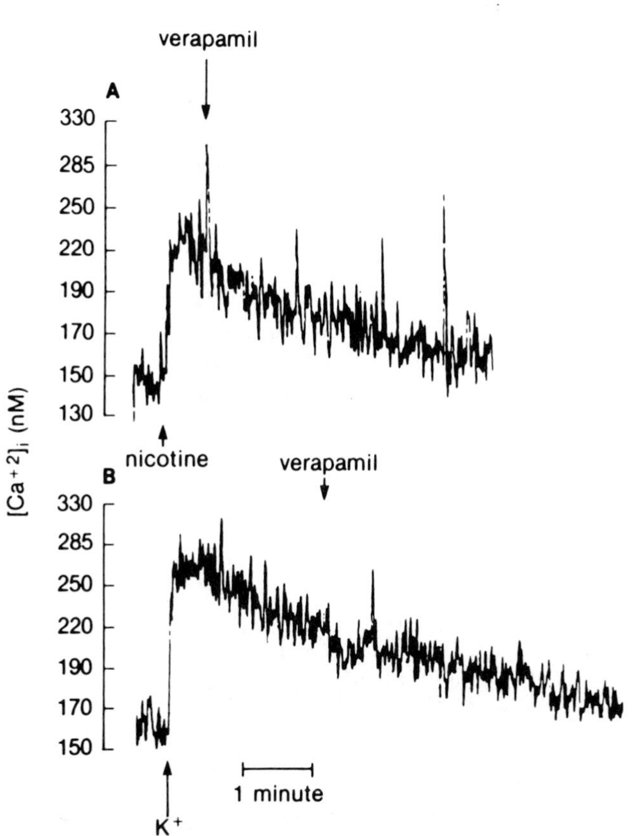

Figure 9. Changes in intracellular free calcium ($[Ca^{2+}]i$) during stimulation with nicotine or high K^+ and consecutive addition of 500 uM verapamil.

These results suggested that during depolarization free calcium may be recruited from intracellular stores. However, under these conditions catecholamines were not secreted. Our interpretation of these results is that calcium influx, rather than elevation in the level of intracellular free calcium, is required to initiate secretion. Indeed, in presence of 2 mM lanthanum (which blocks calcium influx), high K^+ evoked significant rise in intracellular free calcium (Figure 10) and yet catecholamine secretion was inhibited.

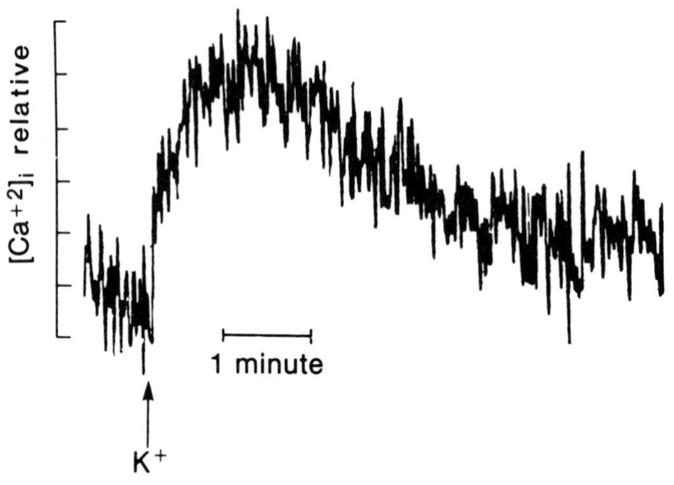

Figure 10. Intracellular free calcium ($[Ca^{2+}]i$) during stimulation with high K^+ in presence of 2 mM $LaCl_3$.

These results support the notion that elevation of intracellular calcium alone is not enough to initiate a secretory response. This hypothesis can be examined quantitatively by comparing the rate of catecholamine secretion, with either measured changes in intracellular free calcium or measured rate of calcium influx. As shown in Figure 11, the best correlation was found between the extent of catecholamine secretion and calcium influx at any given time point, or with any secretagogue. This analysis explains why veratridine induces such an effective secretion, even though the peak level of the intracellular free calcium concentration obtained after veratridine stimulation is relatively low. Veratridine causes a slow but continuous influx of calcium. According to our hypothesis, the cell should thus secrete slowly and continuously, as it indeed does.

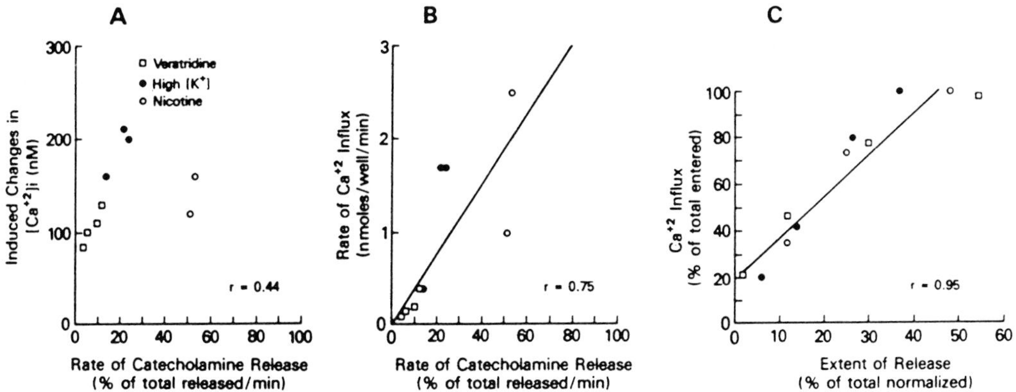

Figure 11. Correlations among rates or extents of catecholamine secretion and changes in intracellular free calcium ($[Ca^{2+}]i$) or rates of calcium influx.

The concept that catecholamine secretion is related to calcium influx, but not to intracellular level of free calcium, suggests that a site on or near the calcium channel must be occupied by calcium in order for optimal release to occur. This interpretation also explains why high K^+-induced catecholamine secretion ceases even at a time when cytosolic calcium levels are still high. It is known that voltage-dependent calcium channels are inactivated a short time after stimulation with high K^+ causing calcium influx to cease. When calcium influx ceases, the high local calcium concentration in the immediate vicinity of the channel would not be long maintained and therefore release would become very slow.

Role of calcium ions in the inactivation of calcium channels

Inactivation of the voltage-dependent calcium channels may possibly be mediated by transported calcium, by analogy to the calcium-dependent K^+ channel in other systems (Sakakibara et al., 1986). Evidence for this is shown in Figure 12, where La^{3+}, a blocker of calcium transport, prevents the voltage-dependent calcium channel from time-dependent inactivation.

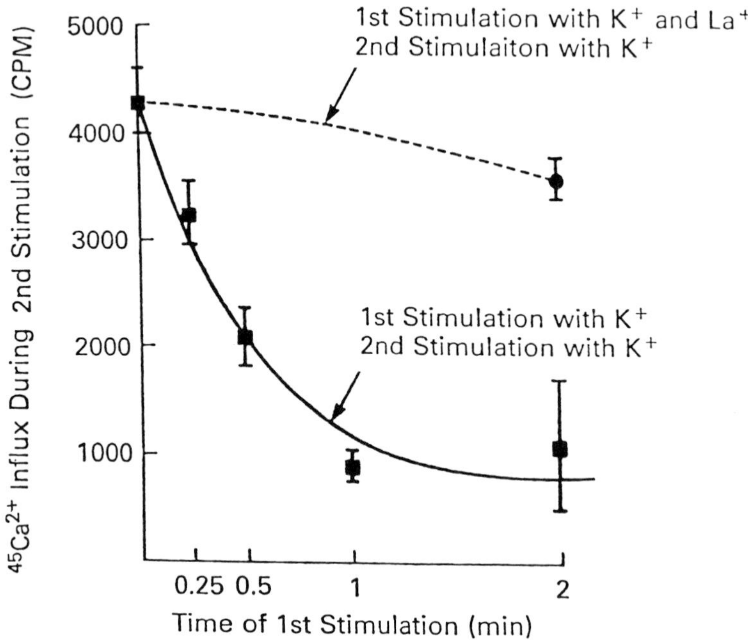

Figure 12. The effect of 2 mM lanthanum on depolarization-induced inactivation of calcium influx.

As expected, a similar protection by La^{3+} or by omission of calcium was also manifest for secretion in response to sequential challenges with high K^+ (Figure 13). As further summarized in Figure 13, when sequential stimulation with nicotine was tested in the presence of calcium, less than 10% of the original response was obtained during the second nicotine stimulation. However, omission of calcium during the primary challenge with nicotine allowed preservation of more than 30% of the response to the second nicotine stimulus (Figure 13). We presume that the fraction of the response lost may have been due to receptor desensitization, which is not calcium-dependent (Sasakawa et al., 1985), but that the remaining activity may have been a consequence of protecting the nicotinic-specific channel from calcium.

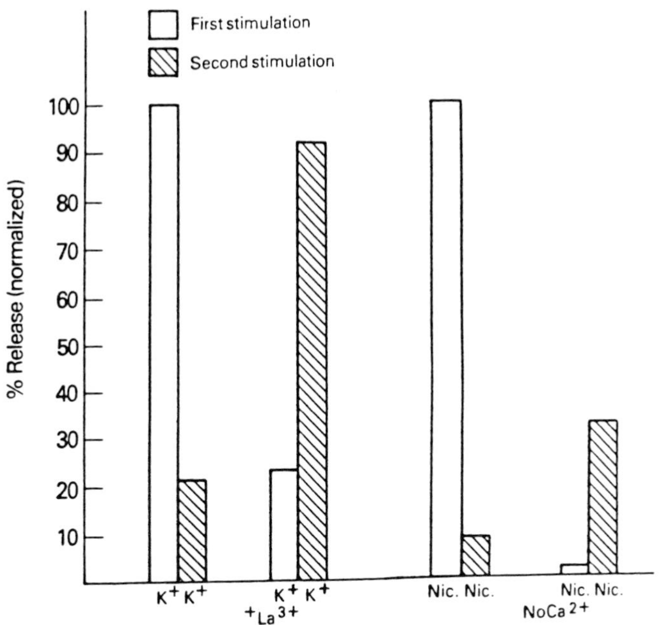

Figure 13: Catecholamine secretion during two repetitive stimulations by either 50 mM KCl or nicotine in presence of 2 mM lanthanum or absence of calcium during the first stimulation.

Inactivation of calcium influx may represent a physiological mechanism to control the rate of secretion. As was mentioned earlier, we observed that a primary challenge with high K^+ blocks a subsequent catecholamine secretion induced by nicotine (Fig 4). However, the reverse is not true. Thus, primary nicotine stimulation does not block subsequent catecholamine secretion induced by high K^+ (Fig 4). These findings could indicate a unidirectional communication between the two separate calcium channels: from the voltage-dependent channel to the receptor-associated channel. This could be related to the ability of chromaffin cells to secrete widely varying amounts of catecholamines (Khalil et al., 1986). For example, under resting conditions, catecholamine secretion is small and is evoked by periodic release of acetylcholine from the splanchnic nerve (Edwards 1982). However, we might anticipate that when the animal is stressed and homeostasis is perturbed, the central nervous system could stimulate the splanchnic nerve more frequently and produce greater depolarization of the chromaffin cell membrane. Increased depolarization would in turn open the voltage-dependent calcium channels, thus promoting additional catecholamine secretion. To permit additional secretion upon depolarization, calcium entry which occurs during activation of the receptor-associated calcium channels must not inactivate the voltage-dependent calcium channels. However, in order to prevent excess amounts of catecholamines from being released, the calcium influx occurring during depolarization might temporarily inactivate both channels. Indeed, our findings are consistent with such a hypothetical mechanism.

SUMMARY

We demonstrated here that calcium influx which occurs via two different channels, and not the rise in intracellular free calcium, initiates catecholamine secretion. The two pathways are: the receptor-associated and the voltage-dependent calcium channels. These two channels are mechanistically distinct but functionally dependent on each other. Our findings suggest that the functional communication between the two separate calcium channels is mediated biochemically by calcium ions which are transported during cell stimulation. This biochemical communication is unidirectional, from the voltage-dependent to the receptor-associated calcium channels. The basis for such unidirectional communication may be the recruitment of calcium from intracellular stores which occurs only during depolarization but not as a result of the activation of the nicotinic receptors.

REFERENCES

Baker, P.F. and Knight, D.E. (1984) TINS 7 (4), 120-126

Coupland, R.E. (1965) J. Anat. 99, 231-254

Edwards, A.V. (1982) J. Physiol. (London) 327, 409-419

Greenberg, A. and Zinder, O. (1982) Cell Tissue Res. 226, 655-665

Heldman, E., Levine, M., Raveh, L. and Pollard, H.B. (1989) J. Biol. Chem. 264 (14), 7914-7920

Kelner,K.L., Levine R.A., Morita, K. and Pollard, H.B. (1985) Neurochem Int. 7, 373-378

Khalil, Z., Marley, P.D. and Livett, B.G. (1986) Endocrinol. 119, 159-167

Pollard, H.B., Orenberg, R., Levine, M., Kelner, K. Morita, K., Levine, R., Forsberg, E., Brocklehurst, K.W., Doung, L., Lelkes, P.I., Heldman, E. and Youdim, M. (1985) in Vitamins and hormones 42 (Aurbach, G.D. and McCormick, D.B. eds), pp. 109-196, Academic Press Inc. Orlando

Sakakibara, M., Alkon, D.L., Delorenzo, R., Boldering, J.R., Neary, J.T. and Heldman, E. (1986) Biophys J. 50, 319-327

Sasakawa, N., Ishii, K. and Kato, R. (1985) Biochem. Biophys. Res. Comm. 133, 147-153

Tsien, R.Y., Pozzan, T. and Rink, T.J. (1982) J. Cell Biol. 94, 325-334

Calcium Channel Modulators in Heart and Smooth Muscle: Basic Mechanisms and Pharmacological Aspects
S. Abraham and G. Amitai, editors
© 1990, VCH, Weinheim/Deerfield Beach, FL and Balaban, Rehovot/Philadelphia

Angiotensin II stimulates catecholamine release in cultured bovine adrenomedullary cells through a cytosolic calcium dependent mechanism

REUVEN ZIMLICHMAN[1], DAVID S. GOLDSTEIN[2], SHULAMIT ZIMLICHMAN[1], ROBIN STULL[2] AND HARRY R. KEISER[2]

[1]*Hypertension Unit and Department of Medicine, Soroka Medical Center and Faculty of Health Sciences, Ben Gurion University, P.O. Box 151, Beer Sheva 84105, Israel, and* [2]*NIH, NHLBI, Bethesda, MD, USA*

INTRODUCTION

Several sites of interaction between the sympathomedullary and renin-angiotensin-aldosterone systems have been proposed. One of these is the adrenal medulla, which contains a large concentrations of receptors for angiotensin II (Healy and Printz, 1984; Israel et al., 1985). Administration of angiotensin II (AII) into laboratory animals elicits increases in plasma levels of epinephrine, the main catecholamine secreted by the adrenal medulla (Rowe and Nasjlettia, 1981; Carroll and Opdyke, 1982). Perfusion of adrenal glands with AII also evokes catecholamine release (Peach, 1971; Feldberg and Lewis, 1965). In vivo end organ perfusion studies cannot separate easily direct from indirect effects of AII on adrenomedullary secretion. The possible direct action of AII on adrenomedullary cells has not been explored in depth. Since AII induces increases in cytosolic ionized calcium ([Ca++]), in smooth muscle cells (Nabika et al., 1985; Alexander et al. 1985), and since catecholamine release from chromaffin cells is calcium dependent (Poisner and Douglas, 1966), we hypothesized that AII would directly increase [Ca++] in cultured adrenomedullary cells.

To ensure that observed changes in [Ca++] were relevant to cellular activation, we also determined whether AII stimulates release of the catecholamines dopamine, norepinephrine, and epinephrine from cultured adrenomedullary cells. We compared the potency of AII and other agonists (nicotine and KCl) in terms of their effects on [Ca++], and we used calcium-free medium and calcium channel blockade to determine whether stimulatory effects of AII depend on transmembrane influx of calcium.

METHODS

Chromaffin cell culture: Primary cell cultures of adrenomedullary cells were prepared as previously described (Levine et al., 1983; Knight and Baker, 1983). Briefly, fresh bovine adrenal glands were perfused three times with balanced salt solution (BSS), then the glands were perfused with 0.2% collagenase (Sigma), incubated for 10 minutes at 37°C, and the procedure repeated three times. The adrenal medullary tissue was excised, minced and filtered through a nylon mesh (Tetco HC-3 160), and shaken with fresh 0.2% collagenase in BSS for 30 minutes at 37°C. The dissociated cells were filtered, washed, and plated either in 250 ml flasks or in 1.8 cm diameter wells in 24 well plates and cultured in minimal Eagle's medium (Gibco) containing 5% fetal calf serum, streptomycin (100 ug/ml), gentamycin (5 ug/ml), cytosine arabinoside (10 ug/ml), and glutamine (2.93 mM).

Catecholamine release: After culture of three days the chromaffin cells were washed with BSS and then incubated in BSS with or without secretagogue. After various incubation periods, the medium was aspirated and filtered using a 0.22 micron Milex (Millipore) filter and stored at -70°C untill assayed for catecholamine content. To measure the catecholamine contents of the cells, the cells were lysed with acetic acid (2%), frozen, and thawed and the lysate filtered through a 0.22 millex filter. Catecholamine contents of the incubation medium and the cellular lysates were assayed using liquid chromatography with electrochemical detection (Goldstein, et al., 1981). The percent release of each of the catecholamines -norepinephrine (NE), epinephrine (E), and dopamine (DA) was calculated from the total NE, E, or DA content of the medium divided by the sum of the content of the medium and the content of the lysate, multiplied by 100.

Determination of cytosolic calcium concentration: Cells were collected from the tissue culture flasks by scraping with a rubber policeman. The cells were then diluted with BSS to a concentration of 4 million cells/ml, washed with BSS (or in some experiments calcium free BSS), 20 uM Quin 2/AM (Calbiochem) was added, and the mixture incubated and rotated for 60 minutes. Quin 2/AM was dissolved in DMSO, the concentration of which was less than 0.1%. After the incubation the cells were centrifuged, washed twice, and resuspended in BSS. The cells were transferred to a quarz cuvette and tested in a thermostatted spectrofluorimeter (Perkin Elmer, LS5) at 37oC during continuous magnetic stirring. Fluorescence was measured using an excitation wave length of 335nm and emission wavelength of 492 nm. Secretagogues or inhibitors were added to the cuvette duting continuous measurements.

Samples of the cells at the same concentration as the test samples were incubated in parallel in DMSO without Quin 2/AM in order to correct the fluorescence of the test samples for light scattering and autofluorescence. In each experiment 50 uM MnCl2 was added to correct for leakage of Quin 2: this leakage was negligible. The cells were lysed with Triton X-100 to determine the maximum fluorescence, and 25 mM Mg-EGTA was then added to determine the minimum fluorescence. Cytosolic calcium concentrations were calculated as previously described (Tsien et al., 1982).

Responses to AII and other agonists: Nicotine, AII, and KCl, were added sequentially to the cuvettes and the effects on Quin 2 fluorescence measured continuously. The doses of nicotine, KCl, and AII were found during pilot studies to elicit maximal increase in [Ca++]. The fluorescence was allowed to return to baseline before addition of the next agonist. KCl was usually the last agonist added, because it induced the longest effect. Additions of the agonists in different sequences did not affect any aspects of the response, except when inactivation of receptors was induced.

In experiments involving measurement of catecholamine secretion, the incubation medium was removed and replaced with medium containing secretagogue. After exposure of the cells to secretagogue for 15 minutes, the medium was aspirated, filtered, frozen, and stored as described above. In some experiments the cells were washed and then incubated in calcium free medium over a 15 minute period before the addition of the agonists, or else the cells were pretreated for 5 minutes with the calcium channel blockers nifedipine or verapamil.

Statistics: Data analysis included dependent-means t-tests. AII mean values were expressed +\- 1 standard deviation in tables and +\- 1 standard error of the mean in figures.

RESULTS

AII caused dose related increases in catecholamine secretion from the cultured adrenomedullary cells (Table 1). The maximal secretory effect was similar to that associated with addition of KCl but smaller than that associated with addition of nicotine (Figure 1), despite similar peak [Ca++] responses to the three agonists (Table 2). All the agonists stimulated approximately

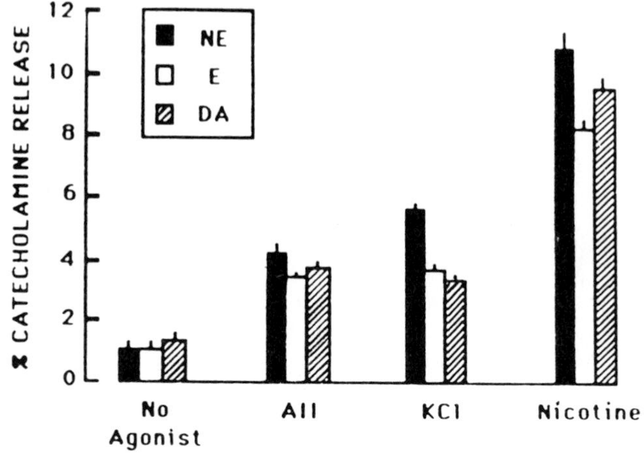

Figure 1. Effects of Angiotensin II (AII), excess extracellular potassium (KCl), or nicotine on release of norepinephrine (NE), epinephrine (E), or dopamine (DA) from cultured bovine adrenomedullary cells. Each mean value is for 14 observations.

proportionate release of norepinephrine, epinephrine, and dopamine. Addition of AII to adrenomedullary cells virtually immediately elicited increases in [Ca++] (Figure 2). This effect lasted several minutes, and then [Ca++] gradually returned to baseline.

The rate of increase in [Ca++] in response to AII was somewhat slower than that observed in response to

Table 1

Effects of Angiotensin II on Catecholamine Release in Cultured Adrenomedullary Cells

	Percent Release (+/- 1 SD) in 15' Incubation		
Dose (M)	**NE**	**E**	**DA**
0	1.1 +/- 0.2	1.1 +/- 0.2	1.4 +/- 0.2
10^{-9}	1.3 +/- 0.6	1.0 +/- 0.5	1.3 +/- 0.6
10^{-7}	3.0 +/- 0.6	2.3 +/- 0.7	1.9 +/- 0.5
10^{-6}	4.3 +/- 0.8	3.7 +/- 0.8	3.9 +/- 1.1
10^{-5}	4.5 +/- 0.5	3.6 +/- 0.6	3.7 +/- 0.7

nicotine or KCl. The subsequent decline in [Ca++] was faster than with KCl but slower than with nicotine. Generally the shape of the response to AII was dome shaped. The peak levels of [Ca++] after AII were similar to those resulting from addition of nicotine or KCl (Table 1). The magnitude of the [Ca++] response to nicotine, compared with the responses to other agonists, seemed to vary from batch to batch of collagenase.

AII caused dose related increases in [Ca++] (Figure

Table 2

Kinetics of Cytosolic Calcium Responses to Angiotensin II, Nicotine, or Potassium in Cultured Adrenomedullary Cells

Condition	Peak Value $[Ca^{++}]i$ (nM)	Time to Peak (sec)	Time to BL (min)
Resting Cells (N=34)	143 +/- 22	---	---
Angiotensin II (1µM, N=17)	313 +/- 43*	20-30	3-7
Nicotine (62 µM, N=14)	321 +/- 41*	10-20	2-4
KCl (50 mM, N=17)	316 +/- 17*	15-25	8-16

Notes: (*) denotes significant difference from resting, p<0.001.
N = number of observations.

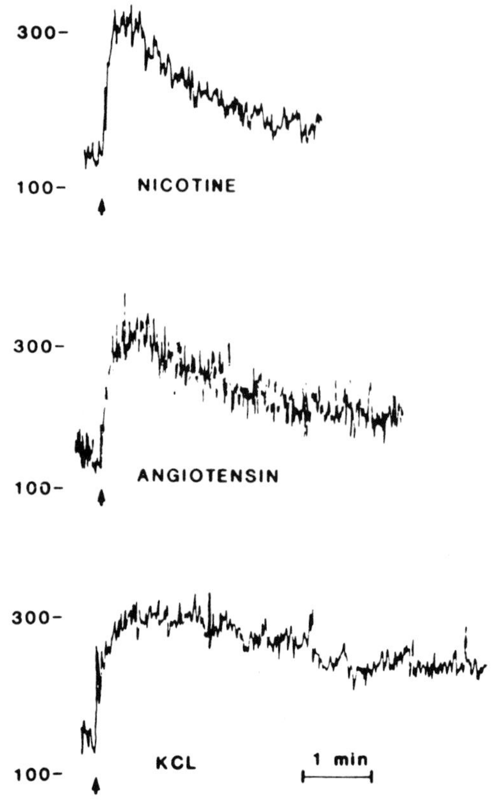

Figure 2. Original tracings for Quin 2 fluorescence upon addition of nicotine, angiotensin II, or KCl to cultured adrenomedullary cells. (Numbers represent calculated cytosolic calcium concentration in nM). All three agonists produced rapid increases in cytosolic calcium, although the pattern of responses differed among the agonists.

3). The minimum AII concentration producing detectable increments in [Ca++] was approximately 10^{-9}M, and the maximum response occured at about 0.5×10^{-6}M.

Addition of nicotine did not affect [Ca++] reponses to AII or to KCl but abolished responses to subsequent addition of nicotine (Figure 4). The inhibition lasted for about 10 minutes, whereas [Ca++] returned to baseline by 4 to 5 minutes after nicotine. Addition of KCl or nicotine did not affect responses to subsequent addition of AII. In contrast, AII attenuated responses to subsequent addition of AII, the attenuation lasting 4-5 mins. AII pre-treatment did not affect responses to KCl or nicotine.

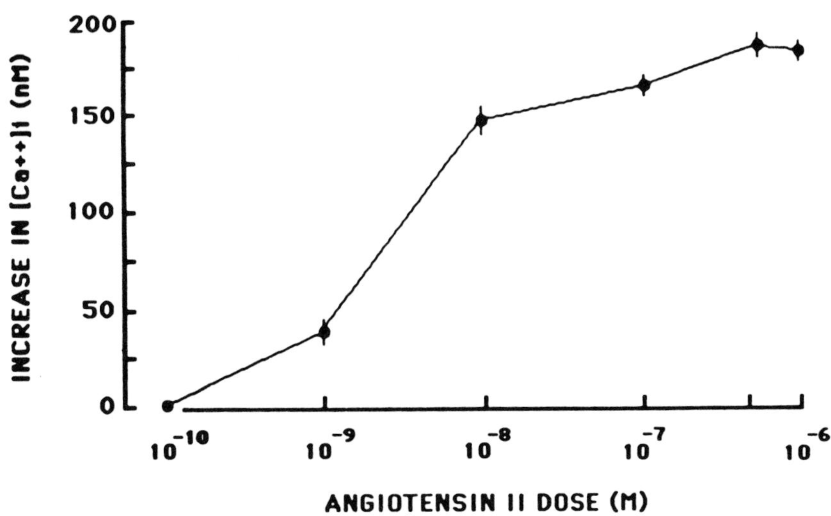

Figure 3. Dose-response relationship for the effect of angiotensin II on cytosolic calcium in cultured bovine adrenomedullary cells. (Each data point is the mean of at least 5 observations. Vertical lines represent 1 standard deviation).

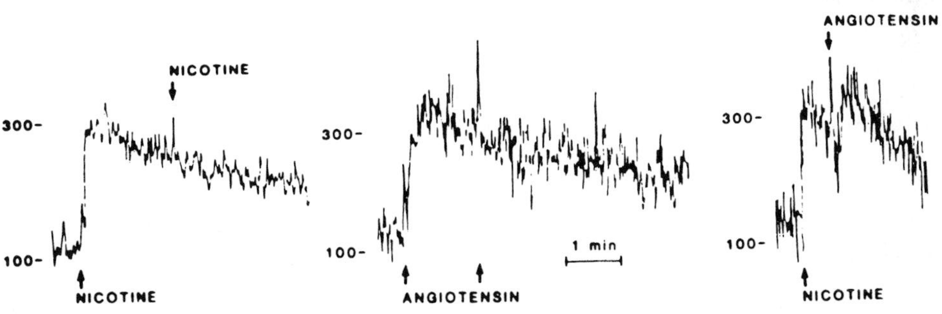

Figure 4. Effects of sequential addition of nicotine or angiotensin II on Quin 2 fluorescence in cultured adrenomedullary cells. (Numbers represent calculated cytosolic calcium concentrations in nM).

Cells suspended in calcium-free medium had markedly attenuated but detectable increases in [Ca++] in response to AII (Figure 5). Addition of verapamil to the cell suspension attenuated responses of [Ca++] to AII as well as to nicotine or KCl (Figure 6). Verapamil at a concentration of 10^{-4}M attenuated and at 10^{-3}M abolished responses of [Ca++] to AII. Similar results were obtained for nifedipine.

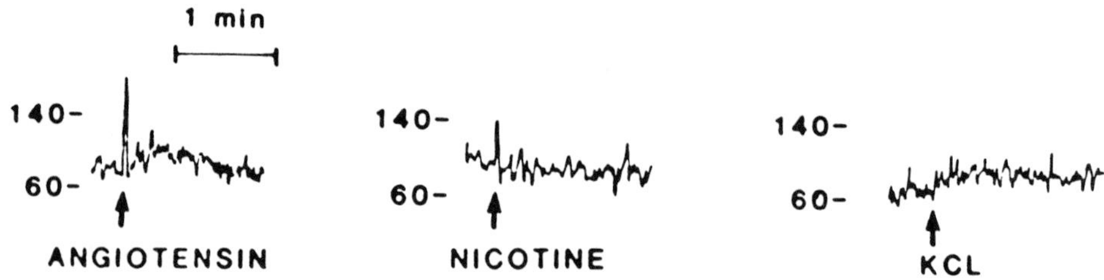

Figure 5. Effects of addition of agonists (angiotensin II, nicotine and potassium chloride) on Quin 2 fluorescence in cultured adrenomedullary cells washed in calcium free medium for 15 mins. (Numbers represent calculated cytosolic calcium concentration in nM).

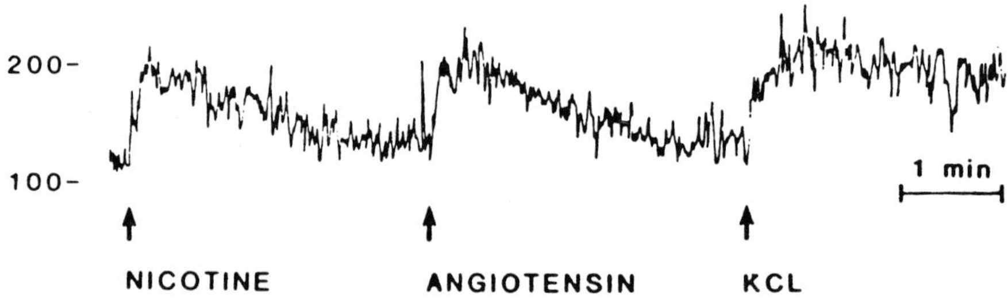

Figure 6. Effects of addition of agonists on Quin 2 fluorescence in cultured adrenomedullary cells incubated in Verapamil for 5 mins. (Numbers represent calculated cytosolic calcium concentration in nM).

DISCUSSION

The results show that AII stimulates adrenomedullary cellular secretion of catecholamines and increases the cytosolic concentration of ionized calcium in cultured adrenomedullary cells. The stimulatory effect of AII on adrenomedullary cellular release of catecholamines generally confirms the in vivo and perfusion studies of others (Rowe and Nasjlettia, 1981; Carroll and Opdyke, 1982; Peach, 1971; Feldberg and Lewis, 1965).

AII elicited Approximately proportional and dose related release of the endogenous catecholamines, norepinephrine, epinephrine, and dopamine. The maximal secretory response to AII was similar in magnitude to that resulting from addition of KCl but smaller than that resulting from nicotine, despite similar increments in [Ca++] in response to these agonists. The results are consistent with an indirect relationship between increments in cytosolic calcium and catecholamine release in adrenomedullary cells. Since at the lowest AII dose, 10^{-9}M, [Ca++] increased but catecholamine release did not, it appears that increments in [Ca++] caused by AII must exceed a threshold amount before cellular secretion is stimulated.

Washing the cells with calcium-free medium decreased levels of [Ca++] and markedly attenuated-but did not abolish-[Ca++] responses to AII. These results indicate that transmembrane movement of calcium ions is important for maintaining basal levels of [Ca++] in adrenomedullary cells and that [Ca++] responses to AII depend predominantly on transmembrane movement of calcium into the cells but also to a minor extent on release of calcium into the cytoplasm from intracellular stores. These findings generally confirm those of Poisner et al (Poisner and Douglas, 1966), who reported that perfusion of cat adrenal glands with calcium-free solution markedly attenuated but did not abolish AII-induced release of catecholamines from the glands. In adrenal glomerulosa cells, AII appears to be able to release calcium from an extra-mitochondrial, intracellular pool (Balla et al., 1985). Since calcium channel blockade with verapamil or nifedipine inhibited [Ca++] responses to AII, it appears that AII can increase [Ca++] in adrenomedullary cells by stimulating transmembrane movement of calcium via calcium specific channels.

Temporary inactivation of cholinergic receptors by stimulating with nicotine did not affect [Ca++] responses to AII. This indicates that the effect of AII on [Ca++] is independent of the cholinergic receptor operated mechanism. Consistent with this suggestion, it was reported (Douglas et al., 1967) that hexamethonium completely blocked nicotine induced depolarization of isolated adrenal chromaffin cells, whereas the depolarization response to AII was unaffected; it was also reported (Feldberg and Lewis, 1965) that hexamethonium prevented acetylcholine-induced release of catecholamines from perfused adrenal glands without affecting responses to AII.

In the present study, the minimum concentration of AII which stimulated catecholamine secretion by cultured adrenomedullary cells was about 10^{-9}M. This is at about two orders of magnitude higher than AII concentrations in antecubital venous blood of humans (Nussberger et al., 1985); the meaning of the present results for normal physiology is therefore unclear. AII in the systemic circulation, however, may not be the only source for AII-induced facilitation of adrenomedullary secretion of catecholamines. It is possible that in vivo concentrations of AII in the adrenal medulla may be higher than in arterial blood, for the following reason. The blood supply to the adrenal gland flows in a corticomedullary direction. Adrenal medullary tissue is therefore exposed to high concentrations of biochemicals released from adrenocortical cells which can affect adrenomedullary catecholamine synthesis (Wurtman and Axelrod, 1966). The work of Mulrow's group has provided evidence for high concentrations of renin in the adrenal cortex which may act locally to generate angiotensin and stimulate aldosterone release (Doi et al., 1984). If the adrenal cortex has the ability to synthesize AII locally, then the adrenomedullary cells may be exposed to much higher concentrations of AII than circulate systemically.

The stimulatory effect of AII on [Ca++] may not be specific to adrenomedullary cells. A calcium-mediated mechanism also appears to participate in AII-stimulated aldosterone secretion by adrenocortical cells (Capponi et al., 1984) and in activation of vascular smooth muscle cells (Nabika et al., 1985; Alexander et al., 1985).

In summary, we report here that incubation of

cultured adrenomedullary cells with AII elicits increases in cytosolic calcium and stimulates catecholamine secretion. The adrenal medulla may be a locus of interaction between two major systems controlling the circulation, the renin-angiotensin system and the sympathoadrenomedullary system.

REFERENCES

Alexander RW, Brock TA, Gimbrone MA, Rittenhouse SE (1985). Angiotensin increases inositol triphosphate and calcium in vascular smooth muscle. Hypertension 7:447-451.

Balla T, Szebeny M, Kanyar B, Spat A (1985). Angiotensin II and FCCP mobilizes calcium from different intracellular pools in adrenal glomerulosa cells; analysis of calcium fluxes. Cell Calcium 6, 327-342.

Capponi AM, Lew PD, Jornot L, Valloton MB (1984). Correlation between cytosolic free calcium and aldosterone production in bovine adrenal glomerulosa cells. Evidence for a difference in the mode of action of angiotensin II and potassium. J Biol Chem 259:8863-8869.

Carrol RG, Opdyke DF (1982). Evolution of angiotensin II-induced catecholamine release. Am J Physiol 243:R65-R69.

Doi Y, Atarashi K, Franco-Saenz R, Mulrow PJ (1984). Effect of changes in sodium or potassium balance, and nephrectomy, on adrenal renin and aldosterone concentrations. Hypertension 6 (2Pt2) I124-I129.

Douglas WW, Kanno T, Sampson SR (1967). Effects of acetylcholine and other medullary secretagogues and antagonists on the membrane potential of adrenal chromaffin cells; An analysis employing techniques of tissue culture. J Physiol 188:107-120.

Feldberg W, Louis GP (1965). Further studies on the effects of peptides on the suprarenal medulla of cats. J Physiol 178:239-251.

Goldstein DS, Feuerstein GZ, Izzo JL, Kopin IJ Keiser HR (1981). Validity and reliability of liquid chromatography with electrochemical detection for measuring plasma levels of norepinephrine and epinephrine in man. Life Sci 28:467-475.

Healy DP, Printz MP (1984). Autoradiographic localization of angiotensin II binding sites in the brain, kidney and adrenal gland of the rat. J Hypertension 2 (suppl. 3):57-61.

Israel A, Niwa M, Plunkett LM, Saavedra JM (1985). High-affinity angiotensin receptors in rat adrenal medulla. Regulatory peptides 11:237-243.

Knight DE, Baker PF (1983). Stimulus secretion coupling in isolated bovine adrenal medullary cells. Q J Exp Physiol 68:123-143.

Levine M, Asher A, Pollard H, Zinder O (1983). ascorbic acid and catecholamine secretion from cultured chromaffin cells. J Biol Chem 258:13111-13115.

Nabika T, Velletri PA, Lovenberg W, Beaven MA (1985). Increase in cytosolic calcium and phosphoinositide metabolism induced by angiotensin II and [Arg]Vasopressin in vascular smooth muscle cells. J Biol Chem 260:4461-4670.

Nussberger J, Brunner DB, Waeber B, Brunner HR (1985). True versus immunoreactive angiotensin II in human plasma. Hypertension 7 (suppl I):I1-I7.

Peach MJ (1971). Adrenal medullary stimulation induced by angiotensin I, angiotensin II, and analogues. Circ Res 28: 29(suppl 2) II107-II117.

Poisner AM, Douglas WW (1966). The need for calcium in adrenomedullary secretion evoked by biogenic amines, polypeptides, and muscarinic agents. Proc Soc Exp Biol Med 123:62-64.

Rowe BP, Nasjlettia A (1981). The effect of angiotensin II infusion on plasma catecholamines in the conscious rabbit. Proc Soc Exp Biol Med 168:110-113.

Tsien RY, Pozzan T, Rink TJ (1982). Calcium homeostasis in intact lymphocytes: cytoplasmic free calcium monitored with a new, intracellularly trapped fluorescent indicator. J Cell Biol 94:325-334.

Wurtman RJ, Axelrod J (1966). Control of enzymatic synthesis of adrenaline in the adrenal medulla by adrenal cortical steroids. J Biol Chem 241:2301-2305.

Calcium Channel Modulators in Heart and Smooth Muscle: Basic Mechanisms and Pharmacological Aspects
S. Abraham and G. Amitai, editors
© 1990, VCH, Weinheim/Deerfield Beach, FL and Balaban, Rehovot/Philadelphia

Quantitative determination of prazosin in plasma using specific antibodies

G. AMITAI[1], Y. SEGALL[2], S. CHAPMAN[1], N. ROTSCHILD[3] AND M. PINTO[3]

[1]Departments of Pharmacology, [2]Organic Chemistry and [3]Microbiology, IIBR, P.O. Box 19, Ness Ziona 70450, Israel

INTRODUCTION

The anti-hypertensive drug prazosin (4-amino 6,7 dimethoxy 2-(furoyl) piprerazine-1-yl quinazoline, Pfizer, USA) reduces peripheral vascular resistance by blocking vascular $alpha_1$ adrenergic receptors (1). Prazosin is a highly specific blocker of the $alpha_1$ adrenergic receptor with a dissociation constant ranging between 50-100 pM in various types of smooth muscle (2,3). Prazosin is an efficacious and safe hypotensive drug except for the orthostatic hypotension effect which usually appears only upon administration of the first dose (4). New analogs of prazosin, alfuzosin and bunazosin which are devoid of this unpleasant side effect has been recently introduced for clinical use (5,6). All the newly developed prazosin analogs contain the 4-amino 6,7-dimethoxy quinazoline moiety. This chemical structure was originally designed by Hess and his colleagues as a molecular combination of two naturally occuring substances, cAMP and papaverine, and was originally designed as a more potent phophodiesterase inhibitor (7,8). Since all the anti-hypertensive drugs which contain 4-amino 6,7-dimethoxy quinazoline in their molecular structure are exceptionally nontoxic and yet display high specificity towards the $alpha_1$ adrenergic receptor, the existence of an endogeneous specific antagonist for this receptor was postulated.

We have, therefore, synthesized a prazosin analog YS-621 (figure 1) which upon conjugation to bovine serum albumin (BSA) becomes an immunogen for producing prazosin-specific polyclonal antibodies. These antibodies may be used as a specific probe for the detection of putative endogeneous prazosin-like compound in human serum and also may serve as a model for the $alpha_1$ adrenergic receptor. In addition, these antibodies are useful for the development of a more sensitive analytical assay for prazosin and its analogs in human serum. Thus, radioimmunoassay (RIA) and enzyme-linked immunosorption assay (ELISA) were developed for the analysis of a series of quinazoline-related anti-hypertensive drugs. The present study summarizes the preparation of prazosin-specific polyclonal antibodies and the delineation of their specificity and capability to detect quantitatively prazosin in human serum.

RESULTS

Synthesis: YS-621 (4-amino 6,7-dimethoxy 2-(4-aminobenzoyl) piperazin-1-yl) quinazoline was prepared by modifications of a previously reported procedure (8) and is summarized in Figure 1A. All intermediates and the final product were characterized by ^1H n.m.r and mass spectrometry and their purity was determined by thin-layer chromatography. YS-621 was coupled to rabbit and bovine serum albumin (RSA and BSA, respectively) via the corresponding diazonium salt in saline using 0.1 M NaOH for maintaining pH 9.0 at 4≥C (figure 1B). The binding level was determined by measuring the relative absorption intensity YS-621/protein of the conjugates at 340 and 280 nm, respectively. The stoichiometry of binding was 16 and 13 YS-621 groups per one BSA or RSA molecule, respectively.

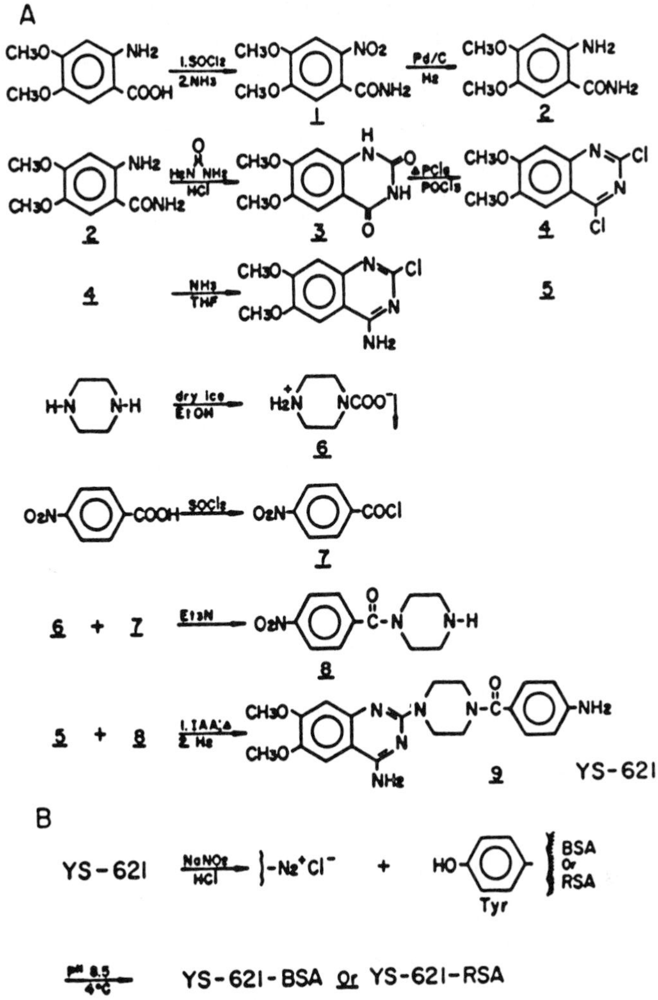

Figure 1: Synthetic pathway of YS-621.

A. Schematic presentation of the synthetic steps.

B. Conjugation of YS-621 with BSA or RSA.

Immunization: BSA-YS-621 conjugate suspended in Freund's Complete Adjuvant (1 mg/ml) was injected first intra-cutaneously and then twice intramuscularly to rabbits on days 0, 10 and 17. Rabbit antiserum was collected on day 24 and screened by immunoprecipitation using RSA-YS-621 conjugate as antigen and YS-621 and prazosin as competing ligands for inhibition of precipitation. The precipitin curves for the rabbit antiserum displayed anti YS-621 immunoglobulin concentration of 0.5 - 0.7 mg/ml. (see figures 2A and 2B).

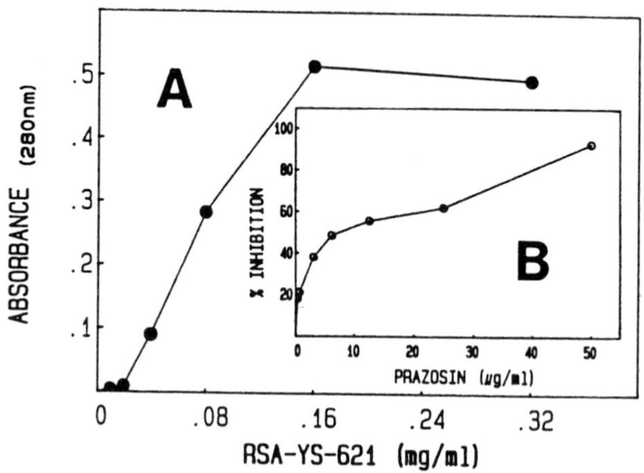

Figure 2: A. Precipitation curve for rabbit anti-BSA-YS-621 antibodies with RSA-YS-621. B. Inhibition of precipitation using prazosin as inhibitor.

IgG separation: Rabbit antiserum was treated with saturated ammonium sulfate solution and spun down at 2500 rpm. The pellet was dialysed extensively with Na phosphate (17.5mM pH 6.3) at 4° C. The IgG was separated from the other serum immunoglobulins on DEAE-cellulose column at pH 6.3 (9). Fractions were collected and the IgG fraction displayed a single band in the Ouchterlony test (10) as compared to multiple bands obtained for the authentic rabbit antiserum (Figure 3).

RIA: The RIA test was carried out according to Hunter (11). Displacement of [^{3}H]prazosin from prazosin polyclonal antibodies (Ab's), which were diluted 1:1000, was measured in the presence of various concentrations of unlabelled prazosin in normal human serum (diluted 1:4). Following a 45 min incubation at 37 C human serum was added as a "filler" and saturated ammonium sulfate solution was added (1:1). The samples were incubated for 16h at 4 C and then spun down at 2500 rpm for 30 min and rinsed twice with cold saline. Residual radioactivity in the pellets represents the amount of [^{3}H]prazosin bound to the Ab's (figure 4). The IC50 value obtained for the displacement of [^{3}H]prazosin (figure 4) is 1.5x10^{-8} M.

Figure 3: Ouchterlony test for the IgG separation.

A. Goat anti rabbit serum.

B. Rabbit anti BSA-621 serum.

C. anti BSA-YS-621 IgG.

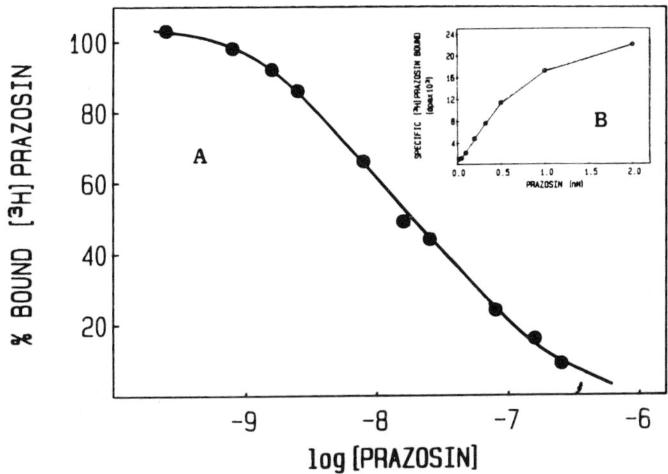

Figure 4: A. Radioimmunoassay curve for prazosin using rabbit anti BSA-YS-621 IgG and [^{3}H]prazosin.

B. Antiserum titration. Maximal prazosin specific binding.

ELISA: ELISA was performed essentially as previously described by Pinto et al (13). Anti-prazosin IgG (20 µg/ml) was bound to a PVP plate and incubated with the antigen (Ag) YS-621-RSA conjugated to horse raddish peroxidase (HRP) (12) (1:64000 dilution). The Ag-Ab binding was carried out in the presence of various prazosin concentrations. Following a 30 min incubation at 37°C the wells were rinsed twice and further incubated with HRP substrate. Percent of inhibition of Ag-Ab binding was calculated from the residual O.D. at 405nm. The IC50 value obtained for prazosin is 1 ng/ml (Figure 5).

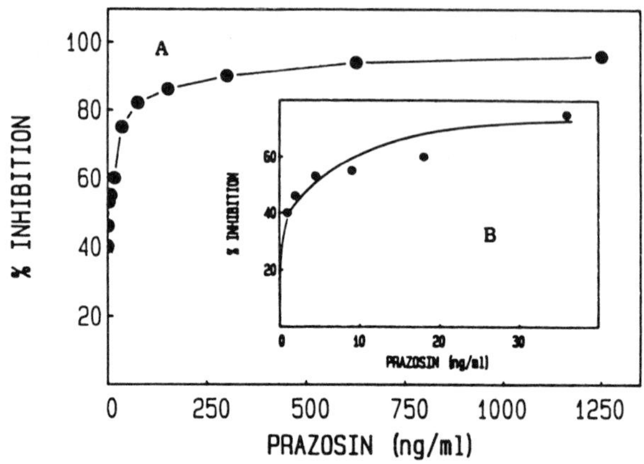

Figure 5: A. Comparative ELISA for anti-BSA-YS-621 antibodies using RSA-YS-621-HRP conjugate as antigen and prazosin as inhibitor. B. Expanded scale for the lower prazosin concentration range.

DISCUSSION

Polyclonal Ab's specific to prazosin (figures 2, 4 and 5) and YS-621 (not shown) were raised in rabbits by using YS-621-BSA as an immunogen. Following three consecutive injections to rabbits we have obtained 0.5-0.7 mg/ml prazosin-apecific immunoglobulins in rabbit antiserum. The IgG separated from this antiserum showed high specificity towards both prazosin and YS-621 as shown by the inhibition test using the ELISA technique. Less than 1 ng/ml of prazosin was detected in human serum by the inhibition of binding shown in Figure 5. Moreover, 0.2-0.5 ng/ml of unlabelled prazosin was detected by the RIA displacement curve (Figure 4). Direct specific binding of [^{3}H]prazosin to its Ab's (inset Figure 4) was saturable at 2 nM prazosin with a detection limit of 10 pg/ml radiolabelled prazosin.

Quantitative multisample analysis of prazosin and other analogs such as alfuzosin and bunazosin can be performed using either RIA or ELISA which were developed in this study. These assays seem to be more sensitive than the previously reported high performance liquid chromatography (HPLC) analysis (14). Furthemore, whereas the HPLC method is suitable only for one specific compound (i.e. prazosin or any of its analog) the RIA and ELISA may be useful for the determination of all antihypertensive drugs which are structurally based on the 4-amino 6,7-dimethoxy quinazoline moiety.

The specificity of these Ab's enables to search for a putative prazosin-like endogeneous ligand in sera of normotensive humans. The prazosin-specific polyclonal Ab's may also be used as a probe for studying the morphology of the alpha1 adrenergic receptor and its ligand binding site.

REFERENCES

1. Brodgen RN, Heel RC, Speight TM, Avery GS, (1977). Prazosin: A review of its pharmacological properties and therapeutic efficacy in hypertension. Drugs 14, 163-197.

2. Amitai G, Brown RD and Taylor P, (1984). The relationship between alpha1 adrenergic receptor occupation and the mobilization of intracellular-calcium. J. Biol. Chem. 259, 12519-12527.

3. Hughes RJ, Boyle MR, Brown RD, Taylor P, and Insel PA, (1982). Characterization of coexisting alpha1 and beta2 adrenergic receptors on a cloned muscle cell line. Mol. Pharmacol. 22, 259-266.

4. Graham RM, Thornell JR, Gain JM, Bagnoli C, Oates HF, and Stokes GS, (1976). Prazosin: The first dose phenomenon. Br. Med. J. 2, 1293-1294.

5. Cavero I, and Roach A, (1980). The pharmacology of prazosin, a novel antihypertensive agent. Life Sci., 27, 1525-1540.

6. Manoury PM, Binet JL, Dumas HP, Lefevre-Borg F, and Cavero I, (1986). Synthesis and antihypertensive activity of a series of 4-amino 6,7-dimethoxy quinazoline derivatives. J. Med. Chem. 29, 1-9.

7. Constantine JW, McShane WK, Scriabine A, and Hess HJ, (1973). In: Hypertension: Mechanism and management. Onesti G, Kim KE, and Mojer JH, Eds. Grune and Stratton, New York NY, Chapter 6 p.429.

8. Althuis, TH and Hess HJ, (1977). Synthesis and identification of the major metabolites of prazosin in dog and rat. J. Med. Chem. 20, 146-156.

9. Levy HB and Sober HA, (1960). A simple chromatographic method for preparation of gamma globulin. Proc. Soc. Exp. Biol. NY 103, 250-252.

10. Ouchterlony O, (1958). Diffusion-in-gel methods for immunological analysis. In: Progress in Allergy, Vol. 5 pp 1-78. Ed. Karlos P,. Basel and New York, Karger.

11. Hunter WM, (1973). Radioimmunoassay. In: Handbook of Experimental Immunology, Ed. Wier DM, Vol 1, Immunochemistry pp. 17.1-17.36 Blackwell Scientific Publications Oxford-Edinburgh.

12. Boorsma DM, and Streefkerk JG, (1979). Periodate or glutaraldehyde for preparing peroxidase conjugates. . Immunol. Meth. 30, 245-255.

13. Pinto M, Herzberg H, Barnea A, and Shenberg E, (1987). Effects of partial hepatectomy on the immune response in mice. Clin. Immunol. and Immunopathol. 42, 123-132.

14. Twomey TM and Hobbs DC (1978). Analysis of prazosin in plasma by a sensitive HPLC fluorescence method. J. Pharm. Sci. 67, 1468-1469.

Calcium Channel Modulators in Heart and Smooth Muscle: Basic Mechanisms and Pharmacological Aspects
S. Abraham and G. Amitai, editors
© 1990, VCH, Weinheim/Deerfield Beach, FL and Balaban, Rehovot/Philadelphia

Purification and labeling of the α_1 adrenergic receptor

JOHN F. REECE[1], R. DALE BROWN[1*], CARIN CAIN[1], GABRIEL AMITAI[1,2], YOFFI SEGALL[2] AND PALMER TAYLOR[1]

[1]Department of Pharmacology, University of California, San Diego, La Jolla, CA, USA 92093 and [2]Israel Institute of Biological Research, Ness Ziona, Israel

SUMMARY

DDT-MF-2 cells, a cell line originating from smooth muscle of hamster vas deferens origin, have been grown under conditions leading to an enrichment of α_1- adrenergic receptors. Addition of dexamethasone immediately prior to saturation of the growth curve arrests cell growth and enhances receptor expression. Hence, conditions are optimized for purification. The receptor has been extensively purified by isolation of a membrane fraction, solubilization in digitonin, affinity chromatography with a conjugated prazosin derivaties and adsorption and elution from a wheat germ agglutinin column. Purification has yielded a solubilized receptor which upon labeling with phenoxybenzamine shows selective protectable labeling of a peptide of 80 kDa. The labeled peptide can be isolated in amounts sufficient for characterization of the site of phenoxybenzamine conjugation.

Introduction

Pharmacology has as an ultimate goal the elucidation of drugs which have absolute specificity toward their target receptors. Unfortunately, a lack of knowledge of structure of the receptor proteins has limited the approaches to rational drug design. Recently, recombinant DNA technology has provided an avenue into the determina-

**Present Address: Department of Pharmacology, University of Illinois, College of Medicine, Chicago, IL 60680, USA*

tion of the primary structures of the ligand-gated channel receptors and the G-protein linked receptors. Moreover, significant progress has been made in identifying the coupling G-proteins which are involved in the activation and inactivation of adenylate cyclase. In contrast, relatively little is known about the proximal events in receptor mediated increases in intracellular Ca2+, either through mobilization of intracellular calcium stores or by opening a membrane Ca^{2+} channel. Such receptors are typified by the alpha-1-adrenergic (α_1) receptor. Although the pharmacologically-labeled hormone-binding subunit of the α_1 receptor has a molecular mass of 80 kDa when analyzed by sodium dodecyl sulfate polyacrylamide gel electrophoresis (SDS-PAGE) (Leeb-Lundberg et al, 1984; Terman et al, 1988), the active receptor complex has been shown to be a possible dimer of 160 kDa in the cell membrane by virtue of radiation inactivation studies (Venter et al, 1984) and as a 145 kDa species in gel filtration studies using digitonin-solubilized receptor preparations (Graham et al, 1982). The binding subunit has been purified and its amino. acid sequence determined by cDNA cloning techniques (Lomasney et al, 1986; Cotecchia et al, 1988), and has been shown to be a 515 amino acid polypeptide of 56 kDa molecular weight with sufficient glycosylation to impede its mobility on SDS-PAGE to 80 kDa (Terman et al, 1988). Physical and biochemical advances such as these have laid the groundwork for future studies on identifying the recognition and coupling sites on the receptor molecule.

Successful purification of the receptor initially required: a) solubilization of the protein in an active form, b) a tissue source with comparatively high expression of the receptor, and c) pharmacologically-specific ligand(s) for affinity purification and receptor identification. Digitonin, a plant glycoside, is the only known detergent capable of solubilizing the receptor with retention of binding properties. A tissue source of receptor-containing material in an easily preparable form is liver, which suffers from large amounts of contaminating protein and low specific activity. The DDT-1 smooth muscle-derived cell line originally described by Cornett and Norris (1982) by virtue of both its high receptor density and its ability to grow in large-scale suspension culture provides an alternative source for α_1-receptors (Lomasney et al, 1986). Of many ligands with appropriate specificity for the receptor, the antagonist prazosin has the requisite chemically modifiable nucleus and specificity in binding for affinity chromatography. Graham et al, (1982) and Lomasney et al, (1986) have pioneered this use of prazosin derivatives for immobilization on affinity resins.

The β-chloroethylamine alkylating agent phenoxybenzamine has long been known to irreversibly block α_1-adrenergic receptors in situ (cf Nickerson, 1957; Venter et al, 1984). In addition, phenoxybenzamine will irreversibly block the reversible binding of prazosin to α_1- receptors in intact cells and membrane fragments (Amitai et al, 1984; Cornett and Norris, 1986). In the present study we report our current progress in α_1-receptor purification and in identifying the site of phenoxybenzamine labeling within the receptor primary structure.

METHODS

Binding Assay

Cell and membrane bound receptor number is determined by incubation with 1.2 nM [^{3}H]prazosin (80 Ci/mmol; New England Nuclear) at 37°C for 30 min in sodium Ringer's solution or STE (100 mM NaCl, 50 mM Tris HCl, 5 mM EDTA, pH 7.4) containing protease inhibitors, (0.1 mM PMSF, 10 μg/ml soybean trypsin inhibitor, 10 μg/ml leupeptin, 10 μg/ml pepstatin, and 0.1 mg/ml bacitracin), respectively. Suspensions are diluted from 0.5 ml to 10 ml with ice cold buffer and immediately filtered onto Whatman GF/B filters, which are rapidly rinsed with two 5 ml aliquots of ice cold buffer. Filters are counted by liquid scintillation. All assays are done in triplicate and nonspecific binding is assayed in 0.1 mM phentolamine. Soluble receptor is assayed in 0.25 ml STE buffer containing 0.05% digitonin and 1.2 nM [^{3}H]prazosin in the presence and absence of 0.5 mM phentolamine for 30 min at 25°C. The samples are cooled to 4°C for 15 min and are applied to identical 2.5 ml Sephadex G-75 columns equilibrated in buffer at 4°C. The excluded fraction containing protein bound tritium is collected for scintillation counting. All assays are done in triplicate.

Membrane Preparation

Cells are harvested by low-speed centrifugation at 4°C and are either resuspended in media containing 10% glycerol for -70°C storage or used immediately. Cells are then washed twice in Ringer's solution. The final pellet is resuspended in lysis buffer (10 mM Tris HCl, 5 mM EDTA, pH 7.4), containing the above protease inhibitors. After incubation on ice for 30 min, the lysed cells are centrifuged at 2000xg for 10 min at 4°C. The pellet is resuspended in an identical volume of lysis buffer and homogenized vigorously with a tight-fitting Dounce pestle, followed by centrifugation at 2000 x g for 10 min at 4°C. The pellet is homogenized and spun twice more. The three supernatants containing suspended membranes are pooled and centrifuged at 20,000 xg for 60 min at 4°C onto a 70% sucrose cushion. Membranes are resuspended at 10-20 mg/ml and frozen at -70°C.

Receptor Solubilization

Membranes are thawed and added to the solubilization buffer (final concentrations: 50% [w/v] glycerol, 100 mM NaCl, 50 mM Tris HCl, pH 7.4, 5 mM EDTA, 1% digitonin, and protease inhibitors) at 0°C to give a final protein concentration of 2 mg/ml. Aliquots of the suspension are gently sonicated on ice at 5 watts for 90 sec on a Branson Heat Systems sonicator using a "micro" tip. Following a 2-fold dilution with buffer lacking digitonin and glycerol and centrifugation at 100,000 x g at 4°C for 3-4 hr, the supernatant is assayed for protein and receptor binding activity and frozen at -70°C.

A55453-Sepharose Affinity Chromatography

The solubilized crude receptor preparation is applied at a 0.5 ml/min flow rate at 4°C. Generally, 2-5 pmol receptor per ml gel is used, and fractions are collected and assayed for receptor concentrations. The column is then washed with two bed volumes on 1 M NaCl, 50 mM Tris/Cl, 5 mM EDTA, pH 7.4, containing 0.05% digitonin, 0.01 mg/ml leupeptin, and 0.1 mg/ml bacitracin, followed by another two volumes of the same buffer containing 100 mM NaCl. Elution is effected by continuing with the above buffer containing 0.5 mM phentolamine (4-5 bed volumes). Receptor is assayed in these fractions by first de-salting the receptor into the above buffer lacking phentolamine using Sephadex G-25, then the binding assay using labelled prazosin is used. The eluted fractions are concentrated to 10 ml by ultrafiltration with an Amicon YM-30 membrane.

Wheat Germ Agglutinin-Sepharose Chromatography

A 1.0 ml column of wheat germ agglutinin Sepharose (WGA-Seph, E-Y Labs) was used to concentrate and further purify the receptor. The column was equilibrated in 100 mM NaCl, 50 mM Tris/Cl, 5 mM EDTA, pH 7.4, containing 0.05% digitonin, 0.01 mg/ml leupeptin, and 0.1 mg/ml bacitracin. The concentrated phentolamine eluate was applied in 0.5 ml aliquots every 20-30 minutes at 4°C. The gel was then washed with 2.5 ml buffer, and the receptor eluted in a 1.0 ml volume with buffer plus 0.3 M N-acetylglucosamine. The receptor was then frozen at -70°C. Typically, a 15-fold receptor enrichment is seen, at about 80% yield. [^{3}H]Phenoxybenzamine was obtained from Amersham Corporation (#TRQ4860) (21 Ci/mmol). Purity was established by thin layer chromatography in hexane:ethanol (85:15) on silica gel plates with fluorescent indicator in the presence and absence of cold phenoxybenzamine, followed by exposure to Kodak X-omat film. Radioactivity comigrated with the cold standard's fluorescent spot, and appeared to be a single molecular species.

Receptor Labeling

Membranes were thawed and diluted to 2 mg/ml in 100 mM NaCl, 50 mM Tris HCl, 5 mM EDTA, pH 7.4 containing protease inhibitors. The membranes were incubated at 37°C for 10 min prior to addition of [^{3}H]POB (final POB concentration is usually 50 nM). Competition experiments utilized 1.0 μM prazosin. The labeling reaction is stopped by the addition of an equal volume of ice cold buffer supplemented with 2 mg/ml BSA followed by centrifugation at 20,000 x g for 30 min at 4°C. The pellet is resuspended in 1 volume of buffer containing 1 mg/ml BSA and briefly homogenized before centrifugation as above. This step is repeated twice, and radioactivity in the supernatants is monitored. The final pellet is resuspended in buffer lacking BSA at 10-20 mg/ml, homogenized, and then assayed for protein and bound radioactivity. Soluble receptor labeling is done essentially as above, but at room temperature. After incubation for 30 min in buffer containing 0.05% digitonin, at protein concentration <1 mg/ml, the free ligand is removed by Sephadex G-25 de-salting, followed by lyophilization and resuspension in SDS-gel buffer.

RESULTS AND DISCUSSION:

DDT1-MF2 cells, a hamster vas deferens smooth muscle-derived cell line, grow in suspension culture to high density and express the adrenergic receptor subtype (Cornett and Norris, 1982). Figure 1 shows the time course of cell growth and receptor expression with and without 10 nM dexamethasone treatment. Hormone addition at day 4, near the end of the log phase of growth, causes an apparent arresting of cell replication, as seen in panel A. There is a concurrent rise in total receptor production (fig. 1B). Whether the induced receptor is identical to its non-induced counterpart has not been shown definitively, although competition studies using [^{3}H]prazosin at various stages of receptor purification have shown no anomalies in binding constant and Hill coefficient (not shown). The receptor also does not show apparent differences in its overall glycosylation state, as evidenced by endo-glycosidase treatment of azidoprazosin labeled receptor preparations and subsequent gel electrophoresis (Terman et al, 1988). Dexamethasone has long been known for its ability to alter gene expression at the transcriptional level, and this is a likely interpretation of the data shown in figs. 1C and 1D, in which receptor expression is enhanced on both a per cell and per protein basis over that in untreated control cultures. Whether similar phenomena are involved in steroid-induced hypertensive states is a matter of speculation. Regardless of its mechanism, it seems clear that dexamethasone induction is useful for increasing both the yield and specific activity of α_1 receptors in DDT1 cells, as a prelude to purification and further analysis.

Affinity chromatography of digitonin-solubilized receptor was effected using a resin made essentially identical to that of Lomasney et al (1986). Briefly, Sepharose 6B is reacted with a long-chain hydrophilic molecule, to give an ether linkage. Glycine is coupled to the linker group, resulting in a free carboxyl group suitable for ligand coupling. We added [^{14}C]-glycine to monitor gel substitution, and determined that the concentration of bound glycine was roughly 1-2 mM in the gel bed. This has allowed us to test the condition of the gel during sample

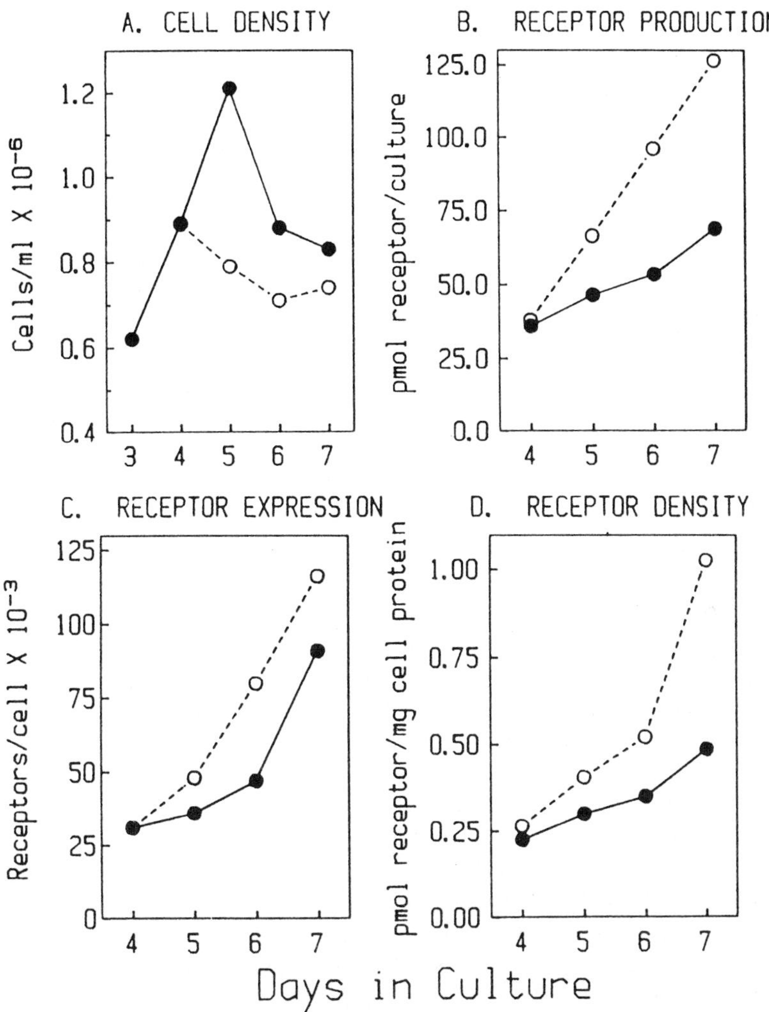

Figure 1: Regulation of α_1-adrenergic receptor expression in DDT-1 cells. Suspension cultures of DDT-1 cells (1000 ml) were grown as described in Methods. Cell number [^{3}H]prazosin binding, and protein content were measured daily. ○, Dexamethasone (1 nM) was added on day 4; ●, uninduced cultures.

throughput, and it was found that little or no glycine was lost during a given elution. The free amino group of the prazosin derivative A55453 (a kind gift of Dr. John S. DeMarinis, Abbott Laboratories) was reacted with the gel using carbodiimide. In our hands the reaction was quite efficient, as measured by absorbance of the gel at 330 nm.

Solubilized receptor preparations were applied to the column as shown in fig. 2. During sample application, the amount of unbound receptor slowly increased. Washing with high salt removed extraneously bound protein. Biospecific elution of the bound receptor was subsequently effected with 0.5 mM phentolamine, resulting in a net 10% yield of the amount applied (Table 1). Roughly 50% of the applied receptor was left unaccounted for. Some remains bound to the resin in a very slowly dissociating state, as evidenced by the final fraction in figure 2, which was collected the following day. Subsequent column runs utilized this effect. It may be that the receptor complex, reportedly a dimer, is relatively unstable after frequent buffer changes and dilutions. Also, it is unstable above room temperature in the solubilized form (not shown). The net purification in our hands was 100-fold from the applied sample, as opposed to the 400- 500-fold reported by Lomasney et al (1986).

After concentration of the phentolamine eluate by Amicon ultrafiltration, the solution was applied to a wheat germ agglutinin Sepharose column. After washing with high salt buffer, the bound glycoproteins were eluted with N-acetyl-glucosamine, resulting in a step yield of 90%, and a final receptor specific activity of 4760 pmol receptor per mg protein, as in Table 1. This corresponds to 38% of theoretical homogeneity, based on an estimated molecular weight of 80 kDa.

As a first step toward characterizing the soluble protein, membrane-bound receptor was labeled with [^{3}H]phenoxybenzamine. Figure 3 shows that unlabeled prazosin competes for an 80 ± 5 kDa species, while not visibly effecting the labeling of the dominant labeled species, at 35 kDa. There is also an abundance of label at the top of the gel. Although biospecifically labeled, it is apparently an aggregate since other experiments using sonication and centrifugation prior to gel loading appear

Table I. Summary of $Alpha_1$-Adrenergic Receptor Purification

STEP	VOLUME (ml)	PROTEIN mg	PROTEIN Recovery (%)	RECEPTOR (pmol)	RECEPTOR Recovery (%)	RECEPTOR Activity (pmol/mg)	RECEPTOR Purification (x-fold)
Cells	16700	2470	100%	956	100%	0.39	1.0
Membranes	57	840	34	550	58	0.66	1.7
Digitonin-Soluble	800	500	20	300	31	0.61	1.6
A55453							
Flow-Through	860	490	20	100	10	0.20	0.5
Salt Wash	170	0.33	0.01	8	0.8	23	60
PTA Wash	180	0.53	0.02	31	3.2	58	150
Amicon Concentrate	10	0.29	0.01	24	2.5	83	210
Wheat Germ Agglutinin							
Flow-Through	11	0.25	0.01	0.4	0.04	1	4
GlcNAc Wash	1	0.005	0.0002	21	2.2	4760	12300

The receptor has been purified to 38% theoretical homogeneity based on an estimated molecular mass of 80 kDa.

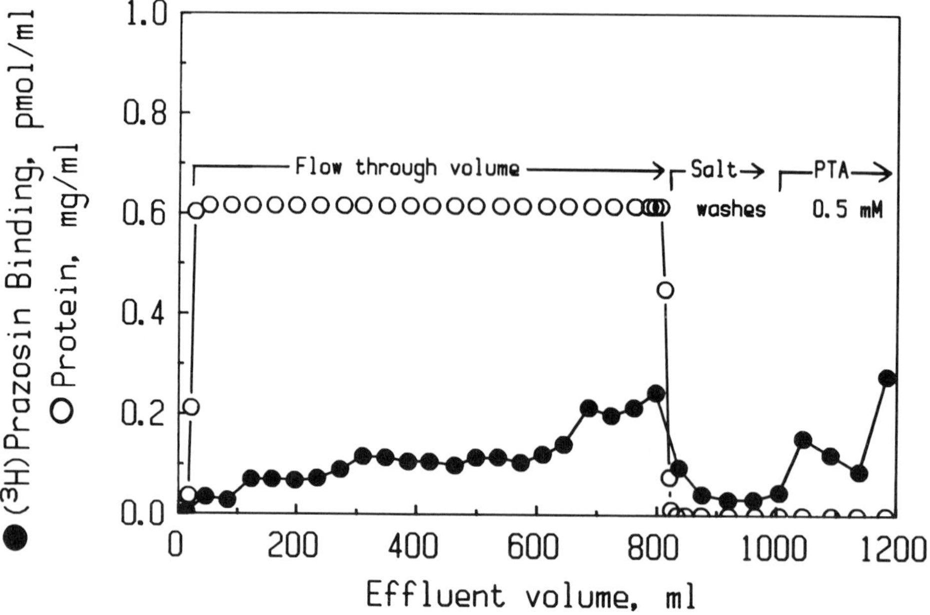

Figure 2: Affinity chromatography of digitonin-solubilized α_1-adrenergic receptor on A55453-Sepharose CL 6B. Soluble receptor (800 ml) was applied to a 40 ml Sepharose CL 6B column conjugated with A55453 at a flow rate of 0.5 ml per min (flow through fraction). The column was washed successively with 2 volumes of 1 M NaCl and 0.1 M NaCl in 50 mM Tris-Cl, 5 mM EDTA, pH 7.4, to remove nonspecifically adsorbed proteins (salt washes). α_1 Receptor was then selectively eluted with 5 column volumes of 0.5 mM phentoamine (PTA wash). Column effluent was monitored for protein concentration and [^{3}H]prazosin binding activity.

to remove this material (not shown). The presence of the lower molecular weight species makes phenoxybenzamine a less attractive choice for routine receptor assays, but retains its utility when purer preparations are used. This pattern of protection of the 80 kDa species is similar to that reported by Cornett and Norris (1986). The labeled receptor band appears relatively diffuse; this is apparently due to its glycosidic nature. Also, similarly diffuse bands are seen when the receptor is photoaffinity labeled using 125I-azidoprazosin (Terman et al, 1988).

We also show in figure 4 that the partially purified preparation described in Table 1 can be specifically labeled with [^{3}H]phenoxybenzamine, yielding a single radiolabeled band of appropriate molecular weight which can be abolished by the addition of unlabeled prazosin. The

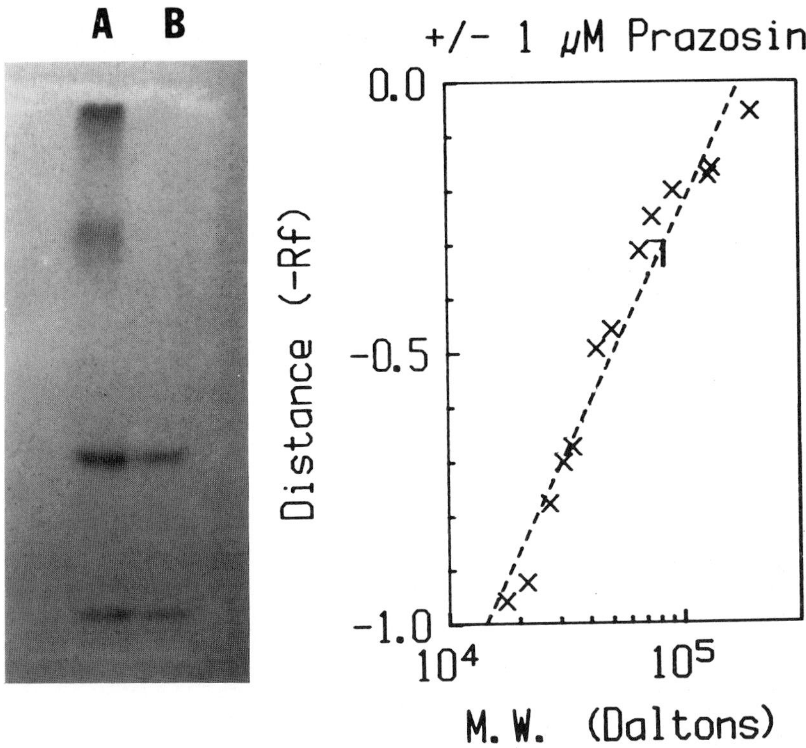

Figure 3: [^{3}H]Phenoxbenzamine (POB) labeling of α_1 receptor-enriched DDT-1 cell membranes. Plasma membranes prepared from DDT-1 cells were reacted with [^{3}H]POB in the presence or absence of 1 μM prazosin. Samples were solubilized and radiolabeled peptides were separated by SDS-PAGE. Positions of the molecular weight markers are shown on the right hand side. A single specifically labeled peptide of 85 kDa was detected similar to that observed from labeling of purified receptor preparations.

protection of a 80 kDa band is consistent with the findings of Graham et al, (1982) using azidoprazosin for labeling.

In conclusion, we have demonstrated the utility of dexamethasone induction of α_1 receptors in DDT1 cells as an adjunct to their subsequent purification by affinity chromatography. Furthermore, we have corroborated studies by Lomasney et al (1984), Venter et al (1984), and Cornett and Norris (1986) detailing α_1-receptor purification and affinity labelling by [^{3}H]phenoxybenzamine. Further studies using these techniques should enable us to identify the phenoxybenzamine binding site within the receptor's primary structure.

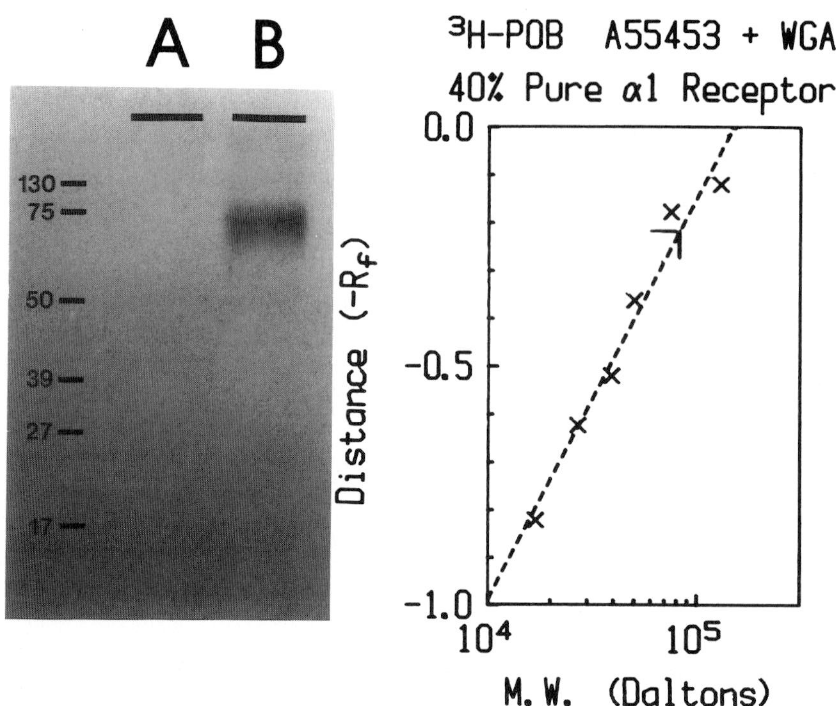

Figure 4: Labeling of solubilized and purified α_1 receptors with [^{3}H]phenoxybenzamine. The purified α_1 receptor fraction obtained from the procedure outlined in Table 1 was incubated with [^{3}H]POB in the presence or absence of 1 μM prazosin. Radiolabeled peptides were separated by SDS-PAGE and visualized by autoradiography. [^{3}H]POB specificity labels a single peptide of 80 kDa.

Supported by USPHS grants GM 36237-04 to PT and GM41470 to RDB.

References

Amitai G, Brown RD, and Taylor P. (1984). The relationship between α_1-adrenergic receptor occupancy and themobilization of intracellular calcium. J Biol Chem 259:12519-12527.

Cornett LE and Norris JS. (1982). Characterization of the α_1-adrenergic receptor subtype in a smooth muscle cell line. J Biol Chem 257:694-697.

Cornett LE and Norris JS. (1986). Affinity labeling of the DDT1 MF-2 cell alpha 1-adrenergic receptor with [3H]phenoxybenzamine. Biochem Pharmacol 35:1663-1669.

Cotecchia S, Schwinn DA, Randall RR, Lefkowitz RJ, Caron MG,and Kobilka BK. (1988). Molecular cloning and expression of the cDNA for the hamster alpha 1-adrenergic receptor. Proc Natl Acad Sci USA 85:7159-7163.

Graham RM, Hess HJ, and Homcy CJ. (1982). Biophysical characterization of the purified α_1-adrenergic receptor and identification of the hormone binding subunit. J Biol Chem 257:15174-15181.

Leeb-Lundberg LMF, Dickinson KEJ, Heald SL, Wikberg JES, Hagen P, DeBernardis JF, Winn M, Arendson DL, Lefkowitz RJ, and Caron MG. (1984). Photoaffinity labeling of mammalian α_1-adrenergic receptors: Identification of the ligand binding subunit with a high affinity radioiodinated probe. J Biol Chem 259:2579-2587.

Lomasney JW, Leeb-Lundberg LM, Cotecchia S, Regan JW, DeBernardis JF, Caron MG, and Lefkowitz RJ. (1986). Mammalian α_1-adrenergic receptor: Purification and characterization of the native receptor ligand binding subunit. J Biol Chem 261:7710-7716.

Nickerson, M. (1957) Nonequilibrium drug antagonism. Pharmacol. Rev. 9:246-259.

Terman BI, Reece JF, Brown RD, and Insel PA. (1988). The oligosaccharide component of α_1-adrenergic receptors from BC3H1 and DDT1 muscle cells: Studies with glycosidases and photoaffinity labeling of intact cells. Biochem J 253:363-370.

Venter JC, Horne P, Eddy B, Greguski R, and Fraser CM. (1984). Alpha 1-adrenergic receptor structure. Mol Pharmacol 26:196-205.

Calcium Channel Modulators in Heart and Smooth Muscle: Basic Mechanisms and Pharmacological Aspects
S. Abraham and G. Amitai, editors
© 1990, VCH, Weinheim/Deerfield Beach, FL and Balaban, Rehovot/Philadelphia

Modulation of the cardiac sodium channel by brevetoxin-3

G. JEGLITSCH[1], K. KRAL[2], H. TRITTHART[1] AND W. SCHREIBMAYER[1]

[1]Institute for Medical Physics and Biophysics, University of Graz, Harrachgasse 21/IV, 8010 Graz and [2]Institute for Zoology, University of Graz, Universitaetsplatz 2, 8010 Graz, Austria

Many ion channels do not exhibit just one open state in accordance to the classical two state - (closed — open) or three state - (closed — open inactivated) models but instead, different open states of the channel exist with different conductivities - so called subconductance states (for review, see Fox (1987)). For voltage dependent ion channels such substate behaviour was first observed by Kazachenko & Geletyuk (1984), Schreibmayer et.al. (1985 a+b) and Schindler & Schreibmayer (1985).
For the sodium channel in electrically excitable membranes, however, only sparse evidence is so far available for the existence of subconductance states (Cachelin et.al. (1983)) and Kunze et.al (1985)). Probably due to short open times of this channel and its small current amplitudes, resolution of these substates has been rather difficult. No mechanistic concepts of channel activities could be derived from these results. Sodium channels exposed to different gating modifiers (predominantly those that prolong the open state of the channel) have been shown to exhibit reduced single channel conductance (Quandt et.al (1982), Yoshii & Narahashi (1984), Chinn & Narahashi (1986), Schreibmayer et.al. 1987)).

If the reduced single channel conductances observed is due to different subconductance states (that are stabilized by the sodium agonist used) then it should be possible to induce flickering of the sodium channel between different open states by using combinations of these modifiers. Results of such experiments are shown in Fig.1 for example.

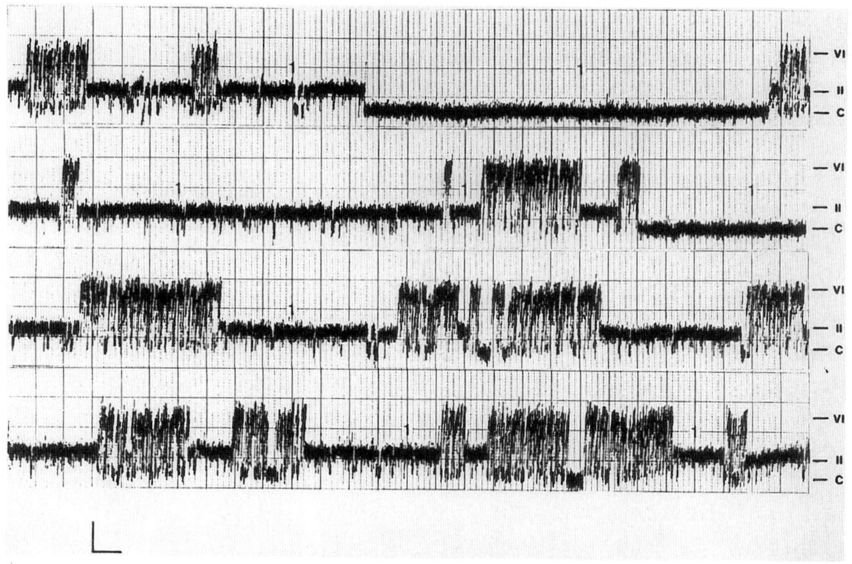

Fig.1 Sodium currents were elicited by repetitive hyper- and depolarization of a cell attached membrane patch. Hyperpolarization potential was 40 mV negative to the resting membrane potential (RP) and depolarization 30 mV positive to it. The extracellular medium contained in mM: NaCl 137; KCl 5.4; $CaCl_2$ 2; $MgCl_2$ 2; Hepes 10 and Glucose 10, buffered with NaOH to pH 7.4. In addition the pipette medium contained 200 µM racemic DPI, 10 µM veratridine and 1 mM $BaCl_2$ to block inward potassium currents. Horizontal bar: 190 ms; vertical bar: 0.5 pA; low pass filtering: 640 Hz.
Selected sodium channel openings showing fluctuations between the closed, open state II and open state VI are shown.

Sodium channels from freshly dissociated rat ventricular cells were modified by mixtures of the sodium channel agonist DPI (4-(3-/4-diphenylmethyl-1-piperazinyl/-2-hydroxy-propoxy)-1H-indole-2-carbonitrile, racemic form) and the plant alcaloid veratridine. Two different substates of the open sodium channel, designated in the figure as VI and II, are stabilized predominantly by using this mixture. With this experimental approach we were able to show (Schreibmayer (1989) and Schreibmayer et.al. (1989)) that:

(i) The sodium channel of this preparation shows six different open states with different conductances. These different conductance states fall into a pattern of six equidistant multiples of 2.3 pS (at 140 mM Na^+ and 2 mM Ca^{2+} on the extracellular side) independent of the agonists and ionic strengths applied (DPI, ATX-II, veratridine and mixtures of them).

(ii) Ion selectivity is conserved for both the different agonists (and mixtures of them) and for the different substates studied.

(iii) Binding of sodium ions to the channel does not depend on channel state - the apparent equilibrium dissociation constant is equal at least for states II and VI.

(iv) These findings are not compatible with the view that the rectangular fluctuations in membrane current - usually observed in patch clamp studies and regarded as openings of single channel molecules - reflect ion passage through only one ion permeation pathway, i.e. per monochannel units. Instead, the alternative view of not one pore but either 6 or 3 pores with normally strictly synchronized gating activity can account for these observations.

In our terminology we call this conceptual molecular entity an "oligochannel" as shown schematically:

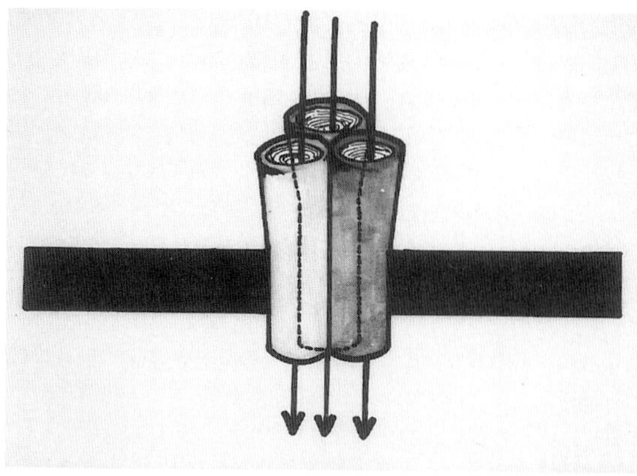

Fig.2 The sodium channel as an oligochannel composed of 3 pores with synchronized gating activities.

Such behaviour, i.e. concerted activity of enzyme protomeres in oligomeric structures has been shown to account for the marked deviations from Michaelis-Menten-kinetics of many enzymes (Monod et.al. (1965), Koshland et.al. (1966)). This behaviour may entail rather dramatic consequences on our understanding of the mechanism of excitation (Hodgkin & Huxley (1956)) and of ligand effects.

One of these dramatic consequences is that our oligopore concept not only predicts multiplicity of ion permeation pathways per channel complex but *also* of receptor domains. Upon state dependent binding of channel modulators (and allosteric interaction) complex kinetics of gating is necessarily the result. With the aid of a novel potent sodium channel agonist, Brevetoxin-3 (Pbtx-3), we were able to observe such complex kinetics for the first time.

Pbtx-3 is isolated from the marine dinoflagellate Ptychodiscus brevis. Blooms of Ptychodiscus brevis are responsible for massive fishkills in the Gulf of Mexico and along the coast of Florida

(red tide). One of the toxic principles isolated from this planctonic organism is the non-protein 11-ring-polyether system Pbtx-3. The lipophylic nature allows Pbtx-3 to pass through biological membranes in contrast to hydrophylic toxins like saxitoxin or Gonyautoxins which are produced by dinoflagellates also. Pbtx-3 acts on a specific receptor binding site (site 5) on the sodium channel protein from nerve (Catterall & Gainer (1985)) and also on the cardiac sodium channel as described here.
Under the influence of the toxin sodium channel inactivation is blocked and sodium channel openings can be observed at maintained depolarization.

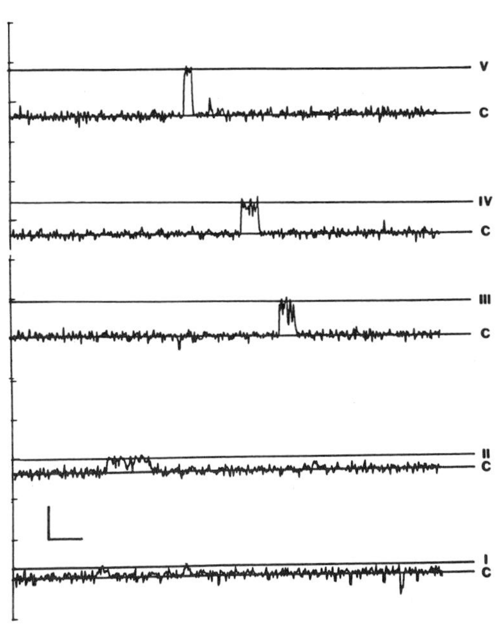

Fig.3 Openings of sodium channels modified by Pbtx-3 in a cell-attached membrane patch. The membrane potential was kept constant. In the lower and in the middle trace the potential was kept 10 mV positive to RP and 20 mV positive to RP for the other traces. The extracellular medium was exactly the same as in Fig.1. The pipette medium contained in addition 2 µM Pbtx-3 and 1 mM $BaCl_2$. Horizontal bar: 50 ms; vertical bar: 1 pA; low pass filtering: 300 Hz. Distinct sodium channel open states are numbered consecutively from I to V.

What makes Pbtx-3 unique among the sodium channel agonists is that it stabilizes up to 5 different open states which can be discerned by their different conductances. Upon opening to one of these five substates the sodium channel does not flicker from this distinct substate to one of the others but usually stays in this state till it closes again. By the use of S-DPI (pure stereoenantiomeric form, agonistic principle), which is known to stabilize state VI of the cardiac sodium channel (Schreibmayer et.al. (1989)), together with Pbtx-3 it is possible to show that the five different observed open states under the action of Pbtx-3 correspond to sodium channel open states I to V in our nomenclature.

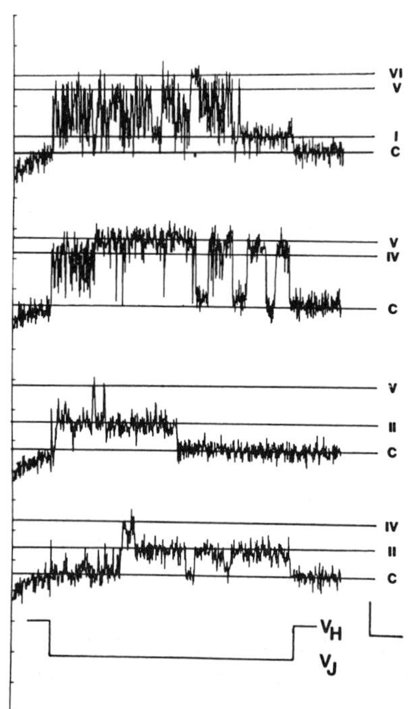

Fig.4 Openings of sodium channel modified by a mixture of S-DPI and Pbtx-3. Openings were elicited by repetitive hyper- and depolarization of a cell-attached membrane patch. Hyperpolarizing potential was 40 mV negative to RP and depolarization was 40 mV positive to RP. The extracellular medium was exactly as stated in Fig.1. In addition the pipette medium contained 50 µM S-DPI, 2 µM Pbtx-3 and 1 mM $BaCl_2$. Horizontal bar: 200 ms; vertical bar: 0.5 pA; low pass filtering: 300 Hz. The open sodium channel now frequently fluctuates between the different open states as indicated.

The inactivation kinetic of these steady state sodium channel openings was investigated in more detail. Sodium channel openings were recorded for several minutes at a membrane potential 20 mV positive to the membrane resting potential. Channel openings were divided arbitrarely into two populations. One with open channel current <

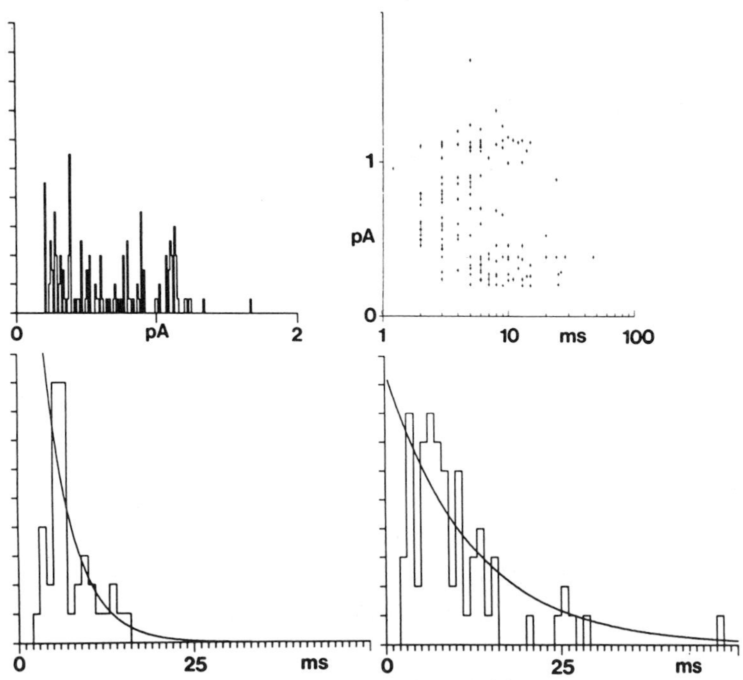

Fig.5 Evaluation of inactivation kinetic of the Pbtx-3 modified sodium channel. Sodium channel openings were registered at maintained depolarization for about 7 min. Same experimental conditions as for Fig.3.
Upper left: Amplitude histogram of detected sodium channel openings. Vertical scaling: 2 counts per division. 150 events were detected. Due to the different channel open states the amplitude scatter between 0.2 pA and 1.7 pA.
Upper right: Amplitude (pA) versus open time (ms) scatter plots of the data shown in the amplitude histogram.
Lower left: Histogram of dwell times in the open states for the opening population with single channel currents > 0.9 pA. Vertical scaling: 1 count per division. 39 events were counted totaly and the time constant was calculated to be 4.1 ms.
Lower right: Histogram of dwell times in the open states for the population with an open channel current < 0.5 pA. 68 events were counted in all and the time constant was calculated to be 11.9 ms.

0.5 pA and another population with an open channel current > 0.9 pA. The population with smaller single channel currents exerted a mean open time of 11.9 ms in contrast to the other population that exerted one of 4.1 ms. This clearly indicates that inactivation kinetic of the larger events differs from that of the smaller events.

What can now be the underlying mechanism for such a complex kinetic behaviour? Let us consider first the gating of a hypothetical homodimeric sodium channel (for the sake of simplicity the number of subunits is reduced from the proposed 6 to only 2. The conclusions remain, however, qualitatively the same). When the transition to the open state of the sodium channel is required prior to inactivation - as is generally acknowledged for this channel protein (for a recent treatment of this topic, see Yue et.al. (1989)) - then what should be observed in most of the patch clamp recordings are complete transitions from the fully closed to the fully open and from this to the fully inactivated state of the channel as exemplified in Fig.6. This is the behaviour of cardiac sodium channels normally observed under physiological conditions. Due to the large differences of the rate constants, k_1 and k_2, the intermediate states should be very shortlived and therefore not visible within the normal resolution of the patch clamp technique. Such conformational transitions of channel protomeres would obey the formalism of Monod et.al. (1965) where the movement of a charged voltage sensor in the transmembrane electrical field replaces the conformation stabilizing binding of a ligand to the protomer.

The situation is completely altered when an additional effector, i.e. a ligand like Pbtx-3, is allowed to bind to the channel protomer and alter its conformation. Assume that the modifier binds to and stabilizes the closed state of the channel protomer. Then the effects of depolarization and ligand binding would be antagonistic.

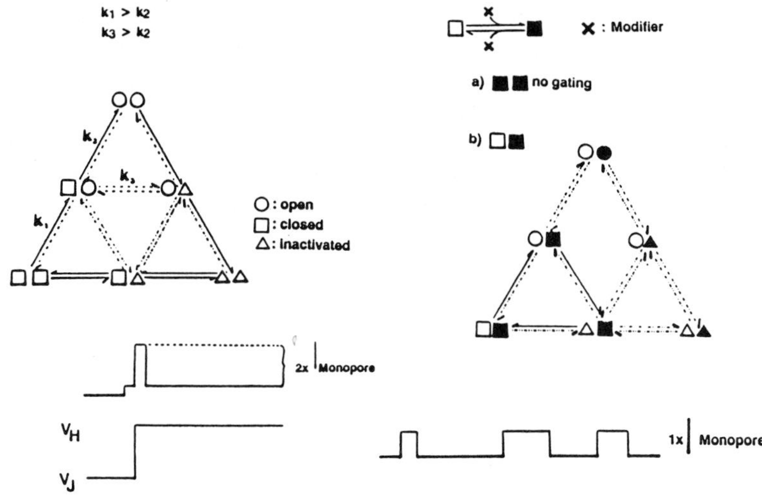

Fig.6: Gating of a hypothetical dimeric channel.
Left: Gating of the unmodified channel. When the intermediate states of the channel are shortlived (compared to the fully closed, open and inactivated state), then the expected single channel trace would contain openings with twice the conductance of a monopore soon after the depolarization. After complete opening of the dimer, the whole complex will inactivate soon. Reopening of the complex at maintained depolarization will (almost always) not occur.
Right: Gating of the modified complex. In case a) two modifiers are bound to the complex, the result will be no gating - the complex is stabilized in the fully closed state. In case b) only one modifier is bound to one protomere of the complex. Upon partial opening (the single channel amplitude will be exactly that of the monopore in this case) full opening of the complex is not possible because of the stabilizing effect of the modifier on the closed state. Instead the unmodified protopore will inactivate and then undergo a conformational change to the closed state again for reasons of symmetry.

Upon binding of Pbtx-3 to both protomeres the whole complex would be stabilized in the closed state and no gating of the sodium channel would result at all. Upon binding of Pbtx-3 to just one of these protomeres, however, partial opening of the complex would be expected. As activation of the whole complex is required for inactivation, however, the open time of this partial open complex should be prolonged - as a transition of the partial opened complex to the inactivated state is energetically unfavourable. Upon partial inactivation the stabilizing effect of Pbtx-3 on the closed state will force the whole complex into the fully closed state - again making reopening at maintained depolarization possible.

As exemplified above the oligopore nature of the cardiac sodium channel brings new dimensions into modelling of drug actions on channel proteins. More detailed insights into possible mechanisms of protochannel coupling will obviously require the direct comparison of protein organization and functional electrical activity of the channel complex. This is being presently approached by a particularly designed reconstitution strategy (Schindler (1989)) and by applying techniques of molecular biology, i.e. cloning, expression and modification of channel subunits (for review see Jan & Jan (1989)) to the cardiac sodium channel.

ACKNOWLEDGEMENTS

We thank Prof. H. Glossmannn (Institute for Biochemical Pharmacology, Innsbruck) for his gift of S-DPI to us, Dr. D. Römer and Dr. E. Rissi (Sandoz AG, Basel) for providing racemic DPI, B. Spreitzer for preparing myocytes and typewriting of the manuscript and L. Patassy for critical reading of the manuscript.
This work was supported by the Austrian Research Fund (S 45) and two travel-fellowships to G.J. and W.S. by the "Amt der Steiermärkischen

Landesregierung, Abt. für Wissenschaft und Forschung".

REFERENCES

Cachelin AB, de Peyer JE, Kokubun S, Reuter H (1983). Sodium channels in cultured cardiac cells. J Physiol 340: 389-401

Catterall WA, Gainer M (1985). Interaction of Brevetoxin A with a new receptor site on the sodium channel. Toxicon 23/3: 497-504

Chinn K, Narahashi T (1986). Stabilization of sodium channel states by Deltamethrin in mouse neuroblastoma cells. J Physiol 380: 191-207

Fox JA (1987). Ion channel subconductance states. J Membr Biol 97: 1-8

Hodgkin AL, Huxley AF (1952). A quantitative description of membrane current and its application to conduction and excitation in nerve. J Physiol 117: 500-544

Jan LY, Jan YN (1989). Voltage-sensitive ion channels. Cell 56: 13-25

Kazachenko VN, Geletyuk IJ (1984). The potential-dependent K^+ channel in molluscan neurones is organized in a cluster of elementary channels. BBA 773: 132-142

Koshland DE, Nemethy G Jr., Filmer D (1966). Comparison of experimental binding data and theoretical models in proteins containing subunits. Biochemistry 5/1: 365-385

Kunze DL, Lacerda AE, Wilson DL, Brown AM (1985). Cardiac Na-currents and the inactivating, reopening, and waiting properties of single cardiac Na-channels. J Gen Physiol 86: 691-719

Monod J, Wymann J, Changeux J (1965). On the nature of allosteric transitions: A plausible model. J Mol Biology 12: 88-118

Quandt FN, Narahashi T (1982). Modification of single Na^+ channels by Batrachotoxin. PNAS 79: 6732-6736

Schindler H, Schreibmayer W (1985). Synchronization of electrically and chemically excitable ion channels. In Changeux JP, Hucho F, Maelicke A, Neumann E (eds.): "Molecular Basis of Nerve Activity", de Gruyter, 387-397

Schindler H (1989). Planar lipid-protein membranes: Strategies of formation and of detecting dependencies of ion transport functions on membrane conditions. In Fleischer S, Fleischer B (eds.): "Methods in Enzymology", Academic Press, Orlando, Florida, 171: 225-253

Schreibmayer W, Hagauer H, Tritthart HA, Schindler H (1985). Potassium channels in rat ventricular cells show voltage-dependent outward rectification and non-linear voltage to current relation. In Changeux JP, Hucho F, Maelicke A, Neumann E (eds.) "Molecular Basis of Nerve Activity", de Gruyter, 145-152

Schreibmayer W, Tritthart HA, Zernig G, Piper HM (1985). Single voltage-dependent and outward rectifying potassium channels in isolated rat heart cells. Eur Biophysics J 11: 259-263

Schreibmayer W, Kazerani H, Tritthart HA (1987). Towards a mechanistic interpretation of the action of Toxin II from anemonia sulcata on the cardiac sodium channel. BBA 901: 273-282

Schreibmayer W (1989). Sublevel characterization of the cardiac sodium channel by combined action of ATX-II, veratridine and DPI 201-106. Arch Pharmacol 339: R1

Schreibmayer W, Tritthart HA, Schindler H (1989). The cardiac sodium channel shows a regular substate pattern indicating synchronized activity of several ion pathways instead of one. BBA submitted

Yoshii M, Narahashi T (1984). Patch clamp analysis of veratridine-induced sodium channels. Biophys J 45: 184a

Yue DT, Lawrence JH, Marban E (1989). Two molecular transitions influence cardiac sodium channel gating. Science 244: 349-352

Excitation-contraction coupling in heart and smooth muscle

Calcium Channel Modulators in Heart and Smooth Muscle: Basic Mechanisms and Pharmacological Aspects
S. Abraham and G. Amitai, editors
© 1990, VCH, Weinheim/Deerfield Beach, FL and Balaban, Rehovot/Philadelphia

Inhibition of sarcoplasmic reticular Ca release in canine ventricular muscle by the Ca channel agonist, BAY K 8644: Are sarcolemmal Ca channels involved?

D. BOSE[1], R. BOUCHARD[2], J.K. SAHA[2], L.V. HRYSHKO[2], T. KOBAYASHI[2], AND T. CHAU[2]

[2]*Departments of Pharmacology and Therapeutics,* [1]*Anesthesia and Internal Medicine, University of Manitoba, Faculty of Medicine, Winnipeg, MB, Canada R3E 0W3*

INTRODUCTION

Ca channel agonists, e.g. BAY K 8644 (Schramm et al, 1983; Thomas et al, 1985a; Hess et al, 1984) increase contractility in rhythmically stimulated cardiac preparations. We have recently reported that this agent, paradoxically, inhibits Ca release from the sarcoplasmic reticulum (SR) in the intact isolated canine ventricular preparation by hastening diastolic 'leak' of this ion (Bose et al, 1986; Hryshko and Bose, 1988; Bouchard et al, 1989; Hryshko et al, 1989a,b). Such a combination of actions is rather unique and lends promise to the prospect of achieving positive inotropy with less intracellular calcium overload. The present report addresses two issues:

a) do other agents, e.g. ryanodine and caffeine, which inhibit the SR, also behave in a manner similar to BAY K 8644 ?
b) is inhibition of Ca release from the SR related to Ca channel agonism? Such a study was necessary because of recent evidence that the SR Ca release site in the SR has structures which exhibit channel-like behavior.

MATERIAL AND METHODS

Thin right ventricular trabecula (usually <1 mm diameter) were obtained from mongrel dogs (5-12 kg) of either sex, anesthetized with pentobarbital (35 mg/kg iv). The preparations were set up for recording isometric tension at optimum length or transmembrane electrical activity using standard KCl (3 M)-filled glass microelectrodes having tip resistances of 8-20 MΩ (Hryshko et al, 1989 a,b). The ambient medium was Krebs Henseleit solution (in mM - NaCl 118; KCl 4.7; KH_2PO_4 1.4; $MgSO_4$ 1.2; $CaCl_2$ 2.5; $NaHCO_3$ 25; dextrose 11) maintained at a temperature of 37°C, as previously described (Hryshko et al, 1989a). pH was kept at 7.4 by bubbling the medium with a mixture of 95% O_2 - 5% CO_2. Rapid cooling contracture was obtained by rapidly cooling the muscle from 37°C to nearly 0°C in less than 1 sec by switching physiological solutions (Hryshko et al, 1989b). In these experiments HEPES (10 mM) was used as a buffer instead of $NaHCO_3$ and the solution was aerated with 100% O_2. The muscles were stimulated with rectangular electrical pulses (width ~ 1 ms; interval 2 s; amplitude 100% above threshold) delivered by a computer-controlled stimulator (Boyechko & Bose, 1984) through bipolar platinum electrodes. For rest-potentiation studies, the normal regular stimulus train was interrupted by randomly varying periods of rests. Scattering of coherent He-Ne laser light by a 300 μm bundle of right ventricular trabecula (Scattered light intensity fluctuation; SLIF) was studied as an index of SR Ca overloading (Lakatta and Lappe, 1981; Bose et al, 1988b; Hryshko et al, 1989b). The optical signal was collected by a photodiode during 30 sec of rest and recorded on a Nicolet digital oscilloscope. The digitized data were transferred to a microcomputer running ASYST (MacMillan Software Co.), to compute the magnitude spectrum of the scattered light intensity fluctuation.

Drugs and Chemicals: Racemic BAY K 8644 and the (-) and (+) enantiomers were kind gifts from Dr. A. Scriabine (Miles Institute of Preclinical Research, USA). Ryanodine was purchased from LVOG, USA. Caffeine was obtained from Sigma Chemicals (USA). Chemicals for making physiological solutions were obtained from Fisher Chemicals (Canada). Drugs were dissolved in deionized distilled water except in the case of BAY K 8644 which was dissolved in ethanol and protected from light during storage and experimentation.

Statistical Analysis: Paired (self-controlled) experiments were analyzed by the Student's paired t-test. Experiments involving more than two groups were subjected to ANOVA followed by comparison with Duncan's multiple range test. All data have been expressed as mean ± SEM. A p value of <0.05 was considered to be statistically significant.

RESULTS

Contractions following rest for 15-480 s showed a transient potentiation which was maximum after rest periods between 60-120 s. In the presence of a racemic mixture of BAY K 8644 (1 μM),

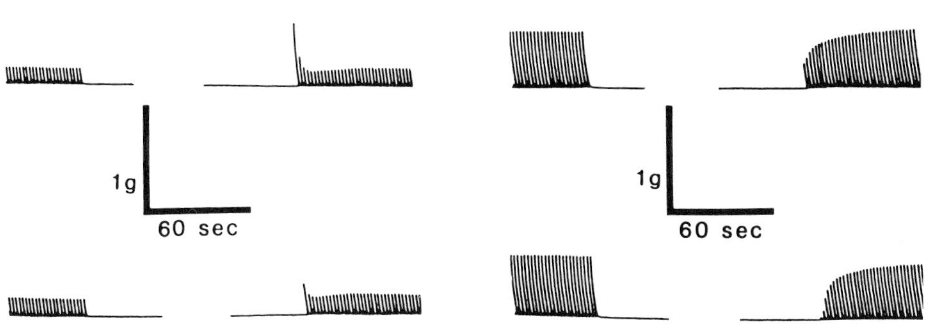

Figure 1. Isometric tension recording from canine trabecula. Control muscle exhibits rest potentiation which is greater at 2 min (top trace) than at 8 min (bottom trace). BAY K 8644 causes rest-depression at both intervals. *Reprinted from Hryshko et al, Am J Physiol 257:H407-H414, 1989.*

the post-rest response was decreased (Figure 1). Inhibition of post-rest response was directly proportional to the rest duration. Inhibition of rest-response was seen even with a BAY K 8644 concentration as low as 0.03 μM. At this concentration the increase in contractility was less than 25%. However the onset of inhibition of the post-rest response was slower than that observed at the higher concentrations.

(-) BAY K 8644 (0.1 μM), a pure Ca channel agonist enantiomer of BAY K 8644 (Schramm and Towart, 1985), also caused inhibition of post-rest response (Figure 2). On the other hand (+) BAY K 8644, a Ca channel antagonist, did not cause a decrease in post-rest response at 0.1 μM concentration, even though it decreased contractility of rhythmically stimulated trabecula (Figure 2). Another Ca channel blocker, nifedipine (0.1 μM), caused the greatest increase in post-rest response. Addition of (+) BAY K 8644, in amounts sufficient to reverse the positive inotropic effect and increased action potential plateau amplitude and duration produced by(-) BAY K 8644 (0.1 mM) , did not reverse the inhibition of post-rest response (Figure 3).

The effect of the racemic mixture of BAY K 8644 on rapid cooling contracture (RCC) was compared with that of caffeine (5 mM) and ryanodine (0.03 μM). RCC was obtained immediately after a train of regular stimuli to assess the loading of the SR by

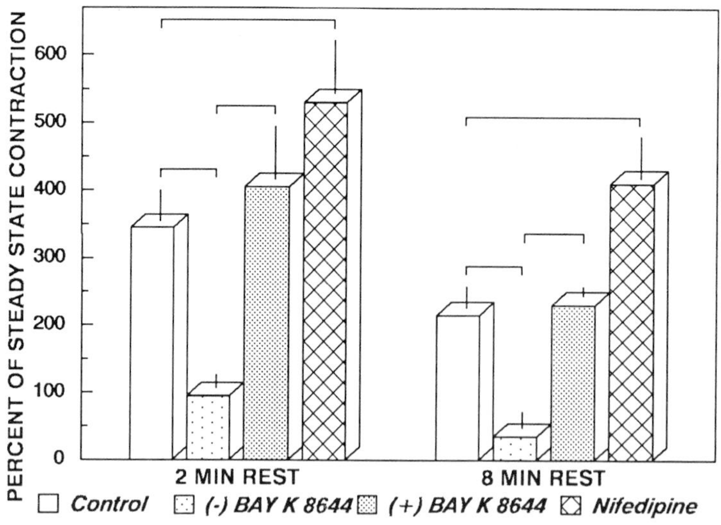

Figure 2. Effect of (-) and (+) BAY K 8644 and nifedipine, all at 0.1 μM concentration, on rest-potentiation. (-) BAY K 8644 (n=6) caused 100% increase in contractile force compared to control (n=16) while (+) BAY K 8644 (n=6) and nifedipine (n=3) reduced force by 50%. Bars joined by a horizontal line are different from each other ($p<0.05$; ANOVA - Duncan's multiple range test).

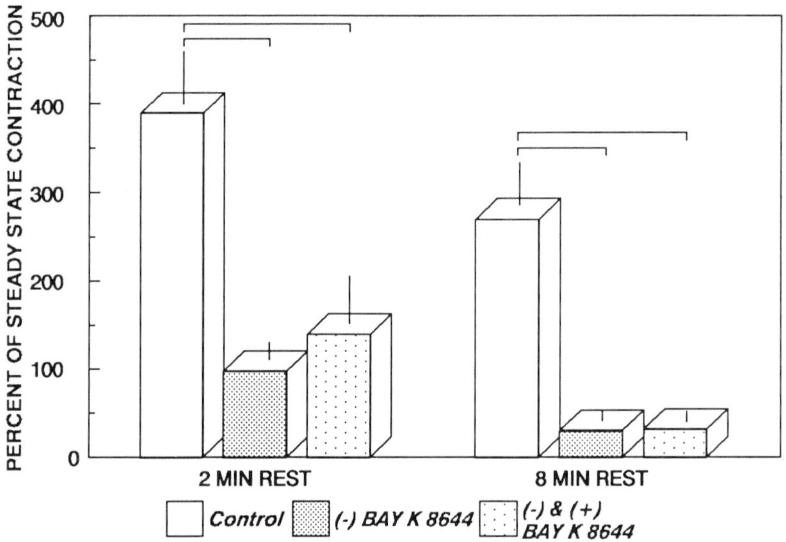

Figure 3. Effect of (-) BAY K 8644 (0.1 μM) on 2 and 8 min post-rest response in 6 experiments. Notice that the depression of the post-response by the Ca channel agonist enantiomer was not reversed by addition of (+) BAY K 8644 in concentrations which couneracted completely the increase in inotropy, action potential duration and plateau amplitude. Horizontal line denotes difference between bars significant at $p<0.05$

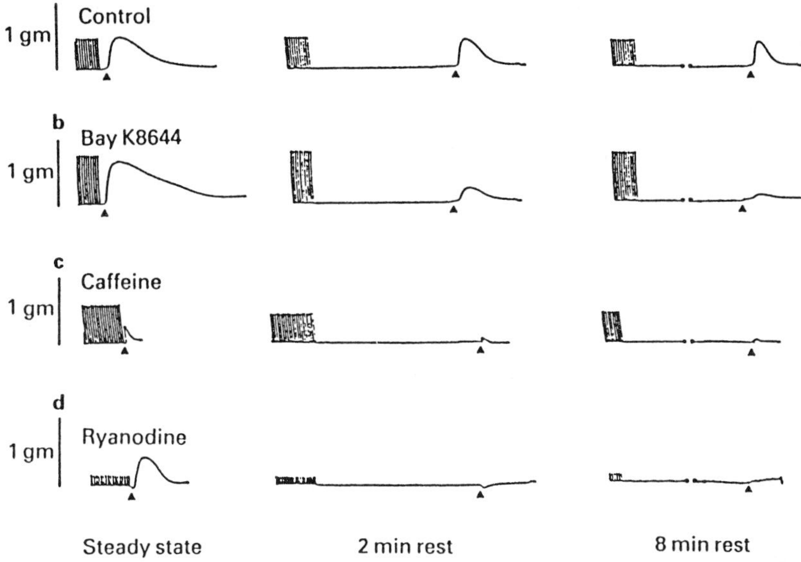

Figure 4. Rapid cooling contracture immediately after a train (steady state) and after 2 and 8 min rest in control muscle and after BAY K 8644 (1 μM), caffeine (5 mM) and ryanodine (0.01 μM). Reprinted from Bouchard et al, Br J Pharmacol 97:1279-1291, 1989.

Ca pump-mediated uptake of the ion from the cytoplasm. RCC was also obtained after 2 and 8 min periods of rest. This was done to measure the diastolic loss of Ca from the SR. As can be seen in Figure 4, RCC was not decreased by BAY K 8644 or ryanodine but was markedly inhibited by caffeine immediately after a regular train. Despite the variable effect of these agents on RCC immediately after stimulation all the agents reduced the amplitude of the post-rest RCC.

Inhibition of Ca release from the SR was also confirmed by examining the effect of BAY K 8644 on SLIF of He-Ne laser light. Figure 5 shows that BAY K 8644 (1 μM) reduced the magnitude of SLIF. This is in contrast to the increase in SLIF seen with ouabagenin (1 μM), which produces inotropy by Na pump inhibition.

DISCUSSION

The main finding reported in this paper is that the Ca channel agonist BAY K 8644, causes inhibition of Ca release from the SR of isolated canine right ventricular trabecula by depleting it during diastole and in this regard it behaves like ryanodine. Previous studies have reported on the ability of the racemic mixture

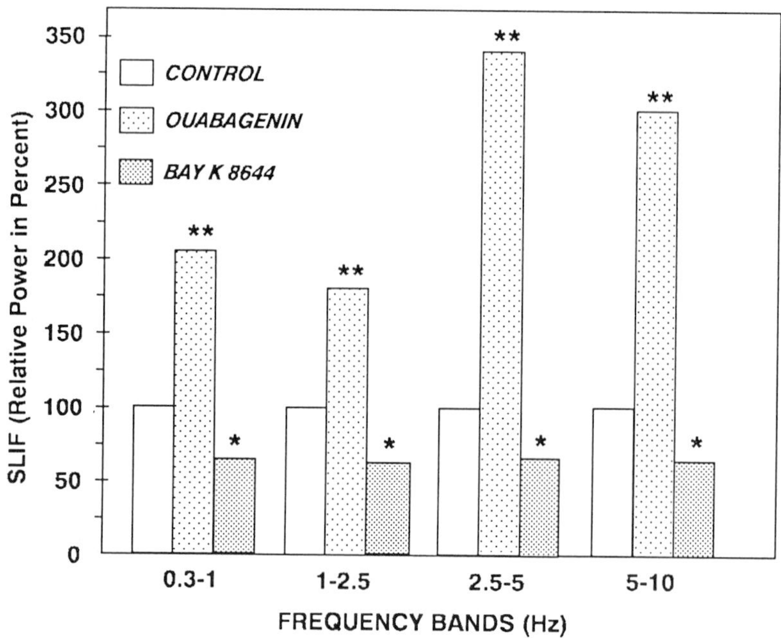

Figure 5. Effect of BAY K 8644 (1 mM) and ouabagenin (1 mM) on SLIF power. The area under the FFT power curve for various frequency bands ater drug treatment have been expressed as a percent of the area under the curve for the same frequency range in the untreated muscle. n=6. **=p<0.01; *=p<0.05 compared to control. Analysis by Student's paired t-test.

of BAY K 8644 to increase the magnitude of the Ca current (Thomas et al, 1985a) due to the increase in the open time of the 'L'-type Ca channels (Hess et al, 1984). A previous study which examined the effect of this agent on skinned muscle fiber (Thomas et al, 1985b) failed to demonstrate any effect on the SR or contractile apparatus. Furthermore, BAY K 8644 also did not have any effect on the Ca handling of isolated SR vesicles (Zorzato, 1985; Kim and Bose, unpublished observations). It is therefore of interest to note that the ability of BAY K 8644 to impair SR function requires an intact cellular preparation. Post-rest potentiation was used to assess changes in the release of Ca from the SR. In a previous study we (Bose et al, 1988a) have suggested that the increased post-rest response can be explained by enhanced Ca release from the SR or a gradual decrease in the refractoriness of the Ca release channels during rest leading to increased release during the post-rest response. The latter possibility, however, is less likely because the time constant of recovery of the Ca release process, as determined by studies on skinned fibers by Fabiato (1985) and mechanical restitution (e.g., Schouten et al, 1987; Bose et al, 1988; Bouchard and Bose, 1989), is relatively fast (~ 1-3 sec). Another process which occurs concomitantly with rest and becomes progressively greater in magnitude with duration of rest is a loss of Ca from the SR. It would appear that the initial loss of Ca from the release site is more than offset by the recirculating fraction of Ca coming from the uptake site. This results in rest potentiation and it peaks around 60-120 s in the canine. Eventually recirculation stops and the Ca 'leak' from the SR predominates leading to rest-depression. Cellular Ca content shows a continuous decrease with rest (Bridge, 1986; Lewartowski et al, 1984).

The concept of compartmentation of Ca within the SR is favored by the difference in the Ca content of the SR estimated by rest-potentiation and RCC. While the magnitude of rest-potentiation is bell-shaped, that of RCC decreases continuously with increasing duration of rest (Bridge, 1986; Bouchard et al, 1989; Hryshko et al, 1989). Thus it seems that RCC better represents total Ca in the SR while post-rest potentiation represents Ca in the SR release compartment. If this interpretation is correct then it can be concluded that BAY K 8644 decreases both the total Ca content of the SR as well as of the release compartment. This conclusion is also supported by the decrease in the magnitude of the scattered light intensity fluctuation signal with BAY K 8644. In the case of other inotropic agents, e.g., ouabagenin (Bose et al, 1988) or high external Ca (Lakatta and Lappe, 1981) there is an increase in the fluctuation of scattered coherent light.

The presence of a large RCC immediately following a regular train of stimulation indicates that the ability of the SR to take up Ca from the myoplasm is not impaired by BAY K 8644 even though it cannot retain it. Similar effects were also seen with ryanodine which is also believed to cause enhanced leakage of Ca from the SR (Hilgeman, 1986a,b; Bose et al, 1988; McLeod and Bers, 1987). Conversely, caffeine blocks RCC immediately after a train of stimuli. This is consistent with caffeine blocking Ca uptake by the SR (Weber and Herz, 1968) as has been previously suggested (Hryshko and Bose, 1989a) (Figure 6).

It is tempting to suggest that the Ca 'leaked' from the SR release compartment is rapidly removed from the myoplasm via the Na-Ca exchange in the sarcolemma. Since the Ca leaving the cell during rest does not cause any obvious contractile activation, either the sarcolemmal Ca extrusion mechanism is very effective or its location is close to the site of release from the SR (i.e. the

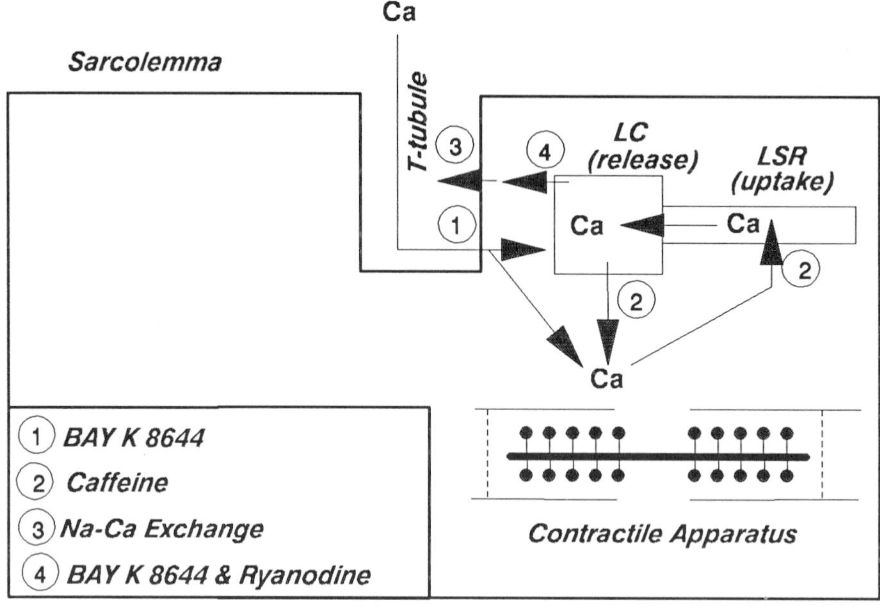

Figure 6. Schematic depiction of Ca movement in the canine ventricular trabecular cell. The proposed sites of action of BAY K 8644, ryanodine and caffeine are shown. 'Leak' of SR Ca during diastole is promoted by BAY K 8644 and ryanodine in a similar fashion while caffeine acts in a different manner (see Discussion).

t-tubule-lateral cisternal apposition) (Figure 6). Whether BAY K 8644 has any direct effect on the Ca extrusion mechanism in the sarcolemma is not known, but its effects can be explained on the possible basis of an action on the Ca release channels in the SR. It is interesting to note that procedures which impair sarcolemmal Ca extrusion mechanisms, e.g. low external Na or Na pump inhibition reduce post-rest depression caused by BAY K 8644 (Bose et al, 1987).

Is the effect of BAY K 8644 on the SR by a mechanism related to the one by which it keeps sarcolemmal Ca channels open? This is unlikely for several reasons. Even though the pure Ca channel agonist enantiomer of BAY K 8644 also causes inhibition of the SR as well as increased contractility, the Ca channel antagonist enantiomer can reverse only its effect on contractility and transmembrane electrical response but not that on the SR. It is therefore reasonable to propose that the sarcolemmal and SR effects of BAY K 8644 are separate. It is interesting to note that a variety of

Ca channel antagonists do not inhibit rest-potentiation (Saha et al, 1989). On the contrary the decrease in contractility of the regularly stimulated beat is accompanied by a relative preservation of rest potentiation. However, there are quantitative differences in the effect of the Ca channel antagonists on rest-potentiation. The rank order of rest-potentiation with the various antagonists is: (+) BAY K 8644 < nitrendipine < nifedipine (Saha et al, 1989). This can be interpreted to mean that nitrendipine and (+) BAY K 8644 are also able to depress SR function, albeit to a lesser extent than (-) BAY K 8644 can while nifedipine has the least effect on the SR. Therefore it is possible that the unique chemical structure of BAY K 8644 confers to it an additional *ryanodine-like* effect, in addition to its ability to increase Ca current (Figure 6). A direct demonstration of any interaction between BAY K 8644 and ryanodine at the SR Ca release site remains to be shown.

The unusual combination of effects of BAY K 8644 on the sarcolemmal Ca channels and SR can be predicted to lead to a very salutory inotropic response, especially at slower heart rates. It should cause a lesser degree of intracellular Ca overload than inotropic agents, e.g., digitalis, which cause inhibition of Ca extrusion (e.g. Na pump inhibition). Intracellular Ca overload is well known to cause triggered automaticity due to oscillatory Ca release from the SR (Kass et al, 1978) and impaired mitochondrial energy production (Vassalle and Lin, 1979). It remains to be seen if these toxic effects can be avoided during Ca channel agonist-induced inotropy. It is unfortunate that the vasoconstrictive effect of Ca channel agonists has prevented their utilization for therapeutic purposes. It may be of interest to see if by combining a vascular selective Ca channel antagonist with a Ca channel agonist, a better therapeutic margin can be obtained.

ACKNOWLEDGEMENTS

This work was supported by grants from the Manitoba Heart Foundation and Medical Research Council of Canada. LVH and RB received Canadian Heart Foundation Studentships.

REFERENCES

Bose D, Hryshko LV, King BW, Chau T. (1988a). Control of interval-force relation in canine ventricular myocardium studied with ryanodine. Br J Pharmacol 95:811-820.

Bose D, Kobayashi T, Bouchard RA, Hryshko LV (1988b). Scattered light intensity fluctuation in the canine ventricular myocardium: Correlation with inotropic drug effect. Canad J Physiol Pharmacol 66:1232-1238.

Bose D, Kobayashi T, Hryshko LV, Chau T. (1986). Depression of canine ventricular sarcoplasmic reticulum by the calcium channel agonist, BAY K 8644. In Dhalla NS, Pierce GN, Beamish RE (Eds): "Heart Function and Metabolism," Boston: Nijhoff, pp 207-220.

Bouchard, RA, Hryshko LV, Saha JK, Bose D et al, (1989). Effects of caffeine and ryanodine on depression of post-rest tension development produced by BAY K 8644 in canine ventricular muscle. Br J Pharmacol 97:1279-1291.

Boyechko G, Bose D (1984). A versatile computer-controlled biological stimulus sequencer. J Pharmacol Methods 12:45-52.

Bridge, JHB. (1986). Relationship between the sarcoplasmic reticulum and sarcolemmal calcium transport revealed by rapidly cooling rabbit ventricular muscle. J Gen Physiol 88:437-473.

Fabiato, A (1985). Time and calcium dependence of activation and inactivation of calcium-induced release of calcium from the sarcoplasmic reticulum of a skinned canine cardiac Purkinje cell. J Gen Physiol 82:247-289.

Hess P, Lansman JB, Tsien RW (1984). Different modes of calcium channel gating behavior favored by dihydropyridine Ca agonists and antagonists. Nature 311:538-544.

Hilgeman, DW (1986a). Extracellular calcium transients and action potential configuration changes related to poststimulatory potentiation in rabbit atrium. J Gen Physiol 87:675-706.

Hilgeman DW (1986b). Extracellular calcium transients at single excitations in rabbit atria measured with tetramethyl murexide. J Gen Physiol 87:707-735.

Hryshko LV, Bose D. (1988). Impairment of Ca release from mammalian ventricular sarcoplasmic reticulum by the calcium channel agonist BAY K 8644. Br J Pharmacol 94:291-292.

Hryshko LV, Bouchard R, Chau T, Bose D. (1989a). Inhibition of rest potentiation in canine ventricular muscle by BAY K 8644: comparison with caffeine. Am J Physiol 257:H399-H406.

Hryshko LV, Kobayashi T, Bose D. (1989b). Possible inhibition of canine ventricular sarcoplasmic reticulum by BAY K 8644. Am J Physiol 257:H407-H414.

Kass RS, Lederer WJ, Tsien RW, Weingart R. (1978). Role of calcium in transient inward currents and aftercontractions induced by strophanthidin in cardiac Purkinje fibers. J Physiol (Lond) 281:178-208.

Lakatta EG, Lappe DL (1981). Diastolic scattered light fluctuations, resting force and twitch force in mammalian cardiac muscle. J. Physiol. (Lond) 315:369-394.

Lewartowski B, Pytkowski B, Janczweski A (1984). Calcium fraction correlating with contractile force of ventricular muscle of guinea-pig heart. Pflug Arch 401:198-203.

MacLeod KT, Bers D (1987). Effects of rest duration and ryanodine on changes of extracellular [Ca] in cardiac muscle from rabbits. Am J Physiol 253:C398-C407.

Saha JK, Hryshko LV, Bouchard RA, Chau T, Bose D (1989). Analysis of the effects of (-) and (+) isomers of the 1,4-dihydropyridine calcium channel agonist, BAY k 8644 on post-rest potentiation in the canine ventricular muscle. Canad J Physiol Pharm *in press.*

Schouten VJA, van Deen JK, de Tombe P, Verveen AA (1987). Force-interval relationship in heart muscle of mammals. A calcium compartment model. Biophys J 51:13-26.

Schramm MG, Thomas G, Towart R, Franckowiak G. (1983). Novel dihydropyridines with positive inotropic action through activation of Ca channels. Nature (Lond) 309:535-537.

Schramm M, Towart R. (1985). Modulation of calcium channel by drugs. Life Science 37:1843-1860.

Thomas G, Chung M, Cohen CJ (1985a). A dihydropyridine (BAY K 8644) that enhances calcium currents in guinea pig and calf myocardial cells: a new type of positive inotropic agent. Circ Res 56:87-96.

Thomas G, Gross R, Pfitzer G, Ruegg JC (1985b). The positive inotropic dihydropyridine BAY K 8644 does not affect calcium sensitivity or calcium release of sarcoplasmic reticulum of skinned cardiac fibers. Naunyn-Schmiedeberg's Arch Pharmacol 328:378-381.

Weber A, Herz R (1968). Relationship between caffeine contracture of intact muscle and the effect of caffeine on reticulum. J Gen Physiol 52:750-759.

Vassalle M, Lin CL (1979). Effect of calcium on strophanthidin-induced electrical and mechanical toxicity in cardiac Purkinje fibers. Am J Physiol 236:H689-H697.

Zorzato F, Volpe P, Salviati G, Margreth A (1985). Ca^{2+} channel agonist BAY K 8644 does not elicit Ca^{2+} release from skeletal muscle sarcoplasmic reticulum. FEBS Lett 186:255-258.

Calcium Channel Modulators in Heart and Smooth Muscle:
asic Mechanisms and Pharmacological Aspects
S. Abraham and G. Amitai, editors
© 1990, VCH, Weinheim/Deerfield Beach, FL and Balaban, Rehovot/Philadelphia

Antigen-stimulated phosphorylation of myosin and histamine secretion from RBL-2H3 cells

ITZHAK PELEG*, RUSSELL I. LUDOWYKE, MICHAEL A. BEAVEN, AND ROBERT S. ADELSTEIN
National Heart, Lung, and Blood Institute, NIH, Bethesda, MD, USA 20892

INTRODUCTION

All eukaryotic cells contain the contractile proteins actin and myosin (for review see Sellers and Adelstein, 1987). The exact function of these proteins is still under study, but they appear to be necessary for cytokinesis (Schroeder, 1973), shape change in platelets (Daniel et al., 1984) and endothelial cell retraction (Wysolmerski and Lagunoff, 1988). Actin is a globular protein with a molecular weight of 42 kDa. It is capable of forming long filamentous polymers with two strands wrapped in a helix. Myosin is a hexamer with a molecular weight of 480 kDa which can form bipolar filaments. In both muscle and nonmuscle cells myosin is composed of two heavy chains (200 kDa each) and two pairs of light chains. The molecular weight of the myosin light chains (MLCs) varies depending on the tissue and species, but appears to be remarkably constant for vertebrate nonmuscle and smooth muscle cells. In these cells myosin is composed of two phosphorylatable or regulatory light chains (20 kDa) and two light chains of 17 kDa, in addition to two heavy chains.

*Present address: Department of Cellular Biochemistry, Hebrew University-Hadassah Medical School, Ein Kerem, Jerusalem, Israel

In vertebrates the regulation of contractile activity in smooth muscle and nonmuscle cells differs from that found in skeletal and cardiac muscle cells (Sellers and Adelstein, 1987). In the case of smooth muscle and nonmuscle cells, phosphorylation of the 20 kDa MLC appears to play a major role in regulating the actin-dependent hydrolysis of MgATP by myosin. Phosphorylation also plays a major role in regulating myosin filament assembly (Scholey et al., 1980).

The enzyme responsible for initiating contraction of actin-myosin filaments in smooth muscle and nonmuscle cells is a substrate specific, Ca-calmodulin-dependent kinase, termed myosin light chain kinase (MLC kinase). The site phosphorylated on the 20 kDa MLC by this enzyme has been identified as being serine-19, a site that is present in both smooth muscle and nonmuscle MLCs. Although phosphorylation of the myosin heavy chain (MHC) has been described in vertebrate nonmuscle cells (Trotter, 1982), it is only recently that enzymes that appear to be responsible for this phosphorylation have been identified both *in situ* and *in vitro* (Ludowyke et al., 1989; Kawamoto et al., 1989).

In addition to being a substrate for MLC kinase, the 20 kDa MLC can also be phosphorylated on serine-1, serine-2 and threonine-9 by protein kinase C (Bengur et al., 1987). The effect of this phosphorylation *in vitro* is two-fold: 1) If myosin has previously been phosphorylated by MLC kinase, phosphorylation by protein kinase C acts to decrease the actin-activated MgATPase activity of this myosin. 2) If myosin is phosphorylated by protein kinase C when it is devoid of phosphate, phosphorylation by protein kinase C decreases the ability of MLC kinase to bind to and hence phosphorylate MLCs (Nishikawa et al., 1984).

To date, there have been few studies on the phosphorylation of myosin under *in situ* conditions by protein kinase C. Most of these studies have used phorbol esters to activate protein kinase C in platelets and nonmuscle cells (see e.g. Kawamoto et al., 1989). To study phosphorylation of myosin by protein kinase C *in situ* using a more physiological approach, we chose the rat basophilic leukemia cell line, RBL-2H3. These cells have receptors for IgE on their surfaces and these receptors can be aggregated by exposing the cells, first to a dinitrophenol-specific IgE and then to a specific antigen, dinitrophenylated bovine serum albumin (DNP-BSA). Aggregation of the IgE receptors on these cells results in two important events: 1) hydrolysis of phosphatidyl inositol bisphosphate with the generation of inositol trisphosphate and diacylglycerol, a known activator of protein kinase C and 2) the release of histamine (Beaven and Ludowyke, 1989). Our purpose was to see if antigenic stimulation of these cells would result in phosphorylation of MHCs and/or MLCs. If phosphorylation of myosin did occur we planned to identify and quantitate the sites phosphorylated in order to see if we could relate them to protein kinase C activation. Finally, we wanted to see if the time course of myosin phosphorylation could be related to the release of histamine from these cells.

RESULTS

Fig. 1 is a flow diagram outlining the procedures used in these experiments. RBL-2H3 cells were first sensitized with IgE (which itself did not alter phosphorylation of myosin) and then labeled with either 32P-orthophosphate or 35S-methionine. The labeled cells were stimulated with the antigen, DNP-BSA and an aliquot of the supernatant was removed for histamine assay. The cells were lysed and labeled cellular myosin was isolated by immunoprecipitation with an IgG antibody raised against human platelet myosin. The purified, washed pellet of IgG and myosin was solubilized in a buffer containing sodium dodecyl sulfate (SDS) for SDS-polyacrylamide gel electrophoresis (PAGE) or in buffer containing urea for 2-dimensional gel electrophoresis.

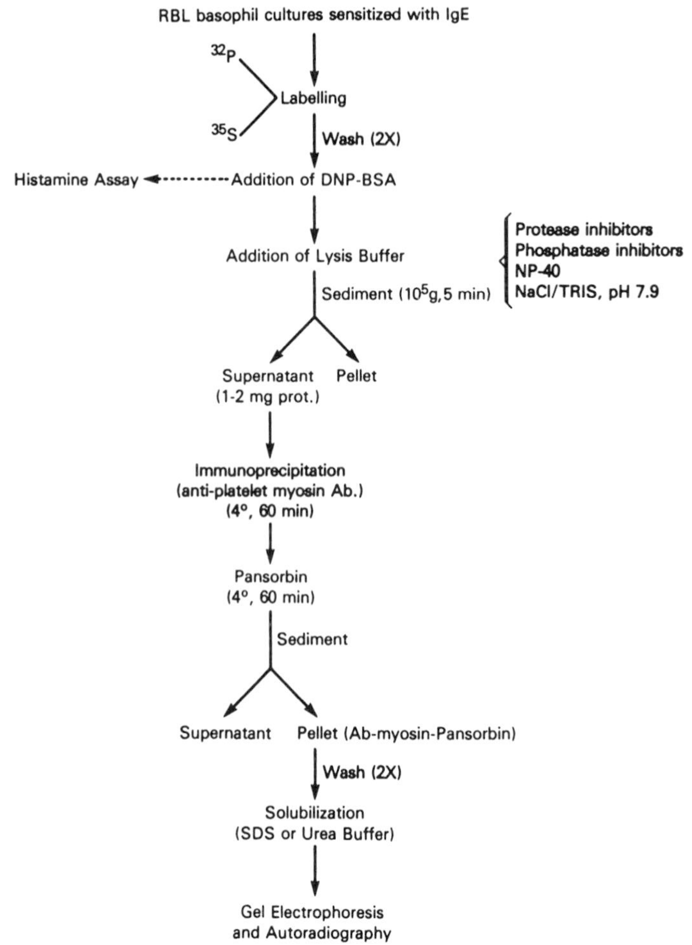

Fig. 1. Flow diagram outlining the procedures used to obtain myosin from RBL-2H3 cells.

Fig. 2A is an autoradiogram illustrating a time course for phosphorylation of the 200 kDa MHC as well as the 20 kDa MLC. The three right lanes of Fig. 2A show control samples that were incubated for the indicated times in the absence of DNP-BSA. These lanes show that prior to antigen stimulation, both the MHC and MLC contain radioactive phosphate. The eight lanes, starting at the left in Fig. 2A, show that the amount of phosphate incorporated into both the MHC and MLC increases following antigen stimulation and reaches a maximum at 10 min after

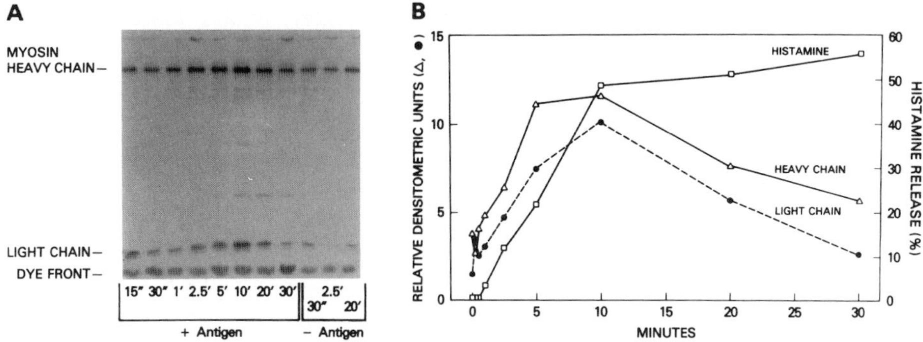

Fig. 2. Autoradiogram (A) and graph (B) showing the time course of MHC and MLC phosphorylation and histamine release. The details used in obtaining the autoradiogram are outlined in the text. In B, the amount of histamine released at each time point was calculated as a percent of total histamine (from Ludowyke et al., 1989).

which it declines. Fig. 2B is a graphic representation of this increase in myosin phosphorylation and also shows the time course of histamine release from these cells. Note that histamine release appears to parallel the time course of myosin phosphorylation.

Our next step was to identify the sites of phosphorylation in both the MLC and the MHC. Our task was made easier by having tryptic peptide maps of the phosphorylated MLCs from both chicken gizzard smooth muscle and human platelet myosin (see e.g. Kawamoto et al., 1989). These maps had been made following in vitro phosphorylation of the MLC with either MLC kinase or protein kinase C. Interestingly, both the chicken and human maps were the same for the same enzyme, showing that these phosphorylatable sites have been conserved between species.

Fig. 3 shows autoradiograms following tryptic hydrolysis of the 20 kDa MLC cut from SDS-polyacrylamide gels similar to those used in obtaining Fig. 2A. Control cells contain a single major phosphorylated tryptic peptide,

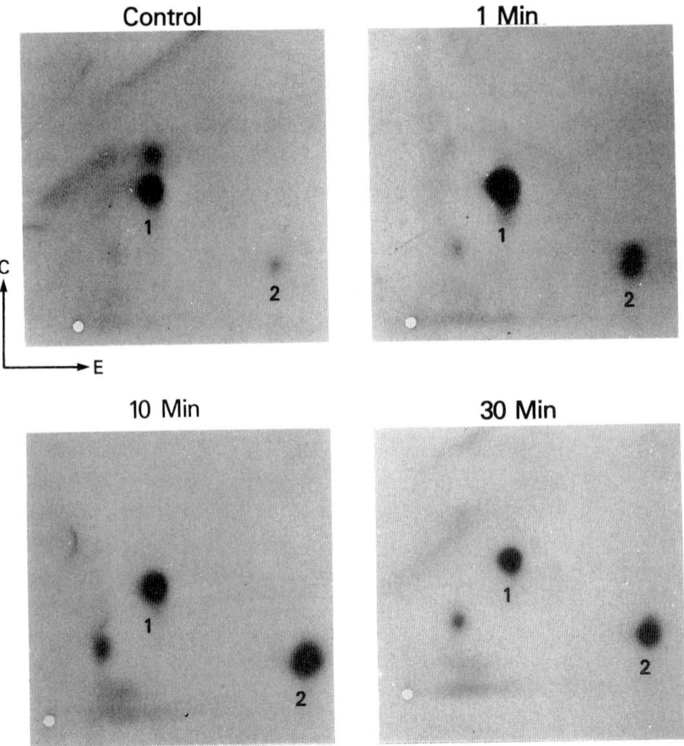

Fig. 3. Autoradiograms of 2-dimensional peptide maps of tryptic digests of 32P-labeled MLCs. Myosin was immunoprecipitated from control cells and cells stimulated with DNP-BSA for the indicated times, analyzed on SDS-PAGE, and the MLCs were subjected to tryptic digestion and peptide mapping. E indicates direction of electrophoresis toward the anode and C indicates direction of chromatography (from Ludowyke et al., 1989).

labeled 1, which can be recognized as the peptide containing serine-19, a site known to be phosphorylated by MLC kinase. (The small spot just above peptide 1 also contained serine-19, but results from a slightly different tryptic cleavage.) The presence of this phosphopeptide shows that, prior to stimulation, RBL-2H3 cells contain MLCs that have been phosphorylated by MLC kinase.

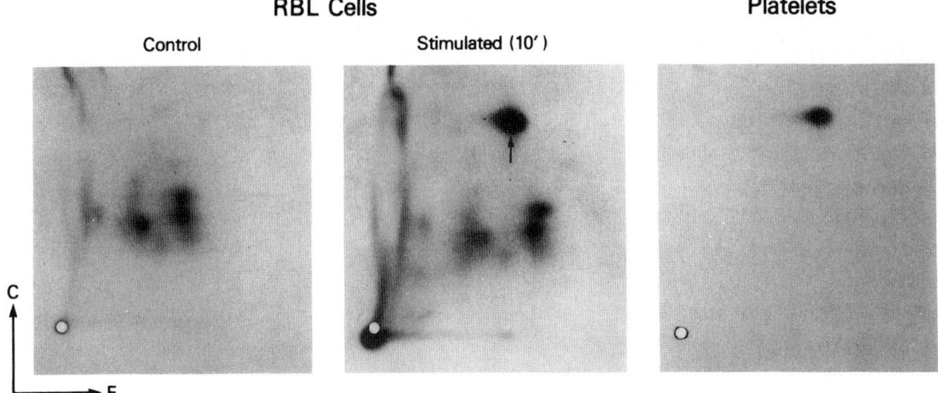

Fig. 4. Autoradiogram of 2-dimensional tryptic peptide maps of 32P-labeled MHCs. RBL-2H3 cell myosin was immunoprecipitated from 32P-labeled unstimulated (control) and stimulated (10 min) cells, analyzed by SDS-PAGE, and the heavy chain subjected to tryptic digestion and peptide mapping. The phosphopeptide produced following tryptic digestion of platelet MHC phosphorylated _in vitro_ by protein kinase C is shown at the right (from Ludowyke et al., 1989).

Following stimulation a second major phosphopeptide, labeled 2, appears. This peptide can be recognized by its location on the 2-dimensional map as containing phosphoserine 1 or 2, sites known to be phosphorylated by protein kinase C from _in vitro_ studies (Bengur et al., 1987).

Mapping of the phosphorylated tryptic peptides was also carried out with the MHC. Following immunoprecipitation of the myosin and migration in SDS-PAGE, the band corresponding to the 200 kDa MHC was cut out of the gel and digested with trypsin. Fig. 4 is an autoradiogram of tryptic peptide maps of the MHC before and 10 min after stimulation of RBL-2H3 cells with DNP-BSA. Prior to stimulation there appear to be 4 major tryptic phosphopeptides. Unlike the MLC phosphopeptides, these peptides and the kinase(s) that catalyze the phosphorylation of the MHC have yet to be identified in nonmuscle cells. Ten minutes after stimulation, the same peptides can still be seen, but now there is a single new phosphopeptide present, indicated with an arrow. Since the time course of increased MLC phosphorylation paralleled that of MHC phosphorylation (see Fig. 2B), we reasoned that, like peptide 2 in Fig. 3, the new heavy chain peptide in Fig. 4 could be due to activation of protein kinase C. To test this idea, we phosphorylated purified human platelet myosin with protein kinase C *in vitro* and digested the phosphorylated heavy chain with trypsin. Fig. 4, right panel, is an autoradiogram of the single heavy chain phosphopeptide obtained following 2-dimensional peptide mapping. This peptide comigrates with the single new phosphopeptide seen in the middle panel after stimulation of RBL-2H3 cells and suggests that this site was phosphorylated *in situ* by protein kinase C.

Having ascertained that the antigen-stimulated increase in MLC and MHC phosphorylation seen in RBL-2H3 cells was due to activation of protein kinase C, we attempted to define the stoichiometry of phosphorylation. To do this we took advantage of the difference in charge between unphosphorylated, monophosphorylated and diphosphorylated forms of the MLC, that allows them to be separated by isoelectric focusing. First we labeled the RBL-2H3 proteins with 35S-methionine. Then we analyzed the stoichiometry of MLC phosphorylation by 2-dimensional protein gel electrophoresis, using isoelectric focusing in the first dimension and SDS-PAGE in the second (Ludowyke et al., 1989). This allowed for direct determination of the extent of MLC phosphorylation for both control and antigen-stimulated cells. Moreover, by comparing 2-dimensional protein gels obtained for 35S-methionine labeled MLCs with those obtained for 32P-labeled MLCs, we were able to determine the specific activity (mol inorganic phosphate/mol MLC) of 32P-labeled MLC and thereby determine the stoichiometry of MHC

TABLE 1. Stoichiometry of Myosin Phosphorylation by Protein Kinase C and Histamine Secretion

	Cell Batch	Stoichiometry MLC	Stoichiometry MHC	Secretion
		mol phosphate/ mol subunit		%
1	Unstimulated	0	0	1
	Stimulated	0.38	0.57	23
2	Unstimulated	0	0	1
	Stimulated	0.7	1.0	44

phosphorylation at the site phosphorylated by protein kinase C.

We quantitated the extent of phosphorylation for both the MLC and MHC for two different batches of cells. Each batch released a given amount of histamine and it was of interest to see if the extent of MHC and MLC phosphorylation correlated with the extent of histamine release. Table I summarizes the stoichiometry of phosphorylation for a batch of cells that secreted 23% of their histamine and a second batch that secreted 44%. Both batches of cells were found to contain 0.4 mols of phosphate/mol MLC prior to stimulation, all of it confined to the site phosphorylated by MLC kinase and this value did not change significantly after stimulation (data not shown). In contrast, the extent of phosphorylation at the site phosphorylated by protein kinase C, rose from 0 mol phosphate/mol MLC to 0.38 mol phosphate/mol MLC in cells secreting 23% of their histamine and to 0.7 mols phosphate/mol MLC in cells secreting 44% of their histamine. A similar scenario is seen for the MHC in that the amount of phosphate at the site phosphorylated by protein kinase C increased from 0 to 0.57 mol phosphate/mol MHC in cells secreting 23% of their histamine and to 1.0 mol phosphate/mol MHC in cells secreting 44%.

DISCUSSION

In these studies we demonstrate that aggregation of the receptors for IgE in RBL-2H3 cells results in phosphorylation of the 20 kDa MLC and 200 kDa MHC by protein kinase C. The extent of phosphorylation of both the MHC and MLC approaches 1 mol/mol in cells secreting 44% of their histamine. Interestingly, the time course of myosin phosphorylation by protein kinase C coincides with histamine release from these cells.

Previous studies have shown that phosphorylation of the 20 kDa MLC at the site phosphorylated by MLC kinase appears to be required for contractile activity in nonmuscle cells (reviewed in Sellers and Adelstein, 1987). In the cells used in this study the 20 kDa MLC was already phosphorylated at this site to a significant extent (0.4 mols/mol MLC) and the amount of phosphate at this site did not appear to change following antigenic stimulation. It would, of course, be useful to know if cells that have no phosphate bound at the MLC kinase site in their 20 kDa MLC are still capable of releasing histamine; by analogy to other nonmuscle cells we predict that they would not. Yet, despite the existence of myosin that is phosphorylated at the MLC kinase site, these cells do not release their histamine until protein kinase C is activated and the MLCs and MHCs become phosphorylated on sites we have shown to be phosphorylated *in vitro* by protein kinase C.

In vitro studies using protein kinase C to phosphorylate the MLCs suggest that protein kinase C acts to oppose the effect of MLC kinase phosphorylation, i.e., to down-regulate contractile activity (see Introduction above).

It is difficult to see how this effect of protein kinase C in phosphorylating the 20 kDa MLC could relate to histamine release other than to down-regulate it. On the other hand, the most novel aspect of our study was the discovery that activation of protein kinase C in these cells results in significant phosphorylation of the MHC. MHC phosphorylation by protein kinase C has not been previously described in nonmuscle cells with the recent exception of platelets, where its function is unknown (Kawamoto et al., 1989). In invertebrate nonmuscle cells MHC phosphorylation results in filament disassembly (reviewed in Warrick and Spudich, 1987) and this is the function we propose for protein kinase C phosphorylation of the RBL-2H3 MHCs in these cells. We envision phosphorylation of the MHCs by protein kinase C as being necessary to dissociate the myosin filaments that lie just inside the membrane (i.e., the cortical myosin filaments) and thereby allow the granules to approach the surface of the cell. Future experiments should allow for testing of this model and for ascertaining the relative importance of this phosphorylation in secretion of histamine from RBL-2H3 cells.

ACKNOWLEDGEMENTS

The authors gratefully acknowledge with admiration the expertise of Catherine Magruder in editing this manuscript.

REFERENCES

Beaven MA, Ludowyke RI (1989). Stimulatory signals for secretion in mast cells and basophils. In Mond JJ, Cambier JC, Wiess A (eds): "Focus on Regulation of Cell Growth and Activation," New York: Raven Press (in press).

Bengur AR, Robinson EA, Appella E, Sellers JR (1987). Sequence of the sites phosphorylated by protein kinase C in the smooth muscle myosin light chain. J Biol Chem 262:7613-7617.

Daniel JL, Molish IR, Rigmaiden M, Stewart G (1984). Evidence for a role of myosin phosphorylation in the initiation of the platelet shape change response. J Biol Chem 259:9826-9831.

Kawamoto S, Bengur AR, Sellers JR, Adelstein RS (1989). _In situ_ phosphorylation of human platelet myosin heavy and light chains by protein kinase C. J Biol Chem 264:2258-2265.

Ludowyke RI, Peleg I, Beaven MA, Adelstein RS (1989). Antigen-induced secretion of histamine and the phosphorylation of myosin by protein kinase C in rat basophilic leukemia cells. J Biol Chem 264 (in press).

Nishikawa M, Sellers JR, Adelstein RS, Hidaka H (1984). Protein kinase C modulates _in vitro_ phosphorylation of the smooth muscle heavy meromyosin by myosin light chain kinase. J Biol Chem 259:8808-8814.

Scholey JM, Taylor KA, Kendrick-Jones J (1980). Regulation of non-muscle myosin assembly by calmodulin-dependent light chain kinase. Nature 287:233-235.

Schroeder TE (1973). Actin in dividing cells: Contractile ring filaments bind heavy meromyosin. Proc Natl Acad Sci USA 70:1688-1692.

Sellers JR, Adelstein RS (1987). Regulation of contractile processes by phosphorylation. In Boyer PD, Krebs EG (eds): "The Enzymes," Orlando: Academic Press, pp 381-418.

Trotter JA (1982). Living macrophages phosphorylate the 20,000 dalton light chains and heavy chains of myosin. Biochem Biophys Res Commun 106:1071-1077.

Warrick HM, Spudich JA (1987). Myosin structure and function in cell motility. Ann Rev Cell Biol 3:379-421.

Wysolmerski RB, Lagunoff D (1988). Inhibition of endothelial cell retraction by ATP depletion. Am J Pathol 132:28-37.

Calcium Channel Modulators in Heart and Smooth Muscle: Basic Mechanisms and Pharmacological Aspects
S. Abraham and G. Amitai, editors
© 1990, VCH, Weinheim/Deerfield Beach, FL and Balaban, Rehovot/Philadelphia

Calcium homeostasis in vascular tissues

CHRISTIAN FRELIN, PAUL VIGNE, AND
CATHERINE VAN RENTERGHEM

Centre de Biochimie du CNRS, Parc Valrose, 06034 Nice, Cedex, France

INTRODUCTION

Calcium ions regulate numerous physiological processes. In muscle cells, including cardiac, skeletal and smooth muscle cells, contractions are triggered by a rise in intracellular Ca^{2+} concentration. Calcium may be released from intracellular stores: the sarcoplasmic reticulum in striated muscles and calciosomes in smooth muscle cells. Calcium also enters cells *via* voltage dependent Ca^{2+} channels and receptor operated channels. Removal of Ca^{2+} from contractile proteins, which induces relaxation in striated muscles, is accomplished either by resequestration of Ca^{2+} into the sarcoplasmic reticulum or by extrusion of Ca^{2+} across the sarcolemma *via* the (Ca^{2+})ATPase or *via* the Na^+/Ca^{2+} exchange system (see Carafoli, 1987 for a recent review).

The temporal aspects of variations in the cytosolic Ca^{2+} levels should be important if these variations serve as an intracellular signal for the cells. It is clear that most of the Ca^{2+} that appears in the cytoplasm of vascular smooth muscle cells that are stimulated with vasoconstricting peptides comes from an intracellular pool called calciosomes (Volpe et al., 1988). Much less is known about other Ca^{2+} transporting systems and about their respective role for shaping intracellular Ca^{2+} signals. This chapter defines the relative activities of the different structures in charge of handling Ca^{2+} across the plasma membrane and of calciosomes of aortic smooth muscle cells. These are receptor-operated channels, voltage-dependent Ca^{2+} channels, the Na^+/Ca^{2+} exchange system and the (Ca^{2+})ATPases of the plasma membrane and of calciosomes.

MOLECULAR PROPERTIES OF THE VASOCONSTRICTING ACTIONS OF ENDOTHELIN AND VASOPRESSIN

Endothelin is a newly discovered peptide that is synthesized by the vascular endothelium and that is one of the most potent vasoconstricting substance known so far (Yanagisawa et al., 1988). It was believed to act as an agonist of voltage-dependent L-type Ca^{2+} channels for a great part of its vasoconstricting action is antagonized by L- type Ca^{2+} channel inhibitors such as dihydropyridines (Yanagisawa et al., 1988). Endothelin is also a potent cardiotonic peptide (Vigne et al., 1989).

The first step in endothelin action on aortic smooth muscle cells is an activation of phospholipase C *via* a *pertussis* toxin-insensitive G protein (Resink et al., 1988, Van Renterghem et al., 1988b, Marsden et al., 1989). Membrane inositol phospholipids are hydrolyzed into inositol polyphosphates and diacylglycerol. Since exogenous diacylglycerol and phorbol esters induce vasoconstriction without any change in intracellular Ca^{2+} concentration (Forder et al., 1985), it is possible that protein kinase C mediates part of the contractile action of endothelin. Inositol trisphosphate is also involved in the action of endothelin through the mobilization of Ca^{2+} from an intracellular pool (Van Renterghem et al., 1988b, Marsden et al., 1989) that is distinct from the caffeine sensitive pool (Kai et al., 1989). The increase in cytosolic Ca^{2+} level activates a Ca^{2+}-dependent K^+ channel in the plasma membrane and leads to a transient hyperpolarization of the membrane. A partial and transient inhibition of voltage-dependent L-type Ca^{2+} channels is also observed possibly as a consequence of the rise in $[Ca^{2+}]_i$ (Van Renterghem et al., 1988a) and of activation of protein kinase C by diacylglycerol (Galizzi et al.,1987). All these events occured within one to two minutes following the application of endothelin. The development of the contraction which usually takes 5-10 minutes is associated with a slow depolarization of the plasma membrane due to the opening of a non selective cationic channel that is permeable to Na^+, K^+ and Ca^{2+}. Although this channel has not been identified using pharmacological probes, it may be identical to the receptor-operated channel that is activated by vasopressin and that has been characterized by $^{45}Ca^{2+}$ flux experiments in the same cell line (Wallnöfer et al., 1987). When the membrane potential reaches the threshold potential for the opening of voltage-dependent Ca^{2+} channels, a spontaneous electrical activity develops. Blockers of voltage-dependent Ca^{2+} channels such as dihydropyridines and verapamil do not prevent the binding of endothelin to its membrane receptor (Hirata et al., 1988). They do not affect the early actions of endothelin, *i.e.* the production of inositol phosphates, the transient rise in intracellular Ca^{2+} concentration, the hyperpolarizing and the depolarizing phases. Dihydropyridines and verapamil only suppress the opening of voltage-dependent Ca^{2+} channels that is consecutive to the depolarization of the plasma membrane (Van Renterghem et al., 1988b).

It seems therefore that, although a great part of the contractile action of endothelin is prevented by inhibitors of Ca^{2+} channels, the action of the peptide on the Ca^{2+} channel is indirect. Another evidence for such an indirect action is the observation that endothelin does not prevent the binding of $[^3H]$PN200-110, of d-cis$[^3H]$diltiazem, of $[^3H]$verapamil and of $[^3H]$fluspirilene to their respective binding sites on the Ca^{2+} channel (Van Renterghem et al., 1988b). An action very similar to that of endothelin has been described for two other vasoconstricting peptides: vasopressin (Van Renterghem et al., 1988a) and bombesin (Lazdunski et al., 1988). One difference is however that endothelin is much less potent than vasopressin (Van Renterghem et al., 1988b) and than angiotensin II

(Araki et al., 1989) for stimulating phospholipase C although its contractile activity is larger and longer lasting. It could be that vasoconstriction is not ultimately determined by the extent of the activation of phospholipase C but rather by the opening of the non selective cationic channel of the plasma membrane. One possibility could be that two distinct G proteins mediate the effects of endothelin. One would act on phospholipase C whereas the other would open the non selective cationic channel in the plasma membrane. It has also been reported that endothelin exerts a potent brochoconstricting action by releasing cyclo-oxygenase products of arachidonic acid metabolism (Payne and Whittle, 1988). A membrane action of metabolites of arachidonic acid has been described in cardiac cells (Kurachi et al., 1989). Whether such compounds mediate some of the actions of endothelin on vascular smooth muscle cells is not known.

THE Na^+/Ca^{2+} EXCHANGE SYSTEM OF AORTIC SMOOTH MUSCLE CELLS IS ACTIVATED BY PROTEIN KINASE C

The presence of a Na^+/Ca^{2+} exchange system in vascular smooth muscles is well known and the system has been postulated to participate in the modulation of vascular tone and peripheral resistance (Blaustein, 1977). The properties of the system in rat aortic smooth muscle cells have been studied using $^{45}Ca^{2+}$ flux experiments and intracellular Ca^{2+} measurements (Vigne et al., 1988a).

Suspending vascular smooth muscle cells into a Na^+ free medium promotes a $^{45}Ca^{2+}$ uptake component that has the following properties (i) it is insensitive to blockers of L-type Ca^{2+} channels, (ii) it is dependent on intracellular Na^+ concentration, (iii) it is inactive in Li^+ loaded cells and (iv) it is increased under depolarizing membrane conditions. Activity was also detected by monitoring changes in the cytosolic Ca^{2+} levels using the dye indicator indo-1. All the experiments that have been performed are consistent with the presence in vascular smooth muscle cells of an electrogenic

Na^+/Ca^{2+} exchange system that has properties similar to those of the system in cardiac cells. The Na^+/Ca^{2+} exchange system of vascular smooth muscle cells is inhibited by derivatives of the diuretic drug amiloride with IC_{50} values at 30, 100 and 100 µM for benzamil, 2,4-dichlorobenzamil and 3,4-dichlorobenzamil respectively. Amiloride by itself is unable to inhibit Na^+/Ca^{2+} exchange activity up to a concentration of 1 mM. Although benzamil and its derivatives do inhibit Na^+/Ca^{2+} exchange activity in vascular smooth muscle cells (Vigne et al., 1988a) and in cardiac cells (Siegl et al., 1984, Lazdunski et al., 1985), they should not be used as specific inhibitors for the system. They inhibit voltage-dependent L-type Ca^{2+} channels at concentrations that are lower than the concentrations needed to inhibit the Na^+/Ca^{2+} exchange system (IC_{50} = 10 µM). Benzamil also inhibits renal epithelium Na^+ channels (IC_{50} = 0.4 µM) (Barbry et al., 1987) and the Na^+/H^+ exchange system (IC_{50} = 0.1 mM) (Vigne et al., 1984). The pharmacological properties of the different targets for amiloride have been reviewed recently (Frelin et al., 1987, 1988).

Activation of protein kinase C with phorbol esters increases the activity of the Na^+/Ca^{2+} exchange system of vascular smooth muscle cells about two-fold (Vigne et al., 1988a). Activation does not result from a change in the membrane potential or of the intracellular Na^+ concentration. It was not observed with phorbol esters that do not activate protein kinase C. It may therefore be a consequence of the phosphorylation of the exchanger protein (or of a protein that is associated to it) by protein kinase C. In embryonic chick cardiac cells phorbol esters have no action on Na^+/Ca^{2+} exchange activity. Two other important ion transporting systems of the plasma membrane of vascular smooth muscle cells have also been reported to be targets for protein kinase C. Phorbol esters stimulate Na^+/H^+ exchange activity (Vigne et al., 1988b), they partially inhibit voltage-dependent Ca^{2+} channels (Galizzi et al., 1987). It seems therefore that the control of ions movements across the plasma membrane of vascular smooth muscle cells is an important function of protein kinase C.

A physiological role for the Na^+/Ca^{2+} exchange activity and its modulation by protein kinase C could be the following. Since under resting conditions, the Na^+/Ca^{2+} exchange system works as a Ca^{2+} extrusion mechanism (Vigne et al., 1988a), an increase in $[Ca^{2+}]_i$, for instance in response to the mobilization of intracellular Ca^{2+} stores by vasoconstricting peptides, would promote net Ca^{2+} efflux. As a consequence, the function of the system could be to reduce the duration of Ca^{2+} transients induced by Ca^{2+} mobilizing hormones such as epinephrine, angiotensin II, vasopressin and endothelin. These hormones also activate protein kinase C *via* the production of diacylglycerol. A protein kinase C mediated activation of the Na^+/Ca^{2+} exchange system could further promote Ca^{2+} efflux and reduce the duration of Ca^{2+} transients. To test this hypothesis, Ca^{2+} transients were induced by vasopressin in A7r5 cells under conditions in which the Na^+/Ca^{2+} exchange system was forced to function as a Ca^{2+} influx mechanism or was rendered inactive by depleting cells of their Na^+. In all cases vasopressin induced Ca^{2+} transients that had the same size and time course as Ca^{2+} transients elicited in cells that have a Na^+/Ca^{2+} exchange system that operated as a Ca^{2+} efflux mechanism. Similar results have been obtained for angiotensin II-induced Ca^{2+} transients in primary cultures of rat aortic smooth muscle cells. These experiments indicate that the Na^+/Ca^{2+} exchange system is not involved in the short-term regulation of cytosolic Ca^{2+} levels of aortic smooth muscle cells and that this function is mainly achieved by the (Ca^{2+})ATPases of the calciosomes and of the plasma membrane. The Na^+/Ca^{2+} exchange system could be involved in the long-term maintenance of intracellular Ca^{2+} levels. The same function has been postulated for the system in cardiac cells (Carafoli, 1987).

Ca^{2+} UPTAKE THROUGH THE PLASMA MEMBRANE CONTROLS THE CALCIUM LOAD OF CALCIOSOMES

Vasopressin and bombesin are two potent activators of phospholipase C in aortic smooth muscle cells of the A7r5 cell line. After stimulation of the cells with vasopressin or bombesin, the intracellular Ca^{2+} level increases rapidly to reach a maximum at 10 seconds and then it decreases to its resting level within 2 minutes. These Ca^{2+} transients are independent of the presence of extracellular Ca^{2+} and arose as a consequence of the mobilization of intracellular Ca^{2+} stores. Although the sizes of Ca^{2+} transients elicited by bombesin and vasopressin were similar, a first application of one of the peptides reduced the response to the second peptide by half (Vigne et al., 1988b). The reason for such a desensitization is that only half of Ca^{2+} that is released by calciosomes into the cytoplasm is pumped back into calciosomes and can be mobilized in response to the application of a second peptide. The remaining Ca^{2+} is pumped outside of the cells as evidenced previously from $^{45}Ca^{2+}$ efflux experiments (Doyle and Ruegg, 1985). The vasopressin- and bombesin-sensitive intracellular Ca^{2+} stores can be depleted of their Ca^{2+} by a 1-2 hour exposure of the cells to a 50 nM Ca^{2+} medium. Depletion occured faster at 4°C than at 25°C. When Ca^{2+}-depleted cells were shifted back to a Ca^{2+}-containing medium, intracellular calcium stores refilled slowly ($t_{1/2}$ = 4 minutes). Addition of (−)D888 to block voltage-dependent Ca^{2+} channels or of Ni^{2+} to block putative receptor-operated Ca^{2+} channels (Hallam and Rink, 1985) slowed the kinetic of refilling of calciosomes. These experiments suggest that the rate limiting step for refilling calciosomes once Ca^{2+} had been mobilized by inositol trisphosphate, is Ca^{2+} influx through the plasma membrane. This was demonstrated in a more direct way by showing that if Ca^{2+} uptake *via* the Na^{+}/Ca^{2+} echange system is promoted, then calciosomes refill almost instantaneously (Vigne et al., 1988b). A direct refilling of calciosomes from the extracellular space as described for acinar cells (Merritt and Rink, 1897) does not exist in aortic smooth muscle cells (Vigne et al., 1988b). The obvious conclusion is that Ca^{2+} efflux out of the cells *via* the (Ca^{2+})ATPase is important for desensitizing vascular smooth muscle cells to the repetitive actions of different vasoconstricting peptide. It also suggests that one role for voltage-dependent Ca^{2+} channels (and for the electrical activity of the plasma membrane) could be to control the importance of the Ca^{2+} stores of calciosomes and therefore the responsiveness of the cells to the action of circulating peptides. A similar role for Ca^{2+} channels has been suggested for cardiac cells (Morad and Cleeman, 1987).

Acknowledgements

This work was supported by the "Centre National de la Recherche Scientifique", the "Association pour la Recherche contre le Cancer" and the "Fondation sur les Maladies Vasculaires". We are grateful to M.-T. Ravier, N. Boyer and C. Roulinat-Bettelheim for expert technical assistance and to Dr. J.-P. Breittmayer for flow cytometic analyses.

REFERENCES

Araki S, Kawahara Y, Kariya K, Sunako M, Fukuzaki H, Takai Y (1989). Stimulation of phospholipase C mediated hydrolysis of phosphoinositides by endothelin in cultured rabbit aortic smooth muscle cells. Biochem Biophys Res Commun 159:1072-1079.

Barbry P, Chassande O, Vigne P, Frelin C, Ellory C, Cragoe EJ Jr, Lazdunski M (1987). Purification and subunit structure of the [^{3}H]phenamil receptor associated with the renal apical Na^+ channel. Proc Natl Acad Sci USA 84:4836-4840.

Blaustein MP (1977). Sodium ions, calcium ions, blood pressure regulation and hypertension: A reassessment of a hypothesis. Am J Physiol 323:C165-C173.

Carafoli E (1987). Intracellular calcium homeostasis. Annu Rev Biochem 56:395-433.

Doyle V, Ruegg UT (1985). Vasopressin induced production of inositol trisphosphate and calcium efflux in a smooth muscle cell line. Biochem Biophys Res Commun 131:469-476.

Forder J, Scriabine A, Rasmussen H (1985). Plasma membrane calcium flux, protein kinase C activation and smooth muscle contraction. J Pharmacol Exptl Ther 235:267-273.

Frelin C, Vigne P, Barbry P, Lazdunski M (1987). Molecular properties of amiloride action and of its Na^+ transporting targets. Kidney Int 32:785-793.

Frelin C, Barbry P, Vigne P, Chassande O, Cragoe EJ Jr, Lazdunski M (1988b). Amiloride and its analogs as tools to inhibit Na^+ transport *via* the Na^+ channel, the Na^+/H^+ antiport and the Na^+/Ca^{2+} exchanger. Biochimie 70:1285-1290.

Galizzi JP, Qar G, Van Renterghem C, Fosset M, Lazdunski M (1987). Regulation of calcium channels in aortic muscle cells by protein kinase C activators (diacylglycerol and phorbol esters) and peptides (vasopressin and bombesin) that stimulate phosphoinositide breakdown. J Biol Chem 262:6947-6950.

Hallam TJ, Rink T (1985). Agonists stimulate divalent cation channels in the plasma membrane of human platelets. FEBS Letters 186:175-179.

Hirata Y, Yoshimi H, Takata S, Watanabe TX, Kumagai S, Nakajima K, Sakakibara S (1988). Cellular mechanism of action by a novel vasoconstritor endothelin in cultured rat vascular smooth muscle cells. Biochem Biophys Res Commun 154:868-875.

Kai H, Kanaide H, Nakamura M (1989). Endothelin sensitive intracellular Ca^{2+} store overlaps with caffeine sensitive one in rat aortic smooth muscle cells in primary cultures. Biochem Biophys Res Commun 158:235-243.

Kuraki Y, Ito H, Sugimoto T, Shimizu T, Miki I, Ui M (1989). Arachidonic acid metabolites as intracellular modulators of the G protein gated cardiac K^+ channel. Nature 337:555-557.

Lazdunski M, Frelin C, Vigne P (1985). The sodium/hydrogen exchange system in cardiac cells: Its biochemical and pharmacological properties and its role in regulating internal concentrations of sodium and internal pH. J Mol Cell Cardiol 17:1029-1042.

Lazdunski M, Romey G, Van Renterghem C (1988). Bombesin modulates the spontaneous electrical activity of rat aortic cells (A7r5 cell line) by an action on three different types of ionic channels. J Physiol (London) 407:100P.

Marsden PA, Danthuluri NR, Brenner BM, Ballermann BJ, Brock TA (1989). Endothelin action on vascular smooth muscle involves inositol trisphosphate and calcium mobilization. Biochem Biophys Res Commun 158:86-93.

Merritt JE, Rink TJ (1987). Regulation of cytosolic free calcium in fura-2 loaded rat parotid acinar cells. J Biol Chem 262:17362-17369.

Morad M, Cleeman L (1987). Role of Ca^{2+} channel in development of tension in heart muscle. J Mol Cell Cardiol 19:527-553.

Payne AN, Whittle BJR (1988). Potent cyclo-oxygenase-mediated bronchoconstrictor effects of endothelin in the guinea pig *in vivo*. Eur J Pharmacol 158:303-304.

Resink TJ, Scott-Burden T, Buhler FR (1988). Endothelin stimulates phospholipase C in cultured vascular smooth muscle cells. Biochem Biophys Res Commun 157:1360-1368.

Siegl PKS, Cragoe EJ Jr, Trumble MJ, Kaczarowski GJ (1984). Inhibition of Na^+/Ca^{2+} exchange in membrane vesicles and papillary muscle preparations from guinea pig heart by analogs of amiloride. Proc Natl Acad Sci USA 81:3238-3242.

Van Renterghem C, Romey G, Lazdunski M (1988a). Vasopressin modulates the spontaneous electrical activity in aortic cells (line A7r5) by acting on three different types of ionic channels. Proc Natl Acad Sci USA 85:9365-9369.

Van Renterghem C, Vigne P, Barhanin J, Schmid-Alliana A, Frelin C, Lazdunski M (1988b). Molecular mechanism of action of the vasoconstrictor peptide endothelin. Biochem Biophys Res Commun 157:977-985.

Vigne P, Frelin C, Cragoe EJ Jr, Lazdunski M (1984). Structure activity relationships of amiloride and certain of its analogues in relation to the blockade of the Na^+/H^+ exchange system. Mol Pharmacol 25:131-136.

Vigne P, Breittmayer JP, Duval D, Frelin C, Lazdunski M (1988a). The Na^+/Ca^{2+} antiporter in aortic smooth muscle cells. Characterization and demonstration of an activation by phorbol esters. J Biol Chem 263:8078-8083.

Vigne P, Breittmayer JP, Lazdunski M, Frelin C (1988b). The regulation of the cytoplasmic free Ca^{2+} concentration in aortic smooth muscle cells (A7r5 line) after stimulation by vasopressin and bombesin. Eur J Biochem 176:47-52.

Vigne P, Breittmayer JP, Frelin C, Lazdunski, M. (1988c). Dual control of the intracellular pH in aortic smooth muscle cells by a cAMP sensitive HCO_3^-/Cl^- antiporter and a protein kinase C sensitive Na^+/H^+ antiporter. J Biol Chem 263:18023-18029.

Vigne P, Lazdunski M, Frelin C (1989). The inotropic effect of endothelin-1 on rat atria involves an hydrolysis of phosphatidylinositol. FEBS Letters (in press).

Volpe P, Krause KH, Hashimoto S, Zorzato F, Pozzan T, Meldolesi J, Lew DP (1988). Calciosomes, a cytoplasmic organelle: The inositol 1,4,5 trisphosphate sensitive Ca2+ store of non muscle cells? Proc Natl Acad Sci USA 85:1091-1095.

Wallnöfer A, Cauvin C, Ruegg UT (1987). Vasopressin increases $^{45}Ca^{2+}$ influx in rat aortic smooth muscle cells. Biochem Biophys Res Commun 148:273-278.

Yanagisawa M, Kurihara H, Kimura S, Tomobe Y, Kobayashi M, Mitsui Y Yazaki Y, Goto K, Masaki T (1988). A novel potent vasoconstrictor peptide produced by vascular endothelial cells. Nature 332:411-415.

Calcium Channel Modulators in Heart and Smooth Muscle:
Basic Mechanisms and Pharmacological Aspects
S. Abraham and G. Amitai, editors
© 1990, VCH, Weinheim/Deerfield Beach, FL and Balaban, Rehovot/Philadelphia

Short and long term effects of antiarrhythmic drugs on cultured cardiomyocytes

GANIA KESSLER-ICEKSON[1], HADASSA SCHLESINGER[1], AVIV MAGER[2], CHAYA BRODIE[3], SAMUEL R. SAMPSON[3], AND MICHAEL BERGMAN[4]

[1]Rogoff Institute and [2]Intensive-care Unit, Beilinson Medical Center, Petah-Tikva; [3]Life-Sciences, Bar-Ilan University, Ramat-Gan; and [4]Internal Medicine C, Hasharon Hospital, Petah-Tikva, Israel

INTRODUCTION

Many medical treatments subject the heart to prolonged administration of antiarrhythmic drugs which interfere, by various mechanisms, with transmembrane ion fluxes. The majority of studies on the action of these drugs elucidate acute events, particularly at the levels of membrane -excitation and of excitation-contraction coupling. However, there might as well be long term effects of antiarrhythmic drug treatments caused by chronic alterations in transmembrane ion fluxes. Role of ion fluxes in the regulation of gene expression has been described in several excitable tissues. Synthesis of c-fos protein in phaeochromocytoma cells is controlled by calcium ions (Morgan & Curran, 1986) and that of acetylcholine-receptor and of sarcoplasmic Na^+-K^+-ATPase, in skeletal muscle cells, is regulated by calcium and sodium ions, respectively (Birnbaum et al, 1980; Wolitsky & Fambrough, 1986). The above examples stimulated our interest in long term effects of antiarrhythmic agents on the expression of myosin isoforms in isolated heart myocytes.

Myosin, the major protein of the thick filament, is a hexamere composed of a pair of heavy chains (200 Kdaltons) and two pairs of light chains (16-27 Kdaltons). The ATP hydrolysing site and the actin binding site are both located on the heavy chains (MHCs) of the molecule so that the functional properties of myosin depend on the primary structure of the MHCs. Two types of MHCs, α and β, are known in the rat ventricular muscle, constructing myosins V1 and V3, respectively. V1 has a higher ATPase activity and it allows a faster muscle contraction than V3. The two MHC genes are organised in tandem on the same chromosome and are expressed in an antithetic fashion. That is, once the transcription of αMHC increases that of βMHC decreases and vice-versa (Mahdavi et al, 1984). Embryonic and neonatal rat heart contain mainly βMHC whereas α MHC prevails in the adult. This pattern is changed under various pathophysiological states such as hypertension, diabetes, hypo- or hyper-thyroidism. Thyroid hormone sustains the expression of αMHC via an interaction of the complex hormone-receptor with regulatory sequences on the αMHC gene (Izumo & Mahdavi, 1988). So far only the mechanism of thyroid hormone action has been resolved at the molecular level, while those of other regulators of MHC expression are still obscure.

We report below preliminary observations on effects of the calcium channel blocker verapamil, and of the class III antiarrhythmic drug amiodarone, on contractile activity and on MHC distribution in cultured heart myocytes. The cells, derived from neonatal rat hearts, were grown in a fully defined medium, free of the neurogenic, haemodynamic or humoral factors that regulate them in-vivo. The phenotypic expression of myocardial features was reflected from the synchronous, rhythmical cell beating that initiated spontaneously one day after cell culture.

RESULTS AND DISCUSSION

A. Verapamil

Verapamil belongs to the diphenylalkylamine class of calcium antagonists which block voltage dependent calcium channels that are open or inactivated. It is therefore most effective where the frequency of channel opening and closing is greatest. Accordingly, verapamil is very useful in treating supraventricular cardiac arrhythmias (Snyder & Reynolds, 1985).

Administration of verapamil to spontaneously beating cultures resulted in the inhibition of beating rate in a dose dependent manner (Fig 1). At 5 μM verapamil, cell contractions were completely arrested.

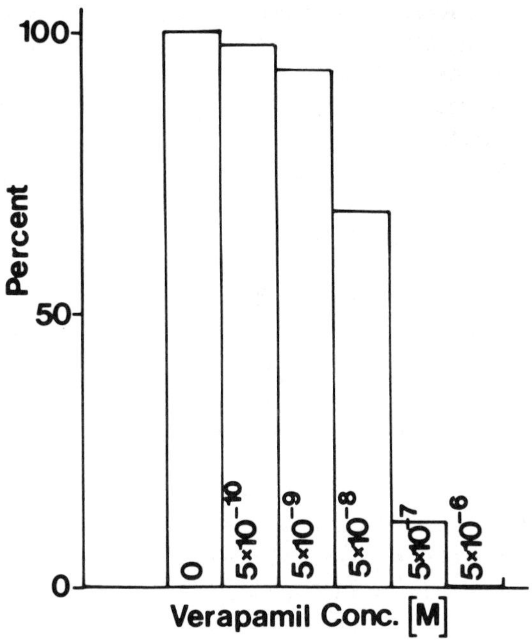

Figure 1. Beating rate of cultures treated by increasing doses of verapamil (% of control), measured one hour after drug administration. The average rate of control cultures was 120 beats per minute. (see also Kessler-Icekson, 1988).

The effect of verapamil on beating rate was very rapid and the cells survived the treatment for several days resuming contraction upon drug removal. Protein synthesis took place in the absence of cell beating and incubation with radiolabeled amino acid allowed the detection of newly formed myosin chains. Alpha - and βMHC were then separated electrophoretically under denaturing conditions and any shift in MHC predmominance was followed by fluorography. In control cultures, βMHC prevailed unless the medium was supplemented with thyroid hormone which induced a transition towards αMHC predominance (Kessler-Icekson, 1988). Chronic (24-48 hours) verapamil treatment enhanced the synthesis of αMHC in hormone depleted cultures and intensified the effect of thyroid hormone when the two effectors were co-administered (Fig 2a lanes V and V/T). The verapamil effect was concentration dependent (Fig 2b) and it diminished when the concentration of extracellular calcium was elevated (Fig. 2a, lane V/Ca).

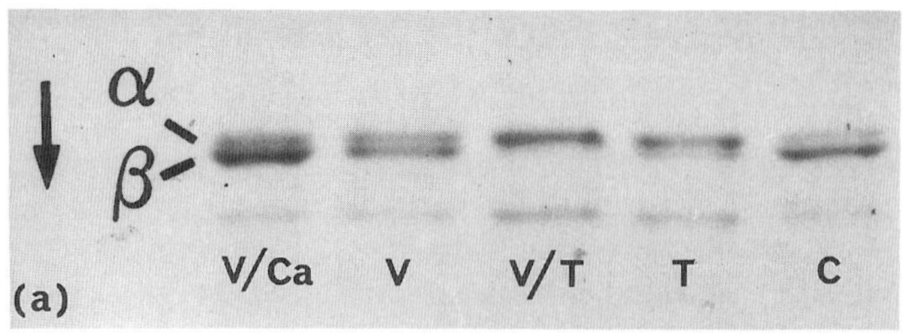

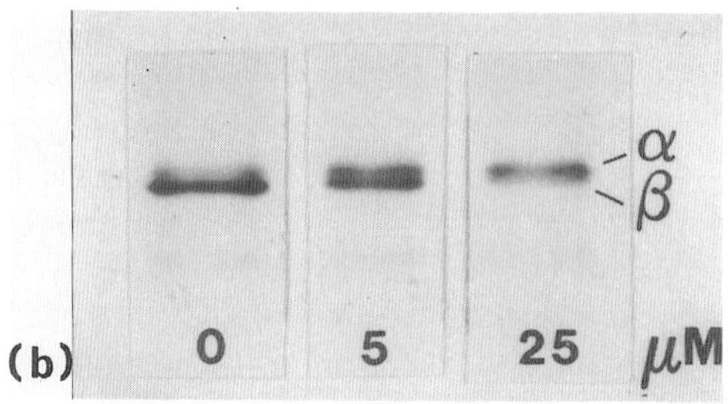

Figure 2. Distribution of α and β MHC in verapamil treated cultures: (a) Autoradiogram of a gel separating ^{35}S-labeled MHCs (3 hour pulse (Kessler-Icekson, 1988)). Cultures were incubated for the previous 48 hours with the following supplements: C-no addition; T-5nM triiodothyronine (T3); V/T - 5nM T3+5μM verapamil; V - 5μM verapamil; V/Ca - 5μM verapamil + 5mM calcium chloride; Arrow=direction of migration. (b) Silver-stain of a gel separating MHCs from cultures grown for two days with 5 or 25 μM verapamil.

It should be mentioned however, that the concentrations required to obtain the shift in MHC distribution were much greater than verapamil therapeutic levels and that different lots of the drug varied in their efficacy to produce the effect. Neither diltiazem nor nifedipine could elicit alterations in MHC predominance.

The mechanism by which verapamil modifies MHC expression is completely unclear. Since it acts on the cell as a calcium channel blocker we assume involvement of calcium ions in the control of isomyosin distribution. A specific role in this regulation may be attributed to the verapamil receptor site since other calcium channel blockers could not mimick the verapamil effect on MHC synthesis. Further experiments are required to clarify these possibilities.

B. Amiodarone

Amiodarone is a diiodinated benzofuran derivative, used for controlling both supraventricular and ventricular arrhythmias, particularly those arrhythmias that are resistant to currently employed drugs (Rotmensch & Belhassen, 1988). Unlike verapamil which is a well characterized calcium antagonist, the antiarrhythmic action of amiodarone is more complex and has not yet been fully elucidated. It inhibits sarcoplasmic Na^{+}-K^{+}-ATPase, sodium and calcium currents, and possibly, sarcoplasmic reticulum Ca^{+2} ATPase (Broekhuysen et al, 1972; Follmer et al, 1987; Nattel et al, 1987; Alkaslasy et al, 1989).

When administered to cultured myocytes amiodarone inhibited the rate of cell beating in a dose dependent manner but a complete arrest of cell contraction was not obtained (Fig 3).

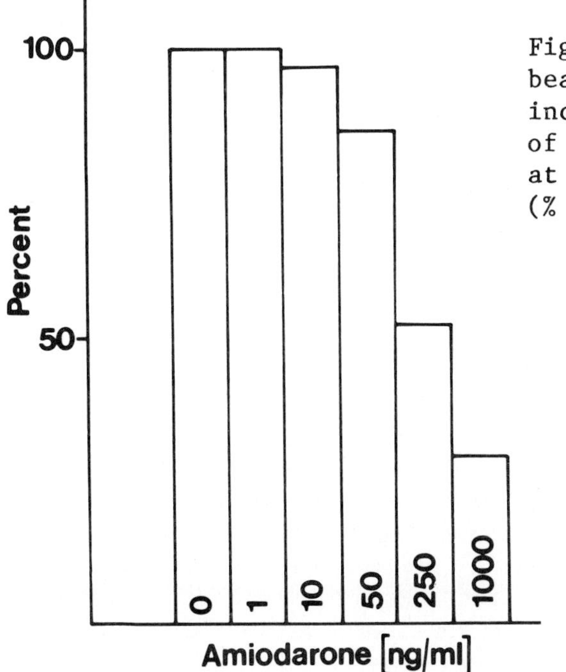

Figure 3. Inhibition of beating rate by increasing concentrations of amiodarone, measured at 2 hours of treatment (% of control).

Registration of membrane electrical activity showed prolongation of action-protential-duration and of the rates of rise and fall of action-potential. Membrane resting potential was reduced as well, within 30 minutes of amiodarone treatment (Fig 4). Furthermore, the activity of Na^+-K^+-ATPase was inhibited by 40-50% in the presence of amiodarone (not shown).

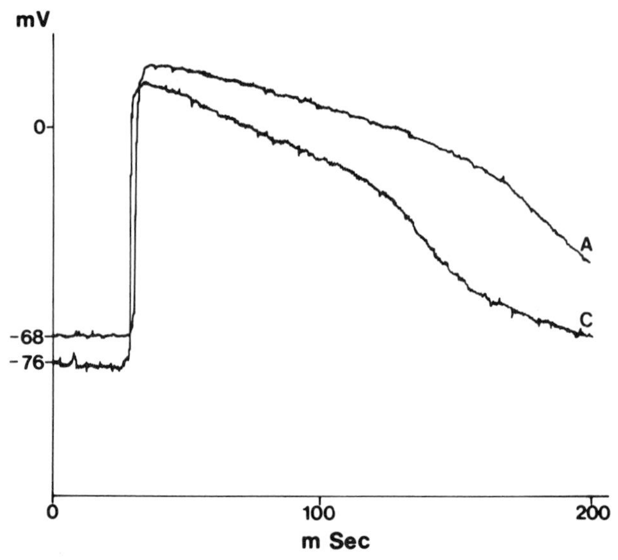

Figure 4. Membrane electrical activity of control (C) and amiodarone-treated (A) cultures. Two representative cells, recorded 30 minutes after drug (1μg/ml) administration (For procedure see Brodie & Sampson, 1988).

Amiodarone did not change the initial pattern of MHC synthesis and βMHC remained the predominant form as long as the expression of αMHC had not been induced by thyroid hormone. If the production of αMHC was first triggered by a 24 hour treatment with thyroid hormone and then the hormone was washed out, the level of α MHC continued to rise with time (Kessler-Icekson, 1988 and Fig 5 lane 1). However, when amiodarone was added upon hormone removal, the synthesis of βMHC reappeared (Fig 5, lane 2). The drug could not prevent thyroid hormone action when the two effectors were co-administered, even in a 100 fold excess of drug to hormone (Fig 5, lane 3).

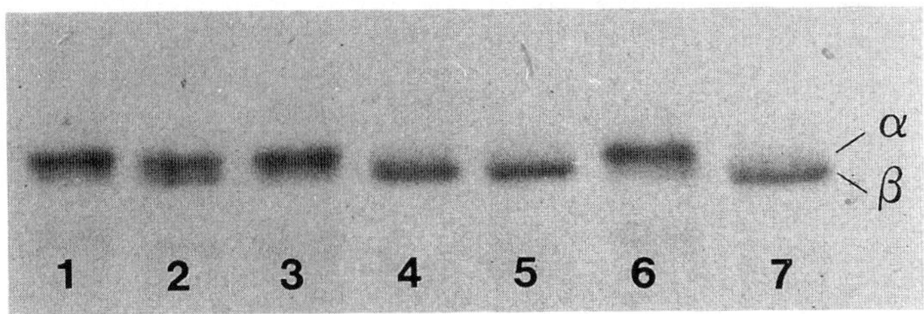

Figure 5. MHC distribution in amiodarone treated cultures.
Myosin labeling and analysis were done as in Fig 2. Experimental groups were the following: (1) 24 hours in T3 followed by 48 hours in hormone deficient medium (HDM). (2) 24 hours in T3 followed by 48 hours in HDM containing amiodarone. (3) 24 hours amiodarone followed by 48 hours of T3+amiodarone. (4) 24 hours in HDM followed by 48 hours in amiodarone. (5) 72 hours in amiodarone. (6) 72 hours in T3. (7) HDM alone. Concentrations were 5nM T3 and 500 ng/ml amiodarone.

It had previously been suggested that amiodarone imposes on the heart a 'hypothyroid-like' state, possibly by interefering with the regulatory action of thyroid hormone within the myocardial cell. This was mainly attributed to structural similarities between the two molecules (the iodinated ring). In receptor binding experiments amiodarone competed very poorly with thyroid hormone (Latham et al, 1987) yet it seemed able to modify the pattern of isomysins in-vivo (Bagchi et al, 1987). A recent publication by Franklyn et al. (1989) reported increased transcription of βMHC mRNA in the hearts of amiodarone-treated hypothyroid rats. Thus, a possible explanation for our results might be that amiodarone supported a shift towards βMHC synthesis by enhancing the displacement of thyroid homone from its receptor. Alternatively, the influence of amiodarone on the pattern of MHC could be mediated, like in the examples mentioned above, by ion fluxes which are greatly susceptible to this drug. Additional experiments are necessary in order to clarify the exact mechanism by which amiodarone modifies isomyosin distribution in the myocardial cell.

To summarize, it appears that the two drugs, verapamil and amiodarone, exert on the myocardial cell short- and long- term actions. The rate of cell beating is attenuated very rapidly by either drug while the pattern of MHC synthesis is affected much slower, only after 24-48 hours. The mechanisms of action of verapamil and amiodarone are certainly different and their effect on MHC distribution is opposite. Verapamil augments the expression of αMHC whereas amiodarone enhances the reappearance of βMHC in thyroid hormone pretreated cells. Considering the functional differences between the two isomyosins it means the opposite physiological implications. Whatever the mechanism of action of these two drugs, at this stage of our study we wish mainly to point out that chronic antiarrhythmic treatments may alter the pattern of myocardial proteins and thus modify myocardial features, a matter we should be aware of.

Acknowledgement: This research was supported, in part, by Grant No. 1518 from the Ministry of Health, the Chief Scientist's Office.

REFERENCES

Alkaslasy L, Shainberg A, Brik H, Rotmensch HH (1989) The effect of amiodarone and its metabolite on Ca^{+2} accumulation in the sacroplasmic reticulum of cultured cardiomyocytes. Abstract. 33rd Oholo Conference. P. 42.

Birnbaum M, Reis MA, Shainberg A (1980). Role of calcium in the regulation of acetylcholin receptor synthesis in cultured mucsle cells. Pfl Arch 385:37-43.

Brodie C, Sampson SR (1988). Characterization of thyroid hormone effect on N-K pump and membrane potential of cultured rat skeletal myotubes. Endocrinology 123:891-897.

Broekhuysen J, Clinet M, Delissee C (1972). Action of amiodarone on guinea pig heart sodium and potassium activated adenosine triphosphatase. Biochem Pharmacol 21:2951-2960.

Follmer CH, Aomine M, Yeh JZ, Singer DH (1987). Amiodarone induced block of sodium current in isolated cardiac cells. J Pharm Exp Ther 243:187-194.

Franklyn JA, Green NK, Gammag MD, Ahlquist JAO, Sheppard MC (1989). Regulation of α and β - myosin heavy chain messenger RNAs in the rat myocardium by amiodarone and by thyroid status. Clinc Sci 76:463-467.

Izumo S, Mahdavi V (1988). Thyroid hormone receptor α isoforms generated by alternative splicing differentially activate myosin HC gene transcription. Nature 334:539-542.

Kessler-Icekson G (1988). Effect of triiodothyronine on cultured rat heart cells: beating rate, myosin subunits and CK-isozymes. J Molec Cell Cardiol 20:649-655.

Latham KR, Sellitti D, Goldstein RE (1987). Interaction of amiodarone and desethylamiodarone with solubilysed nuclear thyroid hormone receptors. J Am Coll Cardiol 9:872-876.

Mahdavi V, Chambers AP, Nadal-Ginard B (1984). Cardiac and myosin heavy chain genes are organized in tandem. Proc Natl Acad Sci USA 81:2626-2630.

Morgan JI, Curran T (1986). Role of ion flux in the control of c-fos expression. Nature 322:552-555.

Nattel S, Talajic M, Quantz M, DeRoode M (1987). Frequence-dependent effect of amiodarone on atrioventricular nodal function and slow-channel action potentials: evidence for calcium-blocking activity. Circulation 76:442-449.

Rotmensch HH, Belhassen B (1988). Amiodarone in the management of cardiac arrhythmias: current concepts. Med Cli North Am 72:321-357.

Snyder SH, Reynolds IJ (1985). Calcium antagonist drugs. New England J Med 313:995-1002.

Wolitzky BA, Fambrough DM (1986). Regulation of the Na^+-K^+-ATPase in cultured chick skeletal muscle: modulation of expression by the demand for ion transport. J Biol Chem 261:9990-9999.

Pharmacodynamic effects of calcium channel modulators and related endogeneous factors

Calcium Channel Modulators in Heart and Smooth Muscle:
Basic Mechanisms and Pharmacological Aspects
S. Abraham and G. Amitai, editors
© 1990, VCH, Weinheim/Deerfield Beach, FL and Balaban, Rehovot/Philadelphia

Growth and metabolism of cultured vascular smooth muscle cells in hypertension

THERESE J. RESINK, TIMOTHY SCOTT-BURDEN, ALFRED W.A. HAHN, AND FRITZ R. BUEHLER

Department of Research, Basel University Hospital, 4031 Basel, Switzerland

INTRODUCTION

Abnormalities of vascular smooth muscle cell (VSMC) differentiation and proliferation are characteristic to both the atherosclerotic plaque and the pathologic alterations of hypertensive blood vessels (Ross, 1981; Schwartz et al. 1986). Raised blood pressure and consequent mechanical load have been ascribed to underly the development of vascular lesions in hypertension (Folkow et al., 1973). However, medial hypertrophy of arterial walls in spontaneously hypertensive rats (SHR) does occur during the prehypertensive stage (Yamori, 1976), while VSMC from SHR possess a greater proliferation capacity in culture, when compared to VSMC from normotensive Wistar Kyoto rats (WKY) (Scott-Burden et al. 1988a). These observations, which are clearly independent of blood pressure, suggest that neural, humoral and genetic factors may also be involved in the pathogenesis of hypertension (Yamori, 1976).

Understanding of the mechanism of control of VSMC proliferation in vivo has been complicated by the vast variety and diverse cellular origin of, and complex interplay between, molecules capable of promoting or inhibiting either proliferation and/or specific growth-related metabolic events.

Nevertheless, much information pertaining to cellular responsive elements and mechanisms involved in VSMC growth control has been gleaned from in vitro studies on cultured VSMC. We have utilized the intrinsic growth differential between VSMC from SHR and WKY as a model system to facilitate assessment of which such processes may actively contribute toward the development of hypertensive vascular abnormalities.

This paper presents an overview of our observations with respect to comparison of a number of proliferative and metabolic events between cultured thoracic aortic smooth muscle cells from SHR and WKY.

PROLIFERATION OF VASCULAR SMOOTH MUSCLE CELLS AND THE EXTRACELLULAR MATRIX

The proliferation of VSMC in culture is dependent on the presence of serum, and we have previously demonstrated that at any given concentration of serum, the capacity for growth is greater in SHR-derived VSMC than in WKY-derived VSMC (Scott-Burden et al, 1988a). However, in addition to the growth stimulatory properties of serum-containing growth factors/mitogens, evidence is accumulating to support an active role for the extracellular matrix in regulating VSMC proliferation (Hynes, 1987; Woods and Couchman, 1988). Once passaged, VSMC undergo a de-differentiative "modulation" process whereby they convert from the functionally contractile to the secretory phenotype (Chamley-Campbell and Campbell, 1981). Amongst the many compounds secreted by such "modulated" VSMC are components which include the glycoproteins, elastin, collagen and the glycosaminoglycans/proteoglycans (Jones et al., 1979; Hay, 1983; Majack and Bornstein, 1987). Compositional perturbation of a given cellular ECM can lead to altered physiological responses of the cells residing within the same (Krieg et al.,1988).

We have observed that VSMC plated onto dishes containing presynthesized ECM (as elaborated by VSMC in previous passage) exhibit an accelerated rate of proliferation when compared with those plated directly onto gelatine-coated dishes (Fig. 1). In the latter case it was apparent that the real onset of VSMC proliferation occurred subsequent to the process of ECM elaboration by these VSMC (Fig. 1; compare enumeration profiles with inset showing [^{3}H] glycine incorporation into matrix material). This apparent relationship between proliferation and ECM-matrix elaboration was true for both SHR- and WKY-derived VSMC, but it was clear that the growth promotional properties of SHR-elaborated ECM were markedly greater than those of WKY-elaborated ECM (Fig. 1). Not only did gelatine-plated SHR-VSMC proliferate more rapidly than their similarly plated normotensive counterparts, but WKY-VSMC plated onto a presynthesized SHR-ECM grew faster than when plated onto a homologously (WKY) presynthesized ECM. The growth rate of SHR-VSMC was, however, not influenced by the source of VSMC (WKY or SHR) - ECM upon which they were plated (Fig. 1).

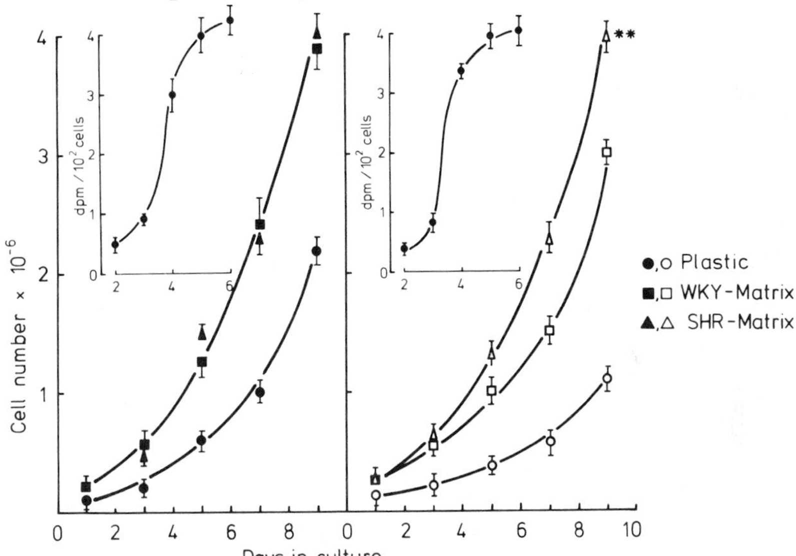

Figure 1. Growth kinetics and matrix elaboration by VSMC: VSMC from SHR (closed symbols) and WKY (open symbols) were plated onto gelatinized plastic ware (● , ○) or onto plasticware coated with ECM elaborated by either SHR-VSMC (▲,△) or WKY-VSMC (■ , □). Cell numbers were determined at the indicated times The inset illustrates total incorporation of [^{3}H]-glycine into ECM's elaborated by the two isolates. * * , $p<0.001$ for greater proliferation of WKY-VSMC on SHR-ECM compared with WKY-ECM.

Determination of total [^{3}H]-glycine incorporated into matrix over 24 hours (insets to Fig. 1) indicated that SHR- and WKY-derived VSMC elaborated comparable gross quantities of matrix material. However, compositional analysis of such matrices revealed a number of differences (SHR vs. WKY) in the overall components (Table 1). Principally, the proportions of glycoprotein/ proteoglycan material and elastin were increased and decreased respectively, in SHR-ECM compared with WKY-ECM. A more detailed analysis of glycosaminoglycan material indicated elevated levels of hyaluronic acid but decreased heparan sulphate/heparin levels in the matrices synthesized by SHR-VSMC (Table 1). The respective growth-stimulatory and growth-inhibitory-properties of these components (Castellot and Karnovsky, 1987; Merrilees, 1987) could partially explain the differential influence between SHR- and WKY-ECM on VSMC proliferation.

Table 1: Compositional analysis of the extracellular matrices elaborated by SHR- and WKY-derived VSMC.

A	SHR	WKY
GLYCOPROTEIN/PROTEOGLYCAN (TRYPSIN)	29.0 ± 3.0	45.5 ± 4.1
ELASTIN (ELASTASE)	19.3 ± 5.0	10.7 ± 1.4
COLLAGEN (COLLAGENASE)	51.8 ± 3.0	43.9 ± 5.3
B		
HYALURONIC ACID	31.8 ± 2.4	20.1 ± 2.1
CHONDROITIN SULPHATE	25.2 ± 2.9	26.4 ± 3.0
HEPARAN SULPHATE/HEPARIN	43.0 ± 3.4	53.5 ± 3.5

Polypeptide and glycosaminoglycan analysis was performed as described previously (Scott-Burden et al, 1986). Data express the proportion of each component relative to total polypeptide (100%) in panel A and to total glycosaminoglycan (100%) in panel B.

SHR-derived VSMC are less susceptible to growth inhibition by both heparin and its homopolysaccharide analogue, pentosan polysulphate (PPS) than are WKY-VSMC (Fig. 2). Additionally, SHR-VSMC possess lower levels of surface heparin receptors (4.5 ± 1.6 pmol/10^6 cells) than their normotensive couterparts (9.1 ± 2.3 pmol/10^6 cells). Such findings (described previously, Resink et al., 1989a) could, in part, account for the observation (Fig. 1) that SHR-VSMC were not growth inhibited by WKY-elaborated ECM in spite of the higher levels of heparin/heparan sulphate in these matrices (vs. SHR-ECM, Table 1).

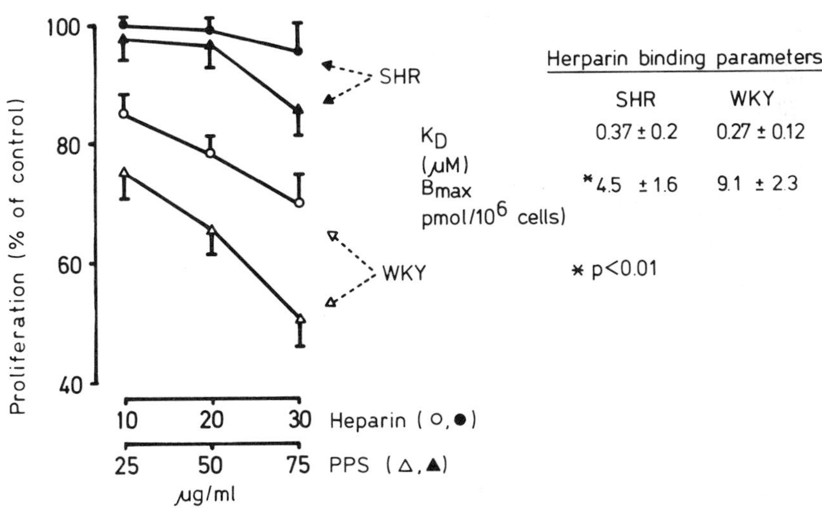

Figure 2. Growth-inhibition of VSMC by heparin and PPS: 14 hours after plating VSMC from SHR (closed symbols) and WKY (open symbols) heparin (●,○) or PPS (▲,△) were added at the doses shown. After 8 days in culture (with routine medium changes ± heparinoids) cell numbers were determined. Data express cell numbers in wells from treated cultures relative to those present (100 %) in control cultures. Parameters for [^{3}H]-heparin binding to intact quiescent VSMC (Resink et al. 1989a) are shown; *, $p<0.01$ indicates significance of difference for B_{max} between the two isolates.

Our observations invoke an active role for the extracellular matrix in the pathophysiology of hypertension. However, the use of physiologically elaborated complex/multi-constituent ECM's complicates interpretation of data, particularly with respect to precise definition of the identity(ies) and mechanisms of action of matrix components influencing VSMC growth. Future studies of the regulation, structure, molecular interactions, function and mechanism of action of matrix molecules are warranted.

GROWTH-RELATED METABOLIC EVENTS

Numerous metabolic processes are stimulated upon exposure of VSMC to growth factors/mitogens including elevation of intracellular Ca^{2+}, intracellular alkalinization, phosphoinositide catabolism and S6-kinase activation (Berridge, 1984; Glaser and Whiteley, 1987; Ives, 1987, Chambard and Pouyssegur, 1986; Hokin, 1985). Alterations in any of these pivotal processes are likely to interfere with co-ordinated control of VSMC function at the level of both proliferation and contraction, with important consequent implications for the pathophysiology of hypertension.

Reactivation of the cellular protein synthetic machinery is a pre-requisite for cells to transit from a quiescent to proliferative state. S6-kinase plays a crucial role in this process of restimulation by promoting the formation of activity translating polysomes (Chambard and Pouyssegur, 1986), although S6-kinase activation per se does not obligatory lead to cell division or elevated incorporation of [^{3}H]-thymidine into DNA (Scott-Burden et al., 1988b,c; 1989a,b). The control of S6-kinase activity itself occurs via phosphorylation/dephosphorylation (Novak-Hofer and Thomas, 1985) and both protein kinase C and receptor tyrosine kinase are believed to be stimulatory in this regard (either independently and/or synergistically). The activation of S6-kinase is intimately associated with intracellular alkalinization as mediated via stimulated Na^+/H^+ exchange (Chambard and Pouyssegur, 1986). Protein kinase C has increasingly been implicated in the activation of Na^+/H^+ exchange (Owen, 1985; Berk et al., 1988a). Therefore, it is not surprising that agonists which promote S6-kinase activation and intracellular alkalinization also stimulate phospholipase C-mediated catabolism of phosphoinositides to release diacylglycerol, the physiological activator of protein kinase C, and inositol trisphosphate, the messenger for release of Ca^{2+} from internal pools (Berridge, 1984; Hokin, 1985). The very same transmembrane signals are elicited in response to vasoactive hormones such as angiotensin II (Resink et al, 1989b; Scott-Burden et al, 1988b; Kawahara et al, 1988). Likewise vasoactive effects have also been reported for growth factors (Berk et al, 1985). Thus, it is imperative that the complex interactions between signal transduction processes be co-ordinated _in vivo_ for regulated maintenance of both VSMC growth and contraction.

We have demonstrated previously that SHR-derived VSMC exhibit enhanced responsiveness to EGF (vs. WKY-VSMC), and partly attributed this differential to increased EGF-receptor capacity on such cells (Scott-Burden et al, 1989a). In subsequent screening experiments it became apparent that amplified signal transduction to agonists might be a gross aberration in SHR-derived VSMC. Increased S6-kinase activation (SHR vs. WKY) in response to a variety of growth factors/mitogens and vasoactive hormones is demonstrated in Fig. 3. It is noteworthy that the extent of S6-kinase activation by either serum (FCS) or phorbol ester (TPA, a diacylglycerol mimetic) was comparable between SHR- and WKY-VSMC (Fig. 3). Such data could imply that the protein kinase-C-dependent pathway for S6-kinase activation is not different between the different VSMC isolates, whereas a differential exists for the protein kinase-C-independent pathway.

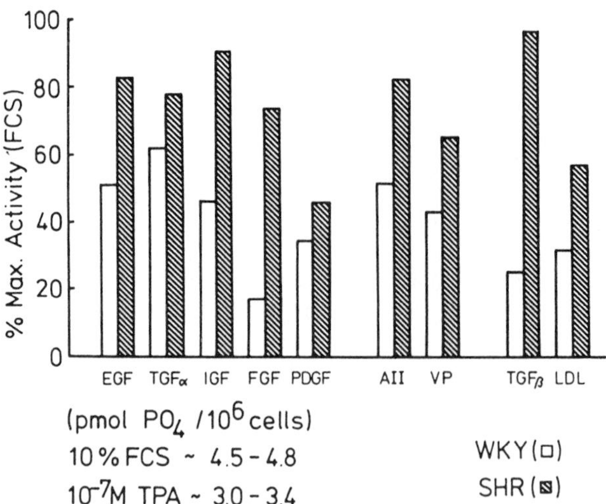

Figure 3. Activation of S6-kinase by in quiescent monolayers of VSMC: VSMC were exposed to the indicated compounds at levels of 5 ng/ml for EGF, TGF (transforming growth factor α/β), IGF (somatomedin C), FGF (fibroblast growth factor), PDGF (platelet-derived growth factor), and at 50 ng/ml for LDL (low density lipoprotein) and at 10^{-7} M for A II (angiotensin II) and VP (vasopressin). Data express the percentage activation of S6-kinase relative to that by 10 % FCS (100 %, absolute values inset).

The alkalinization response to the same spectrum of agonists was also greater for SHR-VSMC compared with their normotensive counterparts (Fig. 4). This differential was also evident for FCS and TPA, which might seem discrepant with data obtained for S6-kinase activation by these agonists. However, it is known that although S6-kinase activation is dependent upon alkalinization, it is most sensitive to conditions which increase intracellular pH by Δ 0.2 units and further alkalinization does not significantly increase polysome formation (Chambard and Pouyssegur, 1986). Amplified alkalinization responsiveness in VSMC need not therefore be necessarily solely dependent on direct activation by protein kinase C, but could also involve an alteration in some process distal to protein kinase C and/or in some moiety of the Na^+/H^+ antiporter itself.

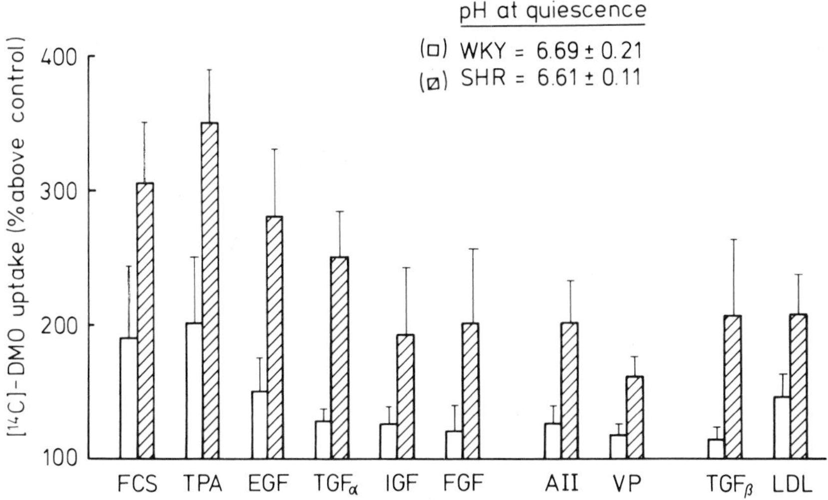

Fig 4: Agonist-induced alkalinization in quiescent VSMC. Quiescent VSMC exposed to the indicated compounds at doses given in Fig. 3. Intracellular pH was determined using 4,4-dimethyl [2-^{14}C]-oxazolidine-2,4-dione (DMO) (Scott-Burden et al., 1989a). pH values at quiescence and without agonist addition are inset; data in the histogram represent extent of alkalinization and express [^{14}C]-DMO uptake relative to that in unstimulated quiescent cells (100 %).

With respect to protein kinase C we have previously found that its activation by FCS or TPA, as assessed by translocation from cytosol to membranes, was comparable between SHR- and WKY-VSMC (Resink et al, 1989b). We additionally demonstrate (Fig. 5) equivalent levels of immunoreactive (80KD) protein kinase C as well as comparable subcellular distribution of phorbol binding between SHR- and WKY-VSMC grown in the continuous presence of FCS. We have also observed comparable (SHR and WKY) phorbol binding to intact VSMC at quiescence (Resink et al, 1989b).

However, our data are insufficient to permit absolute exclusion of a role for protein kinase C in mediating some abnormalities of VSMC in hypertension. Given the existence of multiple isoforms of protein kinase C (Parker and Ullrich, 1987) it is possible that non-parallel relationships exist between protein kinase(s)-C activity and either phorbol binding or immunoreactive peptide. Furthermore, it should be considered that protein kinase C enzymic activity (vis-à-vis it's phorbol binding properties) in SHR- and WKY-VSMC isolates may be different at a given Ca^{2+} and diacylgycerol concentration, and that concentrations of these intracellular cofactors may be different following exposure of such VSMC isolates to agonists.

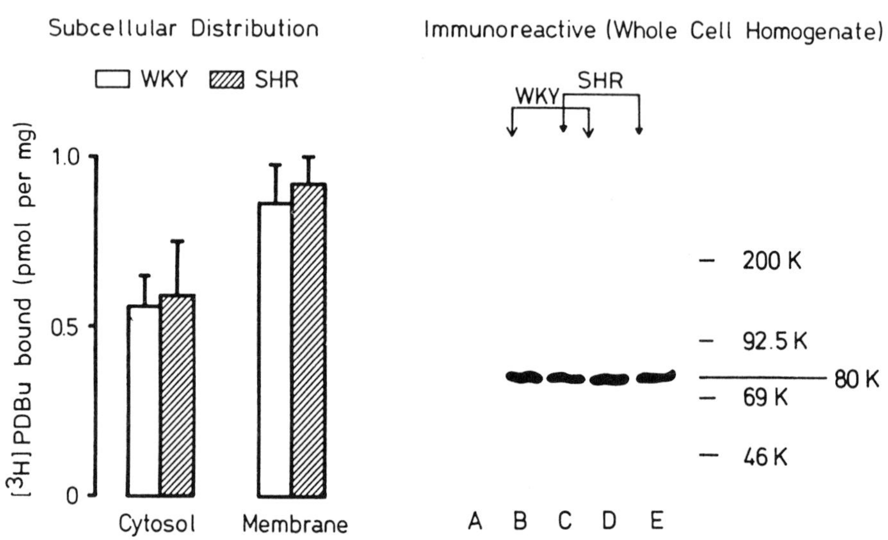

Fig. 5: Protein kinase C: subcellular distribution and immunoreactivity. Binding of [^{3}H]-phorbol dibutyrate ([^{3}H]-PD Bu) was determined (Resink et al, 1989b) in subcellular fractions prepared from VSMC grown continuously in the presence of 10 % FCS. Immunoreactive protein kinase C was measured in whole cell homogenates using monoclonal antibody (clone MC5, Amersham) and [^{125}I]-labelled second antibody. An autoradiogram is presented in which equivalent concentrations of cell protein from two different pairs (SHR vs. WKY) of VSMC isolates were loaded onto SDS-polyacrylamide (10 %) gels.

In assessing the phosphoinositide metabolic response of VSMC to some agonists, we found that compared with WKY-derived VSMC, SHR-VSMC exhibited an amplified catabolism (Fig. 6); the generation of both phospholipase C cleavage products, namely inositol trisphosphate and dianylglycerol was enhanced in the latter VSMC isolates. Available literature abounds with reports of alterations in cellular phosphoinositide signalling in hypertension (reviewed by Heagerty and Ollerenshaw, 1987). However, precise identification of the determinants for such alterations and their functional consequences thereof remain unclear.

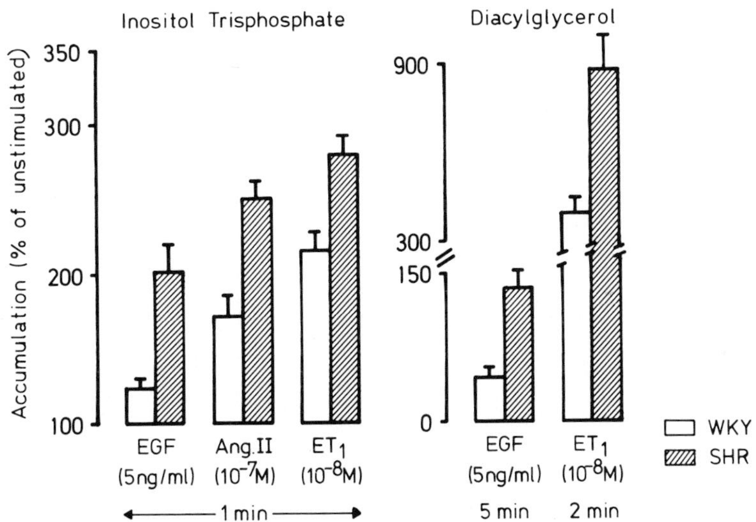

Fig. 6: <u>Activation of phospholipase C is amplified in SHR-VSMC</u>. Quiescent myo-[^{3}H]-inositol-prelabelled or [^{3}H]-arachidonic acid-prelabelled VSMC were exposed to EGF, Ang II or endothelin (ET_1) at levels and for times indicated. Inositol trisphosphate and diacylglycerol accumulation were determined (Resink et al, 1989a,b,c) and data express their increase relative to levels (100 %) in unstimulated VSMC.

MITOGENESIS OR NOT?

It has become increasingly evident that although numerous agonists are capable of stimulating a wide range of metabolic processes characteristic of the proliferative state, many of these agonists per se do not elicit a true mitogenic/proliferative response. For example, with VSMC at quiescence and exposure to epidermal growth factor (EGF) (in the complete absence of serum) it is possible to demonstrate S6-kinase activation (Fig. 3), intracellular alkalinization (Fig. 4), phosphoinositide catabolism via phospholipase C (Fig. 6) and nuclear protooncogene induction (Fig. 7). In spite of all these events, and in the face of their amplification in hypertensive-derived

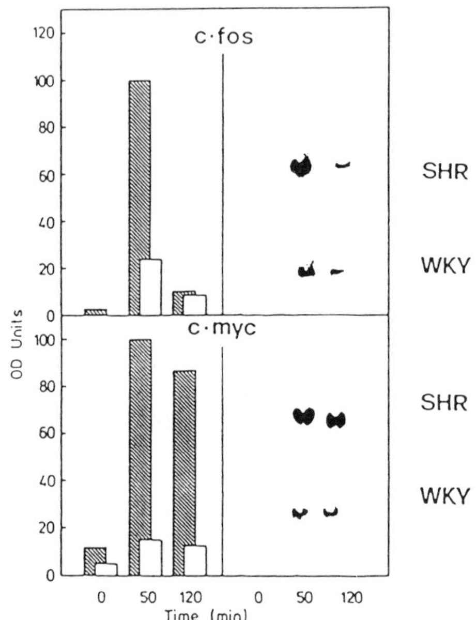

Fig. 7: Activation of c-fos and c-myc by EGF is increased in SHR-VSMC. Quiescent VSMC from SHR (▧) and WKY (□) were exposed to 1 nM EGF for the given times. Northern blot analysis was performed on total RNA (Scott-Burden et al, 1989b) and autoradiographs (examples shown) were subjected to scanner analysis of the 2.2Kb (c-fos) and 2.7Kb (c-myc) bands.

VSMC, a proliferative response did not ensue (Fig. 8). Nevertheless, whereas low serum levels (0.5 %) were sufficient to support growth, the simultaneous addition of EGF and 0.5 % serum produced a marked synergistic effect on the rate of proliferation (Fig. 8). Such synergism may also be found to be operative when using combinations of agonists (and without serum) that are per se ineffectual in eliciting a mitogenic and/or proliferative response. Specifically we have found synergism between EGF and low-density lipoprotein (LDL) (Scott-Burden et al, 1989b) and between angiotensin II and insulin or LDL (unpublished observations).

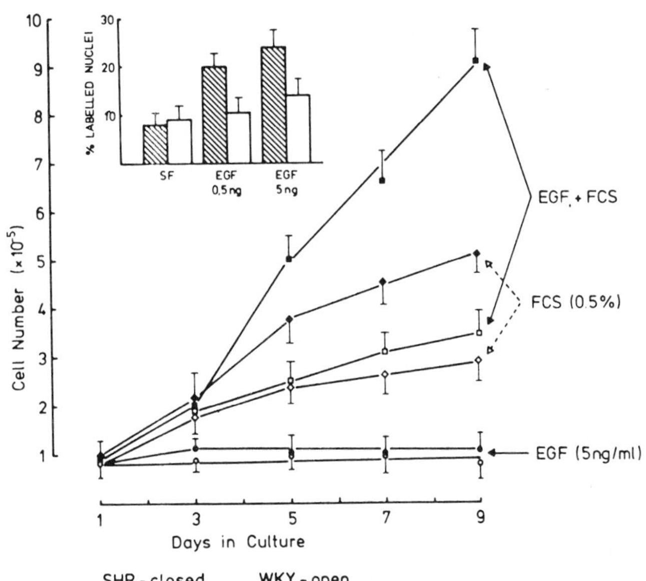

Fig. 8: <u>Growth kinetics of VSMC in response to EGF</u>. 24 hours after seeding and growth in full culture medium, the medium was replaced by serum-free medium containing either EGF (5 ng/ml; ●,○) or FCS (0.5 %,◆,◇) or both EGF and FCS (■,□). Cell numbers were determined at the indicated times (with routine medium changes every 2 days). The inset presents data from nuclear labelling experiments (Scott-Burden et al, 1989a) which expresses the number of labelled nuclei as a percentage of the total nuclei (100 %).

CONCLUSION

A plethora of mechanisms may contribute toward structural and functional abnormalities in the vasculature. The possible determinants of vascular pathology in hypertension are multifaceted and include growth factors/mitogens, vasoactive hormones, extracellular matrix components and cellular/intracellular signal transduction pathways. However, the present paucity of understanding concerning the true complexities of co-ordination and integration between all these "regulatory" components precludes definition of "cause and effect" relationships in hypertension.

Acknowledgements

Sabine Bohnert and Bernadette Libsig are thanked for assistance in preparation of this manuscript. Financial support was provided by the Swiss National Foundation (no.3.827.087).

REFERENCES

Berridge MJ (1984). Oncogenes, inositol lipids and cellular proliferation. Biotechnology 2, 541-546.

Berk BC, Brock TA, Webb RC, Taubman MB, Atkinson WJ, Gimbrone MA, Alexander RW (1985). Epidermal growth factor, a vascular smooth muscle cell mitogen, induces rat aortic contraction. J.Clin.Invest. 75: 1083-1086.

Berk BC, Canessa M, Vallega G, Alexander RW (1988a) agonist mediated changes in intracellular pH: role in vascular smooth muscle cell function. J.Cardiovasc.Pharm. 12 (suppl. 5), S104-S114.

Berk BC, Gordon HM, Uekhstein V, Taubman MB, Alexander RW (1988b). Angiotensin II stimulates protein synthesis in vascular smooth muscle independently of c-fos. Clin.Res. 36(3); 425A.

Castellot JJ Jr., Karnovsky MJ (1987). Heparin and the regulation of growth in the vascular wall. In: Campbell JH, Campbell GR (eds) "Vascular smooth muscle in culture", Florida, CRC Press, vol. 1, pp. 93-115.

Chambard J-C, Pouyssegur J (1986). Intracellular pH controls growth factor-induced ribosomal protein S6 phosphorylation and protein synthesis in the G_0 G_1 transition of fibroblasts. Exptl. Cell Res. 164: 282-294.

Chamley-Campbell JG, Campbell GR (1981). What controls smooth muscle phenotype? Atherosclerosis 40: 347-352.

Folkow B, Hallbäck M, Lundgren Y, Siversson R, Weiss L (1973). Importance of adaptive changes in vascular design for establishment of primary hypertension: studies in man and spontaneously hypertensive rats. Circ Res. 32/33 (suppl 1): 2-6.

Glaser L, Whitely B (1987). Control of ion fluxes by mitogenic polypeptides. Hypertension 10 (suppl 1), 27-31.

Hay ED (1983). Cell biology of the extracellular matrix. New York, Plenum Press.

Heagerty AM, Ollerenshaw JD (1987). The phosphoinositol signalling system and hypertension. 5: 515-524.

Hokin LE (1985). Receptors and phosphoinositide-generated second messengers. Ann. Rev. Biochem. 54: 205-235.

Hynes RV (1987). Integrins: A family of cell surface receptors. Cell 48: 549-554.

Ives HE, Daniel TV (1987). Interrelationship between growth factor-induced pH changes and intracellular Ca^{2+}.

Jones PA, Scott-Burden T, Gevers W (1979). Glycoprotein elastin and collagen secretion by rat smooth muscle cells. Proc. Natd. Acad. Sci. 76: 353-357.

Kawahara Y, Sunako M, Tsuda T, Fukuzaki H, Fukumoto Y, Takai Y (1988). Angiotensin II induces expression of c-fos gene through protein kinase C activation and calcium ion mobilization in cultured vascular smooth muscle cells. Biochem Biophys Res Commun. 150: 52-59.

Krieg T, Hein R, Hatamochi A, Aumilley M (1988). Molecular and clinical aspects of connective tissue. Eur J Clin Invest. 18: 105-123.

Majack RA, Bornstein P (1987). Biosynthesis and modulation of extracellular matrix components by cultured vascular smooth muscle cells. In: Campbell JH, Campbell GR (eds) "Vascular Smooth Muscle in Culture", Florida, CRC Press vol pp. 117-132.

Merrilees MF (1987). Synthesis of glysosaminoglycans and proteoglycans. In: Campbell JH, Campbell GR (eds) "Vascular Smooth Muscle in Culture", Florida, CRC Press vol 1, pp. 133-151.

Novak-Hofer I, Thomas G (1985). Epidermal growth factor-mediated activation of an S6-kinase in Swiss Monse BTB cell. J Biol Chem. 260: 10314-10319.

Owen N (1985). Effect of TPA on ion fluxes and DNA synthesis in vascular smooth muscle. J Cell Biol. 101: 454-459.

Parker PJ, Ullrich A (19879. Protein kinase C: structure and function. Hormones and Cell Regulation. 153: 41-47.

Resink TJ, Scott-Burden T, Bühler FR (1989a). Decreased susceptibility of cultured smooth muscle cells from SHR rats to growth inhibition by heparin. J Cell Physiol. 138: 137-144.

Resink TJ, Scott-Burden T, Baur U, Bürgin M, Bühler FR (1989b). Enhanced responsiveness to angiotensin II in vascular smooth muscle cells from spontaneously hypertensive rats is not associated with alterations in protein kinase C. Hypertension (in press).

Resink TJ, Scott-Burden T, Bühler FR (1989c). Endothelin stimulates phospholipase C in cultured vascular smooth muscle cells. Biochem Biophys Res Commun. 157: 1360-1368.

Ross R (1981). Atherosclerosis: A problem of the biology of arterial wall cells and their interactions with blood components. Atherosclerosis 1: 293-311.

Schwartz SM, Campbell GR, Campbell JH (1986). Replication of smooth muscle cells in vascular disease. Circ Res. 58: 427-443.

Scott-Burden T, Bogenmann E, Jones PA (1986). Effects of complex extracellular matrices on 5-

azacytidine-induced myogenesis. Exptl. Cell Res. 156: 527-535.

Scott-Burden T, Resink TJ, Bühler FR (1988a). Growth regulation in smooth muscle cells from normal and hypertensive rats. J Cardiovasc Pharmacol. 12 (suppl 5). S124-S127.

Scott-Burden T, Resink TJ, Baur U, Burgin M, Bühler FR (1988b). Amiloride sensitive activation of S6-kinase by angiotensin II in cultured vascular smooth muscle cells. Biochem Biophys Res Commun. 151: 583-589.

Scott-Burden T, Resink TJ, Baur U, Bürgin M, Bühler FR (1988c). Activation of S6-kinase in cultured vascular smooth muscle cells by submitogenic levels of thrombospondin. Biochem Biophys Res Commun. 150: 278-286.

Scott-Burden T, Resink TJ, Baur U, Bürgin M, Bühler FR (1989a). EGF responsiveness in smooth muscle cells from hypertensive and normotensive rats. Hypertension 13: 295-305.

Scott-Burden T, Resink TJ, Hahn AP, Bühler FR (1989b). Differential stimulation of growth related metabolism in cultrued smooth muscle cells from SHR and WKY by combinations of EGF and LDL. Biochem Biophys Res Commun. 159: 624-632.

Woods RV, Couchman JR (1988). Focal adhesions and cell-matrix interactions. Collagen Rel Res. 8: 155-182.

Yamori Y (1976). Interaction of neural and non-neural factors in the pathogenesis of spontaneous hypertension. In: Julius A, Esler M, (eds). The nervous system in arterial hypertension. Thomas, Springfield, pp. 17-50.

Calcium Channel Modulators in Heart and Smooth Muscle: Basic Mechanisms and Pharmacological Aspects
S. Abraham and G. Amitai, editors
© 1990, VCH, Weinheim/Deerfield Beach, FL and Balaban, Rehovot/Philadelphia

Calcium antagonists and myocardial protection during ischaemia and reperfusion: The St. Thomas' experience

METIN AVKIRAN AND DAVID J. HEARSE

Cardiovascular Research, The Rayne Institute, St. Thomas' Hospital, London SE1 7EH, United Kingdom

INTRODUCTION

Disturbances in calcium homeostasis have been implicated in the development of cellular injury in the heart during ischaemia and reperfusion. Calcium antagonists may be expected to provide protection against such injury, not only by inhibiting calcium overload through slow calcium channels, but also through other mechanisms such as reduced energy demand and improved myocardial flow distribution. In addition, calcium antagonists may suppress the arrhythmias produced by ischaemia and reperfusion.

The objective of this article is to review some of the studies from the Cardiovascular Research group in the Rayne Institute at St. Thomas' Hospital, which have examined the influence of calcium antagonists on the outcome of myocardial ischaemia and reperfusion. Three experimental preparations were utilised in these studies; transient global ischaemia (as occurs in cardiac surgery), sustained regional ischaemia (as occurs in evolving myocardial infarction) and transient regional ischaemia (a condition which may cause reperfusion arrhythmias).

METHODS

Transient Global Ischaemia

These studies examined the efficacy of the three major calcium antagonists diltiazem, nifedipine and verapamil, as additives to crystalloid cardioplegia (Yamamoto et al., 1983a, 1983b, 1983c).

Isolated hearts from adult rats were perfused in the working mode with modified Krebs Henseleit bicarbonate buffer (37 °C). The left ventricular filling pressure was maintained at 18 cm H_2O and the afterload at 100 cm H_2O. Control measurements of left ventricular function were obtained after 20 min of working perfusion, after which the hearts were perfused for 3 min with the St. Thomas' Hospital cardioplegic solution, either alone or with added diltiazem (0.25, 0.50, 0.75, 1.00 or 1.25 μmol/l), nifedipine (0.05, 0.075, 0.10, 0.15, 0.20, 0.50 or 1.00 μmol/l) or verapamil (0.04, 0.20, 1.10 or 2.20 μmol/l). The hearts were then subjected to a period of normothermic global ischaemia (30 min in the verapamil study, 35 min in the diltiazem and nifedipine studies). Left ventricular function was reassessed after 35 min of reperfusion (the final 20 min of which was in the working mode) and expressed as a percentage of the pre-ischaemic control function. Creatine kinase leakage was determined by measuring the amount of enzyme in the coronary effluent collected during the initial 15 min of reperfusion.

Sustained Regional Ischaemia

These studies investigated the effects of nifedipine and verapamil on the development of myocardial infarction during sustained regional ischaemia (Yellon et al., 1983; Yoshida et al., 1985; Kudoh et al., 1986; Dennis et al., 1986).

Regional myocardial ischaemia was induced in

closed chest anaesthetised greyhounds by coronary artery embolisation. Embolisation was produced by injecting a plastic bead (2.5 mm dia.) into the coronary ostium, using a specially designed cannula inserted through the carotid artery. Radioactive microspheres were administered intraventricularly immediately after embolisation to enable autoradiographic risk zone analysis. Nifedipine and verapamil were administered intravenously as a bolus (16 and 200 µg/kg respectively) followed by continuous infusion (24 and 300 µg/kg/hr respectively) starting 10 min after embolisation. Controls received an equivalent volume of saline. The animals were killed 24 or 48 hours after embolisation and the heart was removed and sectioned into transverse slices. Tissue necrosis was visualised by tetrazolium staining. Infarct and risk zone volumes were determined by planimetric methods and infarct size was expressed as a percentage of the risk zone.

Transient Regional Ischaemia

This study investigated the effects of diltiazem on the vulnerability to arrhythmias induced by transient regional ischaemia and reperfusion (Tosaki et al., 1987).

Isolated hearts from adult rats were perfused in the Langendorff mode with modified Krebs Henseleit bicarbonate buffer (37 °C). Regional ischaemia was induced by tightening a ligature placed around the left anterior descending coronary artery at a point close to its origin. Reperfusion was initiated after 10 min of ischaemia by releasing the ligature. Perfusion with buffer containing diltiazem (0.05, 0.10, 0.50, 1.00 or 5.00 µmol/l) was either started 5 min before ischaemia and maintained throughout the period of 10 min ischaemia and 3 min reperfusion, or started 2 min before reperfusion and maintained during the 3 min of reperfusion. Heart rate and coronary flow were monitored throughout and epicardial ECG recordings were used to analyse the arrhythmias.

RESULTS

Transient Global Ischaemia

Studies on the myocardial protective effects of diltiazem, nifedipine and verapamil, included individually in the St. Thomas' Hospital cardioplegic solution, revealed "bell-shaped" dose-response profiles for all three drugs. Figure 1 shows the effects of the optimum concentrations of the drugs on post-ischaemic recovery of cardiac output and creatine kinase leakage.

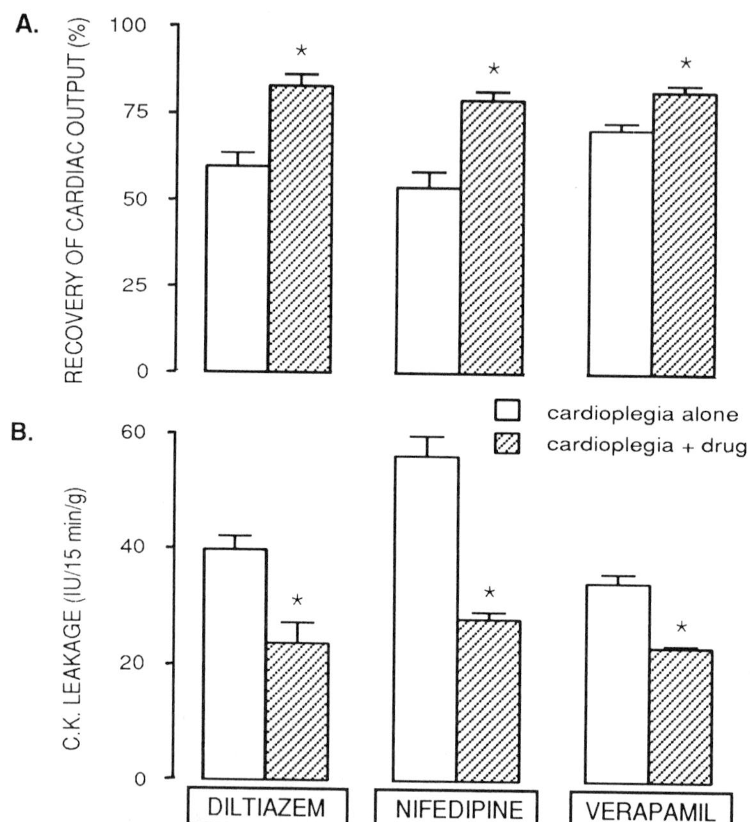

Figure 1. Effects of the optimum concentrations of diltiazem (0.5 µmol/l), nifedipine (0.075 µmol/l) and verapamil (0.2 µmol/l) on **A.** recovery of cardiac output, and **B.** creatine kinase leakage ($n \geq 6$/group, * $p<0.05$ v cardioplegia alone).

Thus, at the optimum concentrations, the post-ischaemic recoveries of cardiac output in the diltiazem, nifedipine and verapamil groups were approximately 40 %, 47 % and 16 % better than the recoveries observed in their respective control (cardioplegia alone) groups. The relatively high recovery in the control group in the verapamil study was due to the shorter duration of ischaemia (30 min compared with 35 min in the diltiazem and nifedipine studies) and may have somewhat masked the additional protective capacity of verapamil. The significant improvements obtained with the calcium antagonists in post-ischaemic recovery of pump function were paralleled by significant reductions in creatine kinase leakage produced by ischaemia and reperfusion.

Interestingly, although each drug was highly effective in studies carried out under normothermic conditions, no additional protection over cardioplegia alone could be demonstrated under hypothermic (20 °C) conditions. This finding suggests that hypothermia and calcium antagonists may share a common mechanism of protection (Fukunami and Hearse, 1984).

Sustained Regional Ischaemia

Figure 2A shows the effects of nifedipine and verapamil on infarct size following regional ischaemia, sustained for 24 hr. Nifedipine significantly reduced the size of the infarct, expressed as a percentage of the risk zone, from 80±4 % in controls (n=9) to 39±5 % (n=10). Verapamil also produced a significant reduction in infarct size from 62±7 % in controls to 18±4 % (n=8/group). When the ischaemic duration was prolonged to 48 hr the reduction in infarct size produced by the drugs was much less, though still statistically significant (figure 2B). Thus after 48 hr of ischaemia, nifedipine reduced infarct size from 86±3 % in controls to 70±5 % (n=9/group) and verapamil from 76±3 % in controls to 51±4 % (n=8/group).

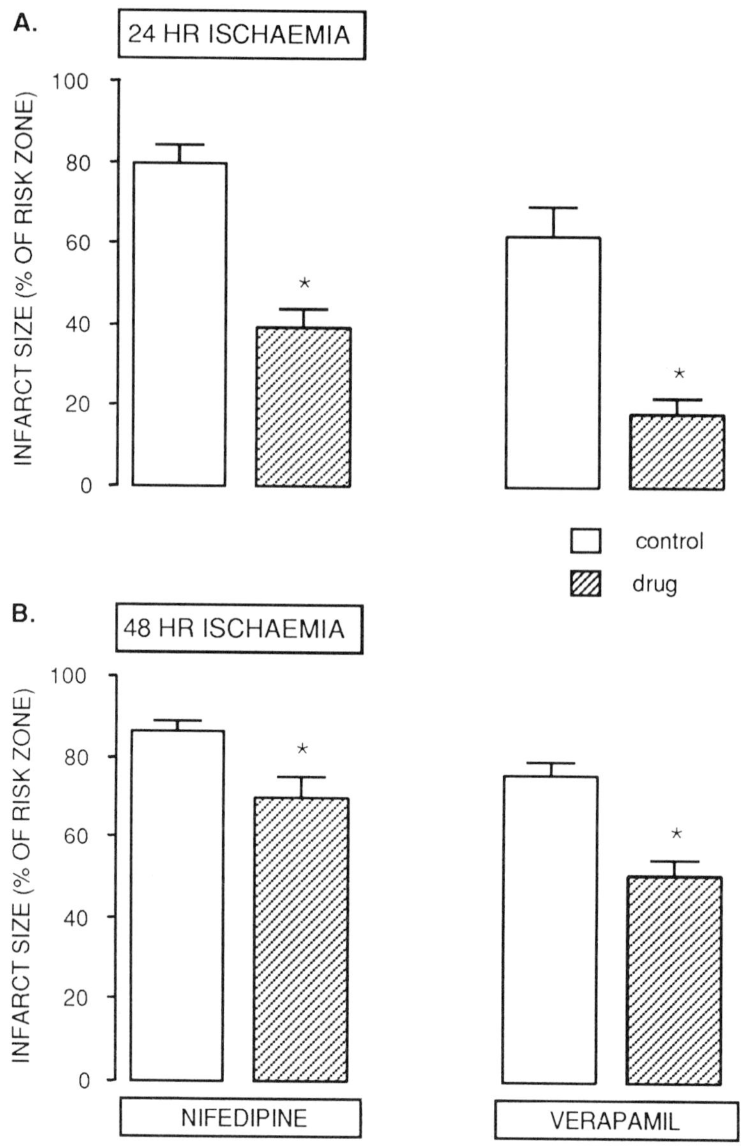

Figure 2. Effects of nifedipine (16 μg/kg bolus + 24 μg/kg/hr infusion i.v.) and verapamil (200 μg/kg bolus + 300 μg/kg/hr infusion i.v.) on myocardial infarct size following **A.** 24 hr, and **B.** 48 hr of regional ischaemia. Controls received an equivalent volume of saline (n=8-10/group, * $p<0.05$ v control).

In the 48 hr nifedipine study (Kudoh et al., 1986) a second microsphere, which did not interfere with the autradiographic risk zone analysis, was given at the end of ischaemia to enable the assessment of temporal flow changes. This study revealed that in tissue destined to necrosis (mainly in the subendocardium) collateral flow at the onset of ischaemia (7±1 % of normal flow) did not increase over the next 48 hr (8±1 %) *but* in tissue which was salvaged (mainly in the subepicardium) flow was greater at the onset of ischaemia (27±3 %) and increased substantially over the next 48 hr (to 65±6 %). Nifedipine did not increase the amount of flow per gram of salvaged tissue but appeared to reduce infarct size by increasing the amount of tissue receiving sufficient flow for salvage.

Transient Regional Ischaemia

The administration of diltiazem prior to the induction of regional ischaemia resulted in a dose-dependent reduction in reperfusion-induced ventricular fibrillation (figure 3A). At concentrations of 0.05, 0.1, 0.5, 1.0 and 5.0 µmol/l (n=12/group) diltiazem reduced the incidence of ventricular fibrillation from the control incidence of 100 % to 91 %, 58 %, 17 %, 0 % and 0 % respectively. Heart rate, measured just prior to the 10 min ischaemic period, was also reduced in a dose-dependent manner to 97±2 %, 91±2 %, 79±2 %, 69±5 % and 51±3 % respectively, from the control value of 340±10 beats/min. These reductions in heart rate persisted until the end of the ischaemic period (figure 3B). Diltiazem also produced dose-dependent increases in total coronary flow prior to ischaemia, during regional ischaemia and during reperfusion. When hearts receiving 0.5 µmol/l diltiazem were paced at 5 Hz during the ischaemic period, the incidence of ventricular fibrillation was increased from 17 % to 91 % (not significantly different from the control incidence of 100 %). There was no significant difference in coronary flow between paced and unpaced diltiazem groups.

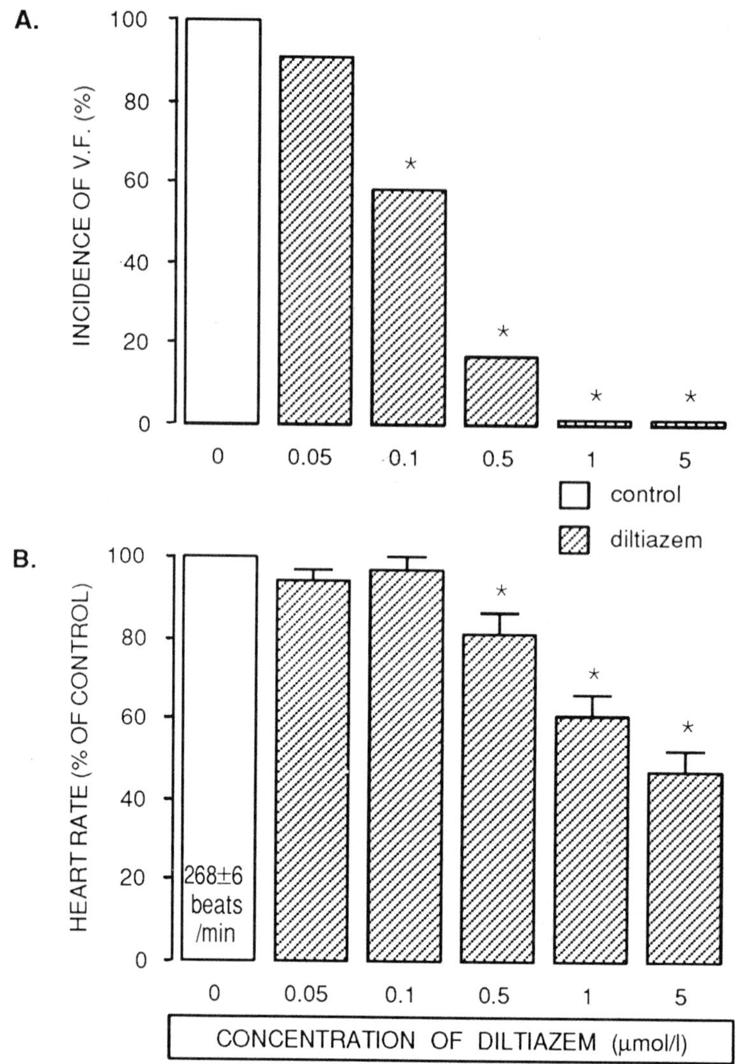

Figure 3. Effects of various concentrations of diltiazem, administered continuously starting 5 min prior to the onset of ischaemia, on **A.** the incidence of reperfusion-induced ventricular fibrillation, and **B.** heart rate at the end of 10 min ischaemia (n=12/group, * $p<0.05$ v control).

When diltiazem administration (0.5 µmol/l) was started 2 min before reperfusion, rather than 5 min before ischaemia, the incidence of reperfusion-induced ventricular fibrillation was 83 %, which was again not significantly different from the control incidence. Diltiazem did not alter heart rate during the 2 min period of ischaemia between the start of its administration and the onset of reperfusion.

DISCUSSION

In the isolated rat heart, under normothermic conditions, all three calcium antagonists tested provided significant additional protection when included in the St. Thomas' Hospital cardioplegic solution, with narrow "bell-shaped" dose-response curves. The reason for the loss of protection observed at high concentrations of the drugs is not known but may be due to residual negative inotropic effects. Although the protective effect was totally lost under hypothermic conditions (20 °C), this would not negate the use of calcium antagonists under the conditions of clinical hypothermia. It is often difficult to maintain adequate, uniform and reproducible myocardial cooling during cardiac surgery and regional "hot spots" are not uncommon. The use of calcium antagonists, which should provide additional protection in these areas, may compensate for inadequate cooling and reduce the reliance upon hypothermia for protection.

The studies with regional ischaemia, sustained for up to 48 hr, have shown that both nifedipine and verapamil can limit infarct size in dogs. However, the smaller reduction in infarct size obtained with both drugs after the longer duration of ischaemia suggests that the protection afforded by calcium antagonism was through an injury-slowing mechanism rather than a permanent limitation of infarct size. It is highly unlikely that any anti-ischaemic intervention can produce a sustained reduction in the ultimate extent of tissue injury, if flow is maintained below a critical threshold. Indeed, the results of clinical trials with calcium antagonists have been discouraging and part of the reason could be the need for relatively early reperfusion in order to achieve a sustained benefit. However recent developments in thrombolytic therapy have made early reperfusion during evolving myocardial infarction a real possibility and may revive interest in the concept of injury-delaying therapy. Thus, injury-delaying drugs such as calcium antagonists may be used to slow the progress of ischaemic injury and prevent the onset of irreversible damage until reperfusion can be achieved (Hearse, 1988).

The studies with diltiazem in the isolated rat heart have revealed a dose-dependent protective effect against reperfusion-induced ventricular fibrillation, when drug administration started prior to the onset of regional ischaemia. The anti-arrhythmic effect of diltiazem was paralleled by a dose-dependent negative chronotropic effect, and all protection was lost if the hearts were paced at normal rate during ischaemia. The protective effect was also lost if drug administration was started just prior to reperfusion. These findings strongly suggest that the anti-arrhythmic effect of diltiazem during reperfusion was a consequence of its protective effects during ischaemia. The anti-ischaemic effect of diltiazem may have been related to its negative inotropic and chronotropic effects or its vasodilatory action. However, pacing the hearts negated the negative chronotropy without altering the vasodilator effect and resulted in the loss of the anti-arrhythmic effect, suggesting the reduction in heart rate to be the most important factor in this model.

In conclusion, the studies reported in this article suggest that calcium antagonists can beneficially influence the outcome of myocardial ischaemia and reperfusion. It is most likely that protection is achieved by *slowing* the rate of injury during ischaemia without limiting its ultimate *extent*.

ACKNOWLEDGEMENTS

The work reported in this article was supported in part by grants from the British Heart Foundation, STRUTH and St. Thomas' Hospital Research Endowments Fund. The authors gratefully acknowledge their colleagues whose work has been included in the article.

REFERENCES

Dennis AR, Downey JM, Hearse DJ (1986). Infarct size limitation with allopurinol and verapamil alone or in combination during 48 h of coronary artery occlusion. Br J Pharmacol 89:737.

Fukunami M, Hearse DJ (1985). Temperature dependency of nifedipine as a protective agent during cardioplegia in the rat. Cardiovasc Res 19:95-103.

Hearse DJ (1988). The protection of the ischemic myocardium; surgical success v clinical failure. Prog Cardiovasc Dis 30:381-402.

Kudoh Y, Hearse DJ, Maxwell MP, Yoshida S, Downey JM, Yellon DM (1986). Calcium antagonists and evolving myocardial infarction; studies of the effects of nifedipine on tissue ATP, collateral flow and infarct size in the closed chest dog. J Mol Cell Cardiol 18 (suppl 4):77-92.

Tosaki A, Szekeres L, Hearse DJ (1987). Diltiazem and the reduction of reperfusion-induced arrhythmias in the rat; protection is secondary to modification of ischemic injury and heart rate. J Mol Cell Cardiol 19:441-451.

Yamamoto F, Manning AS, Braimbridge MV, Hearse DJ (1983a). Calcium antagonists and myocardial protection; diltiazem during cardioplegic arrest. Thorac Cardiovasc Surg 31:369-373.

Yamamoto F, Manning AS, Braimbridge MV, Hearse DJ (1983b). Nifedipine and cardioplegia; rat heart studies with the St Thomas' cardioplegic solution. Cardiovasc Res 17:719-727.

Yamamoto F, Manning AS, Braimbridge MV, Hearse DJ (1983c). Cardioplegia and slow calcium-channel blockers; studies with verapamil. J Thorac

Cardiovasc Surg 86:252-261.

Yellon DM, Hearse DJ, Maxwell MP, Chambers DE, Downey JM (1983). Sustained limitation of myocardial necrosis 24 hours after coronary artery occlusion; verapamil infusion in dogs with small myocardial infarcts. Am J Cardiol 51:1409-1413.

Yoshida S, Downey JM, Friedman, Chambers DE, Hearse DJ, Yellon DM (1985). Nifedipine limits infarct size for 24 hours in closed chest coronary embolized dogs. Bas Res Cardiol 80:76-87.

Calcium Channel Modulators in Heart and Smooth Muscle:
Basic Mechanisms and Pharmacological Aspects
S. Abraham and G. Amitai, editors
© 1990, VCH, Weinheim/Deerfield Beach, FL and Balaban, Rehovot/Philadelphia

Platelet activating factor (PAF) in cardiac and coronary pathophysiology

GIORA FEUERSTEIN

Director, Cardiovascular Pharmacology, Smith Kline & French, Philadelphia, PA, USA 19406

INTRODUCTION

The first evidence for naturally occurring alkyl-phospholipids with ethanolamine moeity was described in egg yolk in 1958; this finding was soon followed by reports on the presence of lipids in other organs such as the heart, liver, brain and blood cells (for review see Snyder, 1987). However, only in the late 1960's were cell-free systems shown to form alkyl ether bond (Snyder et al., 1969). Parallel to the efforts to elucidate the pathways in alkyl and alk-1-enyl (plasmalogen) glycerophospholipid synthesis, biological studies were conducted on renal lipid extracts which exhibited antihypertensive properties; these lipids, later shown to possess similar biological and chromatographic properties as PAF, were termed antihypertensive polar renomedullary lipids (APRL, Muirhead, 1980). In addition, Benveniste et al. (1972) identified a factor released from IgE-stimulated basophils that induced platelet aggregation and histamine release; these investigators first coined the term "Platelet Activating Factor" (PAF) which ultimately was identified as 1-alkyl-2-acety-sn-glycero-3-phosphocholine (Benveniste et al., 1979; Blank et al., 1979; Demopolous et al., 1979) which also defined APRL. The enzymatic steps for the complete biosynthesis and catabolism of PAF were soon completed as well as the proofs for the precise structure of the native PAF isolated from IgE stimulated rabbit basophils (Hanahan, 1980).

Since the discovery of PAF and the availability of pure synthetic material, the research on the biology and pathobiology of PAF has been rapidly expanding. PAF synthesis has been shown not only in basophils, but also in many other blood cells (platelets, monocytes, neutrophils, eosinophils) and endothelial cells (for review see Braquet et al., 1986, 1987, 1989). In addition, PAF was found to be extremely active in numerous systems, organs and cells, and was already associated with several pathological processes such as inflammation, immune responses and shock (for review see Braquet et al., 1986, 1987, 1989).

The following review is aimed at highlighting the pharmacological actions of PAF on the heart, with special attention to the coronary circulation and myocardial function. In addition, this review will emphasize caveats in the current status of the research in this field as well as delineate future research directions needed for better understanding of the mechanisms involved in PAF actions on the heart.

CORONARY CIRCULATION

The coronary effects of PAF have been demonstrated in both _in vitro_ and _in vivo_ models. _In vitro_ studies using guinea pig, rat or rabbit hearts perfused with artificial physiological solutions demonstrated that PAF leads to an increase in coronary vascular resistance (Benveniste et al., 1983; Kenzora et al., 1984; Levi et al., 1984; Saeki et al., 1985; Stahl et al., 1987; Piper and Stewart, 1986, 1987). However, in contrast to the coronary effects of PAF _in vitro_, direct bolus injections of PAF into the coronary circulation of _in situ_, blood perfused heart of the domestic pig produced an initial increase in flow, due to coronary dilation, followed by a decrease in flow due to coronary constriction (Feuerstein et al., 1984). These same observations were later confirmed in the coronary circulation of the intact dog heart (Mehta et al., 1986; Jackson et al., 1986). The vasodilator effect of PAF preceded the vasoconstriction and was invoked by low doses of PAF which did not change systemic or other cardiac variables. The vasodilator phase produced by PAF in coronary vessels of the domestic pig, unlike the constrictor effect, was not blocked by cyclooxygenase inhibitors (e.g., indomethacin) or the cysteinyl-leukotriene antagonist FPL 55712 (Feuerstein et al., 1984). Furthermore, continuous infusion of PAF over several minutes into the coronary circulation of the domestic pig demonstrated that low doses of PAF produce substantial reduction in myocardial contractility as evidenced by over 50% reduction in regional shortening fraction (Figure 1). This reduction in myocardial performance was seen at times where no significant changes could be observed in coronary blood flow, systemic arterial pressure or left ventricular and diastolic pressure (Feuerstein et al., 1985). Furthermore, PAF effect was associated with selective increases in circulating levels of TXB_2 (the stable metabolite of TXA_2), but not in 6-Keto-$PGF_{1\alpha}$ (the stable metabolite of prostacyclin) or leukotriene C_4 immunoreactivity (Feuerstein et al., 1985). Of special interest are observations made by Jackson et al. (1986) indicating that PAF induced coronary dilation is mediated by a platelet derived vasodilator factor. However, larger doses of PAF ultimately result in profound reduction of coronary blood flow, a response largely mediated by TXA_2 (Feuerstein et al., 1984; Ezra et al., 1987; Laurindo et al., 1989), from a clinical point of view, a sudden increase in PAF release (anticipated in systemic anaphylaxis or septicemia) might lead to severe coronary vasoconstriction, ischemia and low cardiac output which are the hallmarks of cardiac anaphylaxis.

MYOCARDIAL FUNCTION

Like many substances which affect systemic hemodynamic variables as well as multiple cardiac functions, the question of whether or not a direct effect on myocardial contractility accounts for the overall reduction in cardiac output is also pertinent to PAF. In fact, profound reduction in cardiac output during intracoronary infusion of PAF, together with reduction of regional "shortening fraction", has been demonstrated in the anesthetized domestic pig at PAF doses which had no substantial effect on systemic blood pressure (Ezra et al., 1987; Goldstein et al., 1986). In the rat, however, PAF reduces cardiac output only at doses which produce severe hypotension (Siren and Feuerstein, 1988). Such data, although suggestive of direct myocardial effects of PAF in some species, are inconclusive since it is impossible to completely isolate the hemodynamic status within different regions of the myocardium from the direct effects of PAF on the myocardium itself. Furthermore, reduction in the force of contraction of in vitro perfused heart preparation (Stahl et al., 1987; Benveniste et al., 1985; Piper and Steward, 1986; Saeki et al., 1985; Levi et al., 1984), even if perfusion pressure is maintained, is not free of similar criticism regarding intramural flow maldistributions. However, evidence obtained from isolated strips of cardiac muscle suggest that PAF can exert a primary negative inotropic effect. For example, Robertson et al. (1988) have shown a direct negative inotropic effect of PAF in the human myocardium. Furthermore, PAF was shown to decrease the force of contraction of electrically paced, non-coronary perfused left atrium and ventricular papillary muscle (Levi et al., 1984; Camussi et al., 1984). In addition, electrophysiological studies have demonstrated that relatively high doses of PAF have direct effects on the canine purkinje fibers (e.g., abnormal automaticity, decreased action potential amplitude) and the guinea pig papillary muscle (e.g., decrease in action potential amplitude and duration); yet, in this latter preparation, a positive inotropic response was observed (Nakaya and Tohse, 1986). A transient inotropic effect in guinea pig papillary muscle was also observed by Camussi et al (1984). Since in such preparations it is possible to exclude neural, humoral, blood cells or hemodynamic factors, a direct myocardial effect is suggested.

Additional evidence in support of the role PAF might have in promoting cardiac dysfunction was obtained by Mickelson et al. (1988) who used a rabbit myocardial ischemia and reperfusion model; PAF infusion at the post-ischemic period exacerbated the hemodynamic status and decreased cardiac performance at doses which had no consistent effect on a normal heart. This data is supported by the studies obtained by Lepran and Lefer (1985) who showed an aggravating effect of PAF infused in a cat myocardial ischemia model, in vivo. Moreover, the production and release of PAF from in vitro perfused hearts has also been demonstrated in response to antigen challenge of sensitized (Camussi et al., 1987) ischemic rabbit heart exposed to ischemia and reperfusion injury, in vitro, (Montrucchio et al., 1989), sensitized and challenged guinea pig (Levi et al., 1984) hearts and the ischemic baboon heart (Annable et al., 1988). More recently, a role for PAF in mediation of the "stunned myocardium" syndrome after ischemia and reperfusion was also suggested by indirect studies utilizing PAF antagonists (Schroeder et al., 1988). Preliminary data was obtained in collaboration with Drs. Goldstein and Ezra (USUHS, Bethesda, Maryland).

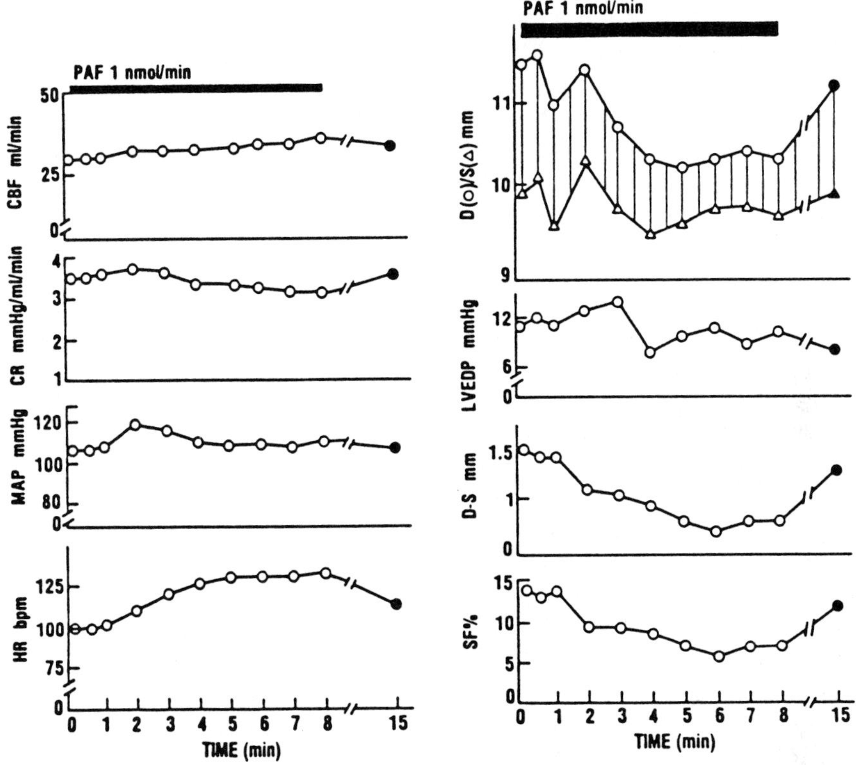

Figure 1. The effect of intracoronary infusion of PAF on systemic and cardiac hemodynamic function.

The horizontal bar denotes the infusion period of 1 nmoles/min of pure synthetic 1-O-hexadecyl-2-acetyl-sn-phosphorylcholine. Open circles denote variables monitored during PAF infusion; closed circles represent responses after the cessation of PAF infusion.

D = diastolic length; S - systolic length
SF = shortening fraction [(D-Si/D]x100

Note the lack of changes in coronary perfusion at time of significant depression (50%) of regional contraction (D-S) and SF.

ELECTROPHYSIOLOGICAL PROPERTIES OF PAF

The data summarized so far strongly suggest that a negative inotropic effect of PAF is one of the most prominent cardiovascular consequences. *In vitro* organ perfusion, isolated muscle preparations and even *in vivo* studies indicate that the negative inotropic effect of PAF is independent of systemic hemodynamic event or coronary blood flow; in many *in vitro* preparations, the effect of PAF was also argued to be independent of secondary mediators such as the eicosanoids (Levi et al., 1984; Robertson et al., 1987). However, the mechanisms of the decrease in cardiac contractility produced by PAF remain, at the present time, unclear.

Tamagro et al. (1985) have demonstrated in the guinea pig papillary muscle preparation that PAF hyperpolarized the resting membrane potential, increased the emplitude and Vmax of the upstroke, shortened the action potential duration and increased the spontaneous activity. However, the most prominent effect of PAF seen at concentrations of 10^{-10}-10^{-11}M was shortening of the action potential duration and increased frequency of spontaneous discharge from the ventricular muscle fibers; most notable was the lack of significant changes in the amplitude and Vmax of the upstroke nor did the resting potential of the membrane change. These data suggest that PAF did not exert any significant effect on the fast inward Na^+ current in the ventricular fibers. Only at higher concentrations ($>10^{-9}$M) did hyperpolarization of the resting membrane potential occur which could explain the increase in upstroke amplitude. Furthermore, these authors demonstrated that PAF restored the electrical responses of the partially depolarized papillary muscle (27 mM K^+), a preparation where generation of slow action potentials is considered to reflect activation of slow Ca^{+2} channels which allows influx of Ca ions (Inui and Inamura, 1986). Further support of PAF activation of slow Ca^{+2} influx was obtained by demonstration that verapamil, an inhibitor of Ca influx, suppressed the electrical responses restored by PAF in the partially depolarized guinea pig papillary muscle. Based on this information, the above authors concluded that PAF induced slow inward Ca currents, a suggestion also compatible with the increase in contractile force at low ($<10^{-9}$M) concentrations of PAF in this preparation. Furthermore, these authors have also shown that PAF inhibited ^{45}Ca efflux from guinea pig papillary muscle. Thus, by both increasing Ca influx through the slow inward Ca currents and reducing Ca^{+2} efflux, PAF increases the intracellular Ca concentration (Ca_i); the increase in Ca_i may explain the shotrening of the action potential duration (APD) by reducing the driving force for Ca ions; these effects could explain the reduction in peak contractile force by 10^{-10}M of PAF; alternatively, high Ca_i may shorten APD by rapid inactivation of slow inward current.

However, Camussi et al. (1984) have presented preliminary data which suggest that PAF may decrease the slow Ca^{+2} current while Robertson et al. (1988) failed to show that PAF affects slow inward Ca^{+2} current because PAF neither affected nor prevented histamine induced restoration of contractile responses in K^+-depolarized guinea pig right ventricular papillary muscle paced at constant rate. However, Robertson et al. (1988) extended their studies to examine the role of the Na^+ current since even small changes in a_{iNa} can produce significant changes in contractile force (Lee, 1985). The latter investigators have shown in the paced cavian ventricular myocardium, utilizing Na^+-selective microelectrodes, that PAF consistently caused a small decrease in a_{iNa}; the decrease in a_{iNa} occurred concomitantly with the peak of negative inotropic action of PAF superfused on the preparation. The negative inotropic effect of PAF was also associated with a concentration dependent decrease in transmembrane action potential duration, but without changes in maximum diastolic potential or action potential amplitude. Because the abbreviation of the action potential would reduce the time available for Ca^{+2} entry into the cell, this would be a possible mechanism whereby the force of contraction is reduced. An alternative explanation for the reduction in action potential duration is that PAF induces an increase in K^+ permeability which thereby decreases a_{iNa}. However, more specific studies using voltage clamp techniques and specific channel (Na^+, K^+, Ca^{+2}) agonist/antagonist application would be needed to clarify these issues.

SUMMARY AND FUTURE DIRECTIONS

Currently all investigations seem to agree that PAF is a potent vasoactive substance in the coronary circulation. The most consistent effect of PAF in *in vitro* systems is coronary constriction; however, *in vivo*, PAF effects on the coronary circulation are more complex. First, low doses of PAF always produce vasodilation by a cyclooxygenase independent mechanism; this vasodilation is blocked by PAF antagonists. Larger doses of PAF also produce coronary vasoconstriction; this phenomenon is sensitive to cyclooxygenase inhibitors and seems to be TXA_2 mediated. The myocardial effects of PAF *in vitro* and *in vivo* are all consistent with a primary negative inotropic effect, although *in vivo* no complete proof to direct myocardial action has been provided. It is important to note that the overall effect of PAF on cardiac output *in vivo* are not only the result of myocardial and coronary derangements, but also due to pulmonary effects and primarily pulmonary hypertension which leads to Cor pulmonale and reduction in pre-load.

The main issues which need to be addressed in the future include: a) source of PAF: is PAF derived from endogenous cardiac sources or exogenous (blood-borne), both are important in producing cardiac pathology; b) which cellular element in the heart produces PAF; c) what are the molecular mechanisms associated with PAF effects on the myocardium; d) what are the temporal and spatial relationships of the functional consequences of ischemic heart disease to PAF.

These issues call for a more sophisticated approach to answer these questions; of particular importance will be electrophysiological studies aimed at elucidating the ion channel(s) involved in PAF effects on myocardial cells. Furthermore, better assay methods should be applied to monitor PAF generation in the heart and elucidate the cellular origin (myocardial cells, endothelium, platelets, white blood cells) of PAF synthetized in the heart. Finally, utilization of specific and potent PAF antagonist to prevent or reverse pathological cardiac functions *in vivo* should be much encouraged.

REFERENCES

Annable CR, McManus LM, Carey KD, Pinckard RN (1988). Isolation of platelet-activating (PAF) from ischemic baboon myocardium. Fed Proc 1271: (abst).

Benveniste J, Henson PM, Cochrane CG (1972). Leukocyte dependent histamine release from rabbit platelets: The role of IgE, basophils and platelet activating factor. J Exp Med 136:1356-1377.

Benveniste J, Chigard M (1985). A role for PAF-acether (platelet-activating factor) in platelet-dependent vascular disease. Circulation 72:713-717.

Benveniste J, Boullet C, Brink C, Labat C (1983). The action of PAF-acether (platelet activating factor) on guinea pig isolated heart preparations. Br J Pharmacol 80:81-83.

Benveniste J, Tence M, Varenne P, Bidault J, Boullet C, Polonsky J (1979). Semi synthese et structure purposee du facteur activant les plaquettes (P.A.F.). PAF-acether un alkyl ether analogue de la lysophosphatidylcholine. C.R. Acad Sci [D], (Paris) 289:1037-1040.

Blank ML, Snyder F, Byers LW, Brooks B, Muirhead EE (1979). Antihypertensive activity of an alkyl ether analog of phosphorylcholine. Biochem Biophys Res Comm 90:1194-1200.

Braquet P, Godfroid JJ (1986). PAF-acether specific binding sites: 2. Design of specific antagonists. TIPS 7:397-403.

Braquet P, Touqui L, Shen TY, Vargaftig BB (1987). Perspectives in platelet-activating factor research. Pharmacol Rev 39:97-145.

Braquet P, Paubert-Braquet M, Koltai M, Bourgain R, Bussolino F, Hosford D (1989). Is there a case for PAF antagonists in the treatment of ischemic states? TIPS 10:23-30.

Camussi GM, Nielsen N, Tetta C, Saunders RN, Nilgrom F (1987). Release of platelet-activating factor from rabbit perfused in vitro by sera with transplantation alloantibodies. Transplantation 44:113-118.

Camussi G, Alloatti G, Montrucchio G, Meda M, Emmanuelli G (1984). Effect of platelet-activating factor on guinea-pig papillary muscle. Experimentia 40:357-361.

Demopolous CA, Pinckard RN, Hanahan DJ (1979). Platelet-activating factor. Evidence for 1-O-alkyl-2-acetyl-sn-glyceryl-3-phosphorylcholine as the active component (a new class of lipid mediators). J Biol Chem 254:9355-9358.

Ezra D, Laurindo FRM, Czaja JF, Snyder F, Goldstein RE, Feuerstein G (1987). Cardiac and coronary consequences of intracoronary platelet activating factor infusion in the domestic pig. Prostaglandins 34:41-57.

Feuerstein G, Ezra D, Ramwell PW, Letts G, Goldstein R (1985b). Platelet-activating factor as a modulator of cardiac and coronary functions. In Bailey JM (ed): "Prostaglandins, Leukotrienes and Lipoxins, "New York: Plenum Press, pp. 301-310.

Feuerstein G, Boyd LH, Ezra D, Goldstein RE (1984). Effect of platelet activating factor on coronary circulation of the domestic pig. Am J Physiol 246:H466-H471.

Goldstein R, Ezra D, Laurindo FRM, Feuerstein G (1986). Coronary and pulmonary vascular effects of leukotrienes and PAF-acether. Pharmacol Res Commun 18:151-162.

Hanahan DJ, Demopolous CA, Liehr J, Pinckard RN (1980). Identification of platelet activating factor isolated from rabbit basophils as acetyl glyceryl ether phosphorylcholine. J Biol Chem 255:5514-5516.

Inui and Imura (1986). Naunyn Schmiedeberg's Arch Pharmacol 294:261.

Jackson CV, Schumacher WA, Kunkel SL, Drisoll EM, Pope T-K, Lucchesi BR (1986). Platelet activating factor release of a platelet-derived coronary artery vasodilator substance in the canine. Circ Res 58:218-229.

Kenzora JL, Perez JE, Bermann SR, Large L (1984). Effects of acetyl glyceryl ether phosphorylcholine (platelet activating factor) on ventricular perload, afterload and contractility in dogs. J Clin Invest 74:1183-1203.

Laurindo FRM, Goldstein RE, Davenport NJ, Ezra D, Feuerstein GZ (1989). Mechanisms of hypotension produced by platelet-activating factor. J Appl Physiol (in press)

Lee CO (1985). 200 years of digitalis: The emerging central role of the sodium ion in the control of cardiac force. Am J Physiol 249:C367-C378.

Lepran I, Lefer AM (1985). Ischemia aggravating effects of platelet-activating factor in acute myocardial ischemia. Basic Res Cardiol 80:135-141.

Levi R, Burke JA, Buo ZG, Hattori Y, Hoppens CM, McManus LM, Hanahan DJ, Pinckard RN (1984). Acetyl glyceryl ether phosphorylcholine (AGEPC). A putative mediator of cardiac anaphylaxis in the guinea pig. Circ Res 54:117-124.

Mehta J, Wargovich T, Nichols WW (1986). Biphasic effects of platelet-activating factor on coronary blood flow in anesthetized dog. Prostag Leukot Med 21:87-95.

Michelson JK, Simpson PJ, Lucchesi BR (1988). Myocardial dysfunction and coronary vasoconstriction induced by platelet-activating factor in post-infarcted isolated heart. J Mol Cell Cardiol 20:547-561.

Montrucchio G, Alloatti G, Tetta C, DeLuca R, Saunders RN, Emanuelli G, Camussi G (1989). Release of platelet activating factor from ischemic reperfused rabbit heart. Am. J. Physiol 256:H1236-H1246.

Muirhead, E (1980). Antihypertensive function of the kidney. Hypertension 2:444-464.

Nakaya H, Tohse N (1986). Electrophysiological effects of acetylglyceryl ether phosphorylcholine on cardiac tissue. Comparison with lysophosphatidylcholine and long chain acyl carnitine. Br J Pharmac 89:749-757.

Piper P, Stewart AG (1987). Antagonism of vasoconstriction induced by platelet activating factor in guinea pig perfused hearts by a selective platelet activating factor receptor antagonist. Br J Pharmac 90:771-783.

Piper P, Stewart AG (1986). Coronary vasoconstriction in the rat isolated perfused heart induced by platelet-activating factor is mediated by leukotriene C_4. Br J Pharmac 881:595-605.

Robertson DA, Wang D-Y, Lee COK, Levi R (1988). Negative inotropic effect of platelet-activating factor: Association with a decrease in intracellular sodium activity. J Pharmacol Exptl 245:124-128.

Robertson DA, Genovese A, Levi R (1987). Negative inotropic effect of platelet-activating factor on human myocardium: A pharmacological study. J Pharmacol Exptl Ther 243:834-839.

Saeki S, Masugi F, Ogihara R, Otsuka A, Kumahara Y, Watanabe K, Tamura K, Aashi H, Kumag A (1985). Effects of 1-0-alkyl-acetyl-glycero-3-phosphorylcholine (platelet activating factor) on cardiac function in perfused guinea pig heart. Life Sci 37:325-330.

Schroeder E, Van Mechelan H, Maldague P, Vuylsteke A, Keyeux A, Rousseau MF, Pouleur H (1988). Increased coronary vascular resistance in the stunned myocardium: Role of the platelet activating factor. Proc 61st Scientific Session of the American Heart Association, II-77 (abst).

Siren A-L, Feuerstein G (1989). Effects of platelet activating factor and its antagonist, BN 52021, on cardiac function and regional blood flow in the conscious rat. Am J Physiol (in press).

Snyder F (1987). Platelet-activating factor and related lipid mediators. Plenum Press, New York.

Snyder F, Malone B, Wykle RL (1969). The biosynthesis of alkyl ether bonds in lipids by a cell free system. Biochem Biophys Res Comm 34:40-47.

Stahl GL, Lefer DJ, Lefer AM (1987). Paf-acether induced cardiac dysfunction in the isolated perfused guinea pig heart. Arch Pharmacol 336:459.

Tamargo J, Tejerina T, Delgado C, Barrigon S (1985). Electrophysiological effects of platelet-activating factor (PAF-acether) in guinea pig papillary muscles. Eur J Pharmacol 109:219-227.

Calcium Channel Modulators in Heart and Smooth Muscle:
Basic Mechanisms and Pharmacological Aspects
S. Abraham and G. Amitai, editors
© 1990, VCH, Weinheim/Deerfield Beach, FL and Balaban, Rehovot/Philadelphia

Interactions between voltage-dependent calcium channels and peripheral benzodiazepine receptor ligands in the heart

GORDON T. BOLGER[1], SHLOMO ABRAHAM[2], NISSIM OZ[2], AND BEN AVI WEISSMAN[2]

[1]*Department of Pharmacology, BioMega Inc. Laval, Quebec, Canada H7S 2G5, and* [2]*Department of Pharmacology, Israel Institute for Biological Research, Ness Ziona 70450, Israel*

INTRODUCTION

High affinity (0.5-2.0 nM), stereospecific and saturable receptors for peripheral benzodiazepine ligands (i.e. Ro 5-4864, PK 11195; Fig. 1) have been identified in a wide variety of mammalian tissues, including cardiac, skeletal, neuronal and intestinal and vascular smooth muscles (Schoemaker et al., 1983; Bolger et al., 1984). These sites, termed peripheral benzodiazepine receptors (PBR) possess properties which are unique from the benzodiazepine receptors characterized in the central nervous system (Squires and Braestrup, 1977; Bolger et al., 1984; Schoemaker et al., 1983; Anholt et al., 1985, 1986). Thermodynamic and pharmacologic studies have concluded that Ro 5-4864 is an agonist and PK 11195 is an antagonist of PBR (Mestre et al., 1984; LeFur et al., 1983). However, some studies have demonstrated that both Ro 5-4864 and PK 11195 can produce similar responses (e.g. Grupp et al., 1987), thus making the differentiation of agonist and antagonist PBR ligands somewhat unclear.

High affinity binding sites for dihydropyridine (DHP) calcium antagonists are present in many but not all tissues containing PBR (Rampe and Triggle, 1986); their largest densities being in skeletal and cardiac muscles and intestinal and vascular smooth muscles (Triggle and Janis, 1984). Biochemical and electrophysiological evidence suggests that DHP calcium antagonists (i.e. nifedipine) block calcium currents while other DHP strucutres act as calcium agonists (i.e. BAY K 8644) (Schramm et al., 1983). DHP binding sites represent a regulatory component of the 'L(long opening)-type' voltage -dependent calcium channel (VDCC) (Triggle and Janis, 1984). Thus, DHP calcium antagonists have been widely employed as markers for 'L-type' VDCC in a variety of tissues.

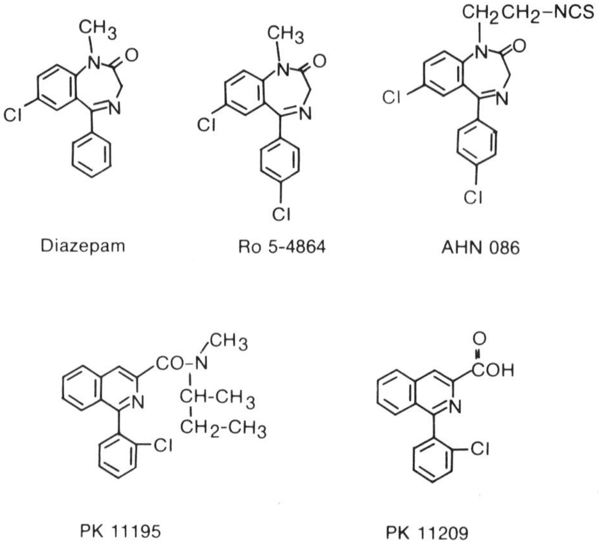

Diazepam Ro 5-4864 AHN 086

PK 11195 PK 11209

Figure 1 Structures of PBR ligands

Mestre et al. (1984) have provided evidence in cardiovascular tissues to suggest that PBR are associated with L-type VDCC and not receptor-operated calcium channels. While such a study implicates an association between PBR and 'L-type' VDCC, other studies have shown that PBR and DHP binding sites are distinct entities suggesting that PBR are not likely associated with 'L-type' VDCC (Bolger et al., 1984; Doble et al., 1985). Furthermore, relatively high (0.1 -10 uM) concentrations of PBR ligands to the <u>in-vitro</u> affinity of PBR are required to alter the function of VDCC (LeFur et al., 1985). Thus, whether a relationship exists between PBR and VDCC remains equivocal.

Recently, the synthesis and biochemical characterization of a specific irreversible inhibitor of radioligand binding to PBR (AHN 086) has been described (Newman et al., 1987, Lueddens et al., 1986; Bolger et al., 1989). In light of the equivocal relationship between PBR and VDCC, we have used AHN 086 as a probe to investigate the effects of PBR occupancy on VDCC function in the absence of high concentrations of PBR ligands. Furthermore, we investigated the pharmacologic effects of Ro 5-4864 and PK 11195 on contractions induced by BAY K 8644 in the spontaneously beating guinea-pig atria, a tissue demonstrating a specific association between PBR and VDCC (Bolger et al., 1989). Finally, the potential interactions between PBR ligands and VDCC were examined *in-vivo* in the rat by measuring changes in the electrocardiogram (ECG) following intracarotid (i.c.) administration of BAY K 8644.

METHODS AND MATERIALS

Isolated Guinea-pig Atria

Male Hartley guinea-pigs (Buckberg, N.Y., 250-300 g) were killed by decapitation. For the preparation of the atria, the heart was rapidly removed and placed in a Tyrode's based salt solution (Bolger et al., 1989). The right and left atria were dissected free from surrounding tissue and mounted with 0.5 g tension to a high compliance strain gauge transducer (Grass Model FT03C, Grass Instruments) in a 20 ml organ bath containing Tyrode's at 37°C. Contractile activity was recorded on a polygraph (Grass Model 7D, Grass Instruments).

In-Vivo Procedure

The procedure for the assessment of the effects of drugs on the ECG (S wave amplitute and ST-segment duration) was as previously reported (Abraham et al., 1987). Briefly, rats were anesthetized with sodium pentobarbitone (25 $mgKg^{-1}$, i.p.) and a catheter was placed in the right femoral artery for monitoring arterial blood pressure (Stoelting pressure transducer). The right carotid artery was cannulated and the cannula pushed to the area of the coronary ostium; this cannula was used for the administration of drugs. ECG (Lead I) was recorded on a HP 7758B electrocardiograph, Hewlett-Packard Instruments); heart rate was derived from the ECG tracings.

Drugs

BAY K 8644 was a gift from Dr. A. Scriabine, Miles Institute for Preclinical Pharmacology, West Haven CT, U.S.A. The following drugs were obtained from the companies named: Ro 5-4864 and clonazepam, Hoffman-LaRoche, Nutely, NJ, U.S.A.; PK 11195 and PK 11209, Pharmuka Laboratories, Genevilliers, France; Nifedipine, Pfizer, Groton CT, U.S.A. AHN 086 was synthesized as previously described (Newman et al., 1987). All other chemicals were obtained from standard commercial sources. BAY K 8644 was dissolved in ethanol and diluted in either saline (0.9% w/v) or distilled water for in-vivo and organ bath experiments, respectively. Ro 5-4864 and PK 11195 were dissolved in a mixture of Emulphor EL-620 (GAF Corporation, New York, U.S.A.) and ethanol (1:1) and diluted in saline for in-vivo experiments. In organ bath experiments all drugs were initially dissolved in ethanol and subsequently diluted in distilled water. The vehicles (diluted ethanol and the diluted mixture of ethanol and Emulphor) had no effect on the activity of the atria, blood pressure heart rate or ECG pattern.

RESULTS AND DISCUSSION

The effects of AHN 086, Ro 5-4864, PK 11195 and clonazepam were evaluated on the rate and tension development of contractions in spontaneously beating guinea-pig atria (Tables 1 and 2). None of the drugs affected the resting tension or the rate of contractions in the dose range of 0.1-10 uM. However, Ro 5-4864 and clonazepam did produce a small negative inotropic effect (5-10%) which was completely reversed by extensive washing of the tissue. Low concentrations (0.5 uM) of AHN 086 potentiated and high concentrations (5-50 uM) inhibited the positive inotropic responses to BAY K 8644 (Table 1 and Fig. 2a). AHN 086(5 uM) inhibited the maximum negative inotropic response to nifedipine (Fig. 2b); these effects of AHN 086 being observed despite extensive washing of the atria. Ro 5-4864 and PK 11195 (10 uM) completely inhibited the positive chronotropic responses to BAY K 8644, while clonazepam and PK 11209 (10 uM), ligands which have a very low affinity for PBR (Rampe and Triggle, 1986; H. Lueddens, personal communication) produced a small (16%) and no inhibition respectively (Table 2); effects which were completely reversed on extensive washing of the tissue. Ro 5-4864 potentiated and PK 11195 inhibited the positive inotropic responses to BAY K 8644 (Table 2). Ro 5-4864 and PK 11195 did not alter the negative inotropic and chronotropic effects of nifedipine.

While complex, these results suggest that PBR ligands can modulate VDCC. Quantitative differences between the effects of Ro 5-4864 and PK 11195 may be due to their different physical and chemical properties and interaction with different domains of PBR (Lueddens et al., 1986; LeFur et al., 1983). The inability of PBR ligands to significantly alter VDCC in contrast to their effects on activation of VDCC by BAY K 8644 strongly suggest that they are modulating the DHP binding site to evoke changes in the function of VDCC. The inability of PBR ligands to affect the activity of nifedipine suggest that the 'activated' state of VDCC may be more susceptable to modulation by PBR ligands.

AHN 086 was found to be a specific acylator of PBR even at high (micromolar) concentrations (Bolger et al., 1989). While AHN 086 produced effects similar to Ro 5-4864 (e.g. potentiation of responses to BAY K 8644) it also produced different effects to those of Ro 5-4864 in the atrium (e.g. high concentrations inhibited the positive ionotropic response to BAY K 8644 and the negative chronotropic response to nifedipine and none of the concentrations affected the chronotropic response to BAY K 8644). Quanatitave differences between the effects of Ro 5-4864 and AHN 086 may be due to the irreversible occupancy of PBR by AHN 086.

TABLE 1. The effect of AHN 086 on the inotropic response to BAY K 8644 in guinea-pig atria

Concentration of AHN 086 (uM)	%-BAY K 8644 Response
0 (control)	100
0.5	188 ± 22*
2.0	107 ± 44
5.0	57 ± 4*
50.0	50 ± 3*

Guinea-pig atria were incubated for 10 min. in the presence of varying concentrations of AHN 086. All tissues (n=3-4) were washed for 1 hr. in Tyrode's and subsequently challenged with 1 uM BAY K 8644. *Significantly different from control, $p < 0.05$, Wilcoxon's assigned rank-test.

This clearly appears to be the case for high concentrations of AHN 086 which significantly blocked PBR and evoked a 'down regulation' of DHP binding sites (Bolger et al., 1989) thereby providing a mechanistic interpretation for the inhibition of DHP calcium agonist and antagonist responses. Differences in the ability of AHN 086 and Ro 5-4864 to alter the chronotropic response to BAY K 8644 may be due to an interaction of these PBR ligands with 'T(transient opening)-type' VDCC which predominantly control chronotropy (Kass et al., 1984).

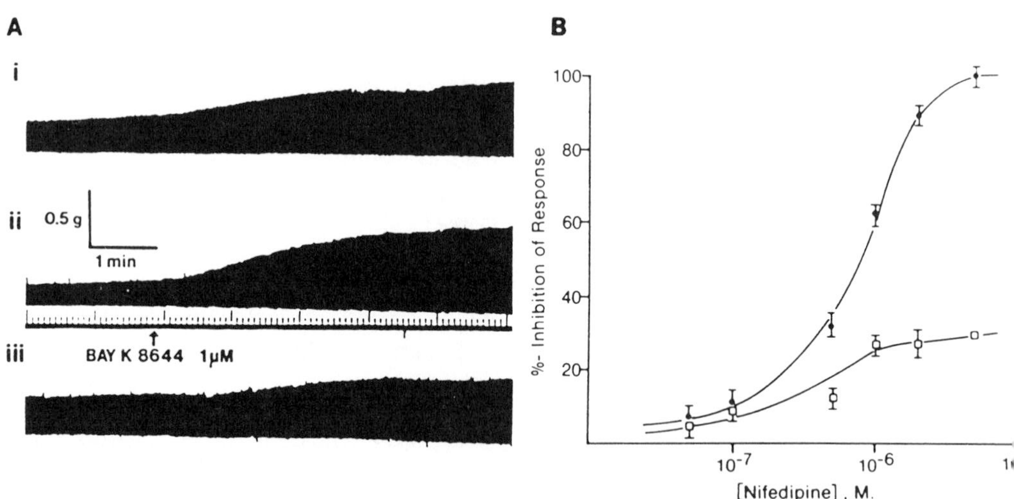

Figure 2 A) The effects of BAY K 8644 in (i) drug free guinea-pig atria and those treated with (ii) 0.5 uM AHN 086 and (iii) 5.0 uM AHN 086 for 10 min. followed by extensive washing. B) The effects of nifedipine on the contractility of guinea-pig atria in naive (●) and those tissues treated with 5 uM AHN 086 (□) (n=4-6).

Of paramount importance is the question, are PBR mediating the interactions of PBR ligands with DHP binding sites at VDCC? The rather small effect of clonazepam and the inability of PK 11209 to affect the response of the atria to BAY K 8644 suggest that PBR are indeed mediating the effects of PBR ligands. This would apply to both the ionotropic and chronotropic effects of PBR ligands, although for the chronotropic effects it is possible that 'atypical' PBR (Basile et al., 1986) which are non-alkylatable (in light of the inactivity of AHN 086) are associated with this response.

It has been proposed that PBR are associated with DHP binding sites in aortic smooth muscle, where PK 11195 blocked BAY K 8644 mediated contractions (Mestre et al., 1984). Even though the highest densities of PBR exist in mitrochondria (Anholt et al., 1986), it is plausible that a subpopulation of PBR are linked with 'L-type' VDCC. Further support for

TABLE 2. Effects of PBR ligands on the chonotropic and inotropic responses of guinea-pig atria to BAY K 8644.

Drug (uM)	Chronotropy (beats/min)	Inotropy[a] (%-Of BAY K 8644 response)
Control	174 ± 6	
BAY K 8644 (1)	228 ± 12	100
Ro 5-4864 (10)	150 ± 12	
PK 11195 (10)	150 ± 12	
Clonazepam (10)	168 ± 12	
PK 11209 (10)	150 ± 31	
BAY K 8644 plus:		
Ro 5-4864 (10)	162 ± 6	163 ± 34**
PK 11195 (10)	114 ± 18	41 ± 20**
Clonazepam (10)	192 ± 6	125 ± 12
PK 11209 (10)	223 ± 16	125 ± 10

[a]The results were obtained over a 3 s measurement period. Nifedipine and other drugs were introduced to the tissue bath 5 and 10 min before the test period respectively. In each experiment the inotropic response to BAY K 8644 for drug treated tissues was normalized to that in a drug free tissue. *Significantly different, $P < 0.05$ from control, unpaired Student's t-test. **Significantly different, $p < 0.05$, from control, Wilcoxon's Assigned Ranks test.

such a concept arises from the 'down-regulation' of DHP binding sites in guinea-pig atria following irreversible occupancy of PBR with AHN 086 (Bolger et al., 1989). In addition, Doble et al., (1985) have suggested that PBR and DHP binding sites are both present within the same compartment on canine cardiac sarcolemma. If PBR are indeed associated with DHP binding sites, then only a small subpopulation would be involved since there is only one fifteenth the number of DHP binding sites as there are PBR on atrial membranes (Bolger et al, 1989). It is likely for this reason that even though low concentrations of AHN 086 (e.g. 0.5 uM) potentiated the inotropic responses to BAY K 8644, no decernable decrease in the number of PBR was noted (Bolger et al., 1989).

The concentrations of PBR ligands to elicit an effect in the atrium were in the 0.5-50 uM range. Nifedipine inhibited contractile activity with an IC_{50} of 8.0 x 10^{-7}M, 10,000 fold greater than the affinity of atrial DHP binding sites (Bolger et al., 1989). Large differences between the pharmacologic potency of DHPs and the affinity of DHP calcium antagonist binding sites have been attributed to the large differences between the receptor environment in isolated tissues and broken cell preparations (Triggle and Janis, 1984). If similar differences exist for PBR, then PBR ligands would possess a pharmacologic activity at concentrations 10,000 fold the affinity of PBR in the atrium (1-3 nM, Bolger et al., 1989), which is 10-30 uM. Alternate explanations for the high concentrations of PBR ligands required for activity at VDCC have been proposed, such as the coupling of 'low affinity' (micromolar) PBR to VDCC (Basile et al., 1986).

Given the effects of PBR ligands in isolated tissues, it was of interest to study their effects on the interaction of BAY K 8644 with VDCC <u>in-vivo</u>. A recently developed model of BAY K 8644-induced ischemia has been described (Abraham et al., 1987). In this model the effects of BAY K 8644 and its interactions with PBR ligands were assessed on the ECG in the rat. BAY K 8644 at 5 and 10 $ugkg^{-1}$ (intracarotidly, i.c.), produced a transient pressor response (30-40 mm Hg) accompanied by ECG ST-segment alterations of longer duration (60-120 s vs. 5-20 s). BAY K 8644 caused an elevation of the S wave at 5 $ugkg^{-1}$ and either depression or elevation of the ST-segment following the administration of 10 $ugkg^{-1}$ (Abraham et al., 1987).

At doses up to 1 $mgkg^{-1}$ (i.c.) both Ro 5-4864 and PK 11195 had no effect on either blood pressure or the ECG pattern of anesthetized rats. Ro 5-4864, injected 1 min prior to BAY K 8644 (5 and 10 $ugkg^{-1}$) produced a dose dependent elevation of the ST-segment (ED_{50} values of 221 and 243 $ugkg^{-1}$; Fig. 3a). In contrast PK 11195 inhibited the BAY K 8644-induced changes in the ST segment profile (IC_{50} values of 52 and 58 $ugkg^{-1}$; Fig 3b). The effect was noted at doses as low as 25 $ugkg^{-1}$ and complete blockade was observed at about 100 and 250 $ugkg^{-1}$ BAY K 8644 respectively. PK 11195 also abolished the combined effect of Ro 5-4864 and BAY K 8644.

In general, these findings suggest that PBR ligands can modulate 'L-type' VDCC in-vivo. More specifically, the results presented herein indicate a similar interaction between PBR and VDCC in the spontaneously beating guinea-pig atria and in the BAY K 8644 induced changes in the ECG in rats. Ro 5-4864 both increased the positive inotropic response and ST-segment of the ECG following BAY K 8644. In contrast, PK 11195 blocked the responses to BAY K 8644. The correspondence between in-vitro and in-vivo data strengthens the premise the PBR are important physiologic regulators of VDCC in the cardiovasculature.

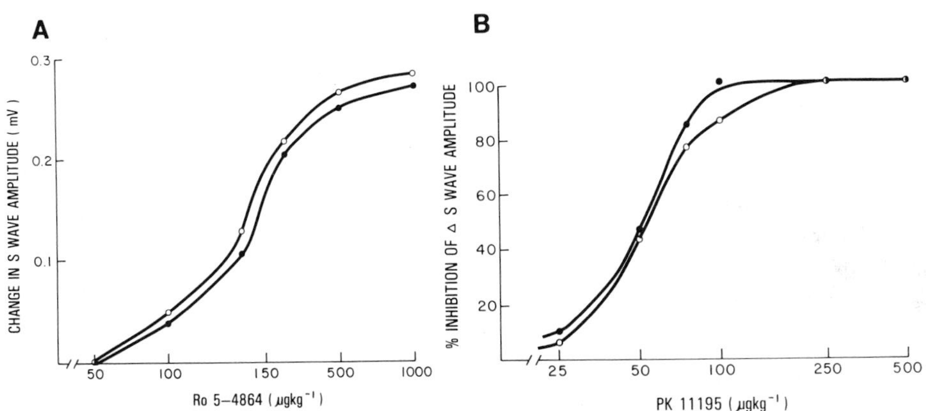

Figure 3. A) Ro 5-4864 elicited increases of ST-segment alteration induced by 5 (○) and 10 (●) $ugkg^{-1}$ BAY K 8644. The results are presented as the mean change of the BAY K 8644-induced S wave alteration in mV (n=3-4).
B) Inhibition of BAY K 8644-elcited change in S wave by PK 11195. PK 11195 was administered 1 min before the channel activator and alterations in ST-segment were monitored 30 s later. The results are presented as the mean percent inhibition of 5 (○) or 10 (●) $ugkg^{-1}$ BAY K 8644 induced changes (n=3-4).

Recently, it was demonstrated that PBR ligands modulated the activity of VDCC in the heart, but not in intestinal smooth muscle (Bolger et al., 1989). Thus, selectivity is an added consideration to the complex interaction between VDCC and PBR ligands; an interaction which clearly needs further experimentation to determine its mechanistic basis. However, the possibility that PBR ligands are cardioselective and may interact with VDCC via an interaction with DHP binding sites raises important possibilities for their future development as clinically useful cardiovascular drugs.

ACKNOWLEDGEMENTS

The authors wish to acknowledge financial support from the Medical Research Council of Canada and the Faculty of Medicine, Memorial University of Newfoundland, St. John's, Newfoundland, Canada.

REFERENCES

Abraham S, Amitai G, Oz N, Weissman BA (1987). (±)-BAY K 8644 induced changes in the ECG pattern of the rat and their blockade by antianginal drugs. Brit J Pharmacol 92:603-608.

Anholt RRH, DeSouza EB, Oster-Granite ML, Snyder SH (1985). Peripheral-type benzodiazepine receptors: Autoradiographic localization in whole-body sections of neonatal rats. J Pharmacol Exp Ther 233:517-526.

Anholt RRH, Pederson PL, DeSouza EB, Snyder SH (1986). The peripheral benzodiazepine receptor: Localization to the mitochondrial outer membrane. J Biol Chem 261: 576-583.

Basile AS, Klien DS, Sklonick P (1986). Characterization of benzodiazepine receptors in the bovine pineal gland: Evidence for the presence of an atypical binding site. Mol Brain Res 1:127-135.

Bolger GT, Weissman BA, Lueddens HWM, Basile AS, Mantione CR, Barrett JE, Witkin JM, Paul SM, Skolnick P (1985). Late evolutionary appearance of "peripheral-type" binding sites for benzodiazepines. Brain Res 338:366-370.

Bolger GT, Newman AH, Rice KC, Lueddens HWM, Basile AS, Skolnick P. Characterization of the effects of AHN 086, an irreversible ligand of peripheral benzodiazepine receptors on contraction in guinea-pig atria and ileal longitudinal smooth muscle. Can J Physiol Pharmacol 67:126-134.

Braestrup C, Squires RF (1977). Specific benzodiazepine receptors in rat brain characterized by high-affinity [^{3}H]diazepam binding. Proc Natl Acad Sci (USA) 74:3805-3809.

Dolbe A, Benevides J, Ferris O, Bertrand P, Meneager J, Vaucher N, Burgevin M-C, Uzan A, Gueremy C, LeFur G (1985). Dihydropyridine and peripheral type benzodiazepine binding sites: Subcellular distribution and molecular size determination. Eur J Pharmacol 119:153-167.

Grupp IL, French JF, Matlib MA (1987). Benzodiazepine Ro 5-4864 increases coronary flow. Eur J Pharmacol 143:143-147.

Kass RS, Sanguinetti MC, Bennett PB, Coplin BE Krafte DS (1984). Voltage-dependent modulation of cardiac Ca-channels by dihydropyridines. In Fleckenstein A, Van Breemen R, Hoffmeister F (eds): "Cardiovascular Effects of Dihydropyridine type calcium antagonists and agonists", Berlin Heildelberg: Springer-Verlag pp 198-215.

LeFur G, Vacher N, Perrier MI, Flamier A, Benevides J, Renault C, Dubroeucq MC, Gueremy C, Uzan A (1983). Differentiation between two ligands for peripheral benzodiazepine binding sites, [^{3}H]Ro 5-4864 and [^{3}H]PK 11195, by thermodynamic studies. Life Sci 33:449-457.

LeFur G, Mestre M, Carriot T, Belein C, Renault C, Dubroeucq M-C, Gueremy C, Uzan A (1985). Pharmacology of peripheral type benzodiazepine receptors in the heart. In Lal H, LaBella F, Lane J (eds): "Endocoids", New York: Alan R. Liss Inc. pp 175-186.

Lueddens HWM, Newman AH, Rice KC, Skolnick P (1986). AHN 086: An irreversible ligand of "peripheral" benzodiazepine receptors. Mol Pharmacol 29:540-545.

Mestre M, Carriot T, Belin C, Uzan A, Renault C. Dubroecq MC, Guermy C, LeFur G (1984). Electrophysiological and pharmacological characterization of peripheral benzodiazepine receptors in a guinea-pig heart preparation. Life Sci 35:953-962.

Newman AH, Lueddens WMH, Skolnick P, Rice KC (1987). Novel irreversible ligands specific for "peripheral" type benzodiazepine receptors: (+)-1-(2-chlorophenyl)-N-(1-methylpropyl)-N-(2-isothiocyanoethyl)-3-isoquinolinecarboxamide (AHN 070), AHN 070 R(+)- and S(-)-enantiomers and 1-(2-isothiocyanatoethyl)-7-chloro-1,3-dihydro-5-(4-chlorophenyl)-2H-1,4-benzodiazepine-2-one (AHN 086). J Med Chem 30:1901-1905.

Rampe D, Triggle DJ (1986). Benzodiazepines and calcium channel function. Trends Pharm Sci 11:461-464.

Schoemaker H, Boles RG, Horst D, Yamamura HI (1983). Specific high affinity binding sites for [^{3}H]Ro 5-4864 in

rat brain and kidney. J Pharm Exp Ther 225:61-69.
Schramm MG, Thomas G, Towart R, Franckowiak G (1983). Novel dihydropyridines with positive inotropic action through activation of Ca^{2+} channels. Nature 303:535-537.
Triggle DJ, Janis RA (1984). The 1,4-dihydropryidine receptor: a regulatory component of the Ca^{2+} channel. J Cardiovasc Pharmacol 6:S949-S955.

Calcium Channel Modulators in Heart and Smooth Muscle: Basic Mechanisms and Pharmacological Aspects
S. Abraham and G. Amitai, editors
© 1990, VCH, Weinheim/Deerfield Beach, FL and Balaban, Rehovot/Philadelphia

Characterization of partially purified peripheral benzodiazepine binding inhibitor from human cerebrospinal fluid (CSF): Its arrhythmogenic effects in the rat

YESHAYAHU KATZ[1], NAHUM ALLON[2], SHLOMO ABRAHAM[2], NISSIM OZ[2] AND MOSHE GAVISH[1]

[1]Rappaport Family Institute for Research in the Medical Sciences and Department of Pharmacology, Faculty of Medicine, Technion-Israel Institute of Technology, Haifa and [2]Department of Pharmacology, Israel Institute for Biological Research, P.O. Box 19, Ness Ziona, Israel

INTRODUCTION

Benzodiazepines (BZs), such as diazepam (Valium), are a class of drugs widely used clinically as anxiolytics, hypnotics, anticonvulsants, and muscle relaxants. These drugs bind specifically and with high affinity to central-type BZ receptors (CBR), which are located in the central nervous system (CNS) (Mohler and Okada, 1977; Squires and Braestrup, 1977). CBR are part of a macromolecular protein complex which contains sites for the main inhibitory brain neurotransmitter, τ-aminobutyric acid (GABA) (Tallman et al., 1978), the chloride ionophore (Mackerer and Kochman, 1978. Costa et al., 1979), barbiturates (Johonston and Willow, 1982), and probably steroids (Harrison and Simmonds, 1984).

In addition to CBR, several BZs bind to specific sites in peripheral tissues, such as kidney (Regan et al., 1981), heart (Taniguchi et al., 1982), adrenal gland (De Souza et al., 1985), and platelets (Wang et al., 1980), as well as to glial cells within the CNS (Schoemaker et al., 1981). These sites are referred to as "peripheral-type BZ receptors" (PBR). The BZ ligand Ro 5-4864 and the non-BZ ligand PK 11195 (an isoquinoline carboxamide derivative), bind to PBR with high affinity and are considered to be agonist and antagonist, respectively, of PBR (Le Fur et al., 1983a,b). Clonazepam, which binds with high affinity to CBR is relatively ineffective at PBR. Diazepam shows high affinity to CBR and moderate affinity to PBR (Squires and Braestrup, 1977. Taniguchi et al., 1982). PBR differ from CBR not only in their drug specificity, but also with respect to their distribution in the body, their cellular localization and their coupling to ion channels. PBR are located throughout most organs, whereas CBR are present only in the CNS. PBR are located on the outer mitochondrial membrane (Anholt et al., 1986), while CBR are part of the plasma membrane (Bosmann et al., 1978)]. PBR are coupled to calcium channels (Bender and Hertz, 1985) whereas CBR are coupled to chloride channels (Costa et al., 1979).

BZs AND THE HEART

In dog open-chest preparation, intravenous administration of diazepam increases coronary blood flow and cardiac output, while reducing blood pressure, heart rate, and myocardial contractile force (Clanchan and Marshall, 1979). At the same time, arterial coronary sinus oxygen differences narrow and myocardial oxygen consumption decreases. These observations suggest that diazepam encourages oxygen conservation and increases oxygen delivery and thus may benefit a patient with coronary insufficiency. BZs potentiate the coronary vasodilatory action of adenosine in anesthetized dogs. Pharmacological doses of diazepam, which increase blood flow, augment adenosine response and duration of action (Clanchan and Marshall, 1980). Therapeutic concentrations of diazepam potentiate the effect of adenosine on isolated guinea pig atria (Moritoki et al., 1985). This potentiation may be due to inhibition of an adenosine uptake site, since dipyridamole, an adenosine uptake site blocker, interacts with PBR, as previously shown in guinea pig and rat heart preparations (Davis and Huston, 1981). Dipyridamole is at least as potent as diazepam in displacing the specific binding of [^{3}H]Ro 5-4864 from PBR of dog heart (Doble et al., 1985).

PBR AND THE HEART

Although the exact biological function of PBR is not yet clear, a strong correlation exists between affinity of BZs to PBR and their physiologic effects on the heart. In addition to the aforementioned actions of diazepam on the various cardiac parameters, Ro 5-4864, the PBR specific ligand, decreases in a dose-dependent manner the duration of intracellular action potential and the contractility in guinea pig papillary muscle. Diazepam is less effective, and clonazepam, the CBR specific ligand, is completely ineffective (Mestre et al., 1984). The effect of Ro 5-4864 is GABA-independent and is antagonized by PK 11195 (which has no effect by itself), but not by Ro 15-1788, a selective antagonist of CBR.

Studies on the regional distribution of PBR in the rat heart indicated that their highest densities are present in the left ventricle and, in decreasing order, the right ventricle, left atrium,and right atrium (Davis and Huston, 1981). Using autoradiographic techniques and utilizing $[^3H]$Ro 5-4864 as a selective labeling ligand, the highest grain density of PBR was located in the left ventricle (Gehlert et al., 1985).

$[^3H]$ PK 14105, a nitrophenyl derivative of PK 11195, can photoaffinity label PBR of various rat tissues including the heart (Doble et al., 1987, Skowronski et al., 1988). Heart rat mitochondrial fraction was photoaffinity labeled by $[^3H]$ PK 14105 and the molecular weight (MW) of the photoaffinity labeled protein was 18 KDa, as detemined by sodium dodecyl sulfate polyacrylamide gel electrophoresis.

Ro 5-4864 induced coronary vasodilation and increased coronary flow in isolated retrograde-perfused Langendorff rat heart model, without affecting heart rate or force of contraction. Clonazepam, had no relaxant effect (Grupp et al., 1987). PK 11195, although increasing coronary flow, failed to block the effect of Ro 5-4864. KCl-induced contraction of rat aortic rings was inhibited by micromolar concentrations of both Ro 5-4864 and PK 11195, suggesting that this inhibitory effect is related to low-affinity PBR (French and Matlib, 1988) rather than to the high-affinity PBR. Recently, we demonstrated high and low affinity sites for Ro 5-4864 binding in various tissues of rat, but PK 11195 labeled only high affinity PBR (Awad and Gavish 1987). Thus the mechanism of the inhibition of KCl induced contraction by Ro 5-4864 may be explained by binding of Ro 5-4864 to low affinity sites. However, the fact that PK 11195 labels only high affinity PBR (nanomolar affinity) can not explain its effect at μM concentrations. We also found that PBR treated with the detergent Triton X-100 exhibited a diverse sensitivity, when either $[^3H]$Ro 5-4864 or $[^3H]$PK 11195 were used as radioligands. It is therefore proposed that the so-called high-affinity PBR are composed of two domains, a BZ binding site and an isoquinoline binding site, which are not identical (Awad and Gavish, 1988; Benavides et al., 1984).

PBR AND CALCIUM CHANNELS

One line of evidence suggests that PBR modulate calcium channels in the heart (Cantor et al., 1984). The agonist Ro 5-4864 possesses calcium channel-blocking properties, which are antagonized by PK 11195, the putative PBR antagonist. PK 11195 not only antagonizes the calcium channel-blocking actions of Ro 5-4864, but also abolishes calcium channel blockade produced by nifedipine, verapamil, diltiazem, and BAY K-8644 (Mestre et al., 1985b).

As calcium channel blockers prevent the accumulation of calcium ions in the cytosol of cardiac cells, thus reducing the earliest pathological consequences of ischemia, BZs can protect against immediate cardiac ischemic damage (Mestre et al., 1985a). PK 11195 prevents both early and delayed ventricular arrhythmias induced by 20 min of ischemia, and ventricular fibrillation following reperfusion in anesthetized dogs (Mestre et al., 1985a). Unlike classical calcium channel antagonists (Van Zwieten et al., 1983), PK 11195 is devoid of α-adrenergic blocking (Bergey et al., 1984) and β-adrenergic properties (Bergey et al., 1984). Therefore, peripheral BZ ligands may represent a novel treatment for angina and cardiac ischemia, which are the main causes of morbidity and mortality in the Western world (Levy, 1982).

Nevertheless, subcellular distribution and molecular size determination indicate that, although PBR and voltage-sensitive calcium chananels in the heart do not interact directly or allosterically, they are packaged in the same membrane compartment and hence may influence each other (Doble et al., 1985).

ISOLATION OF ENDOGENOUS LIGAND FOR PBR

In order to attain better understanding of the physiological and pharmacological functions of PBR, we attempted to identify and isolate an endogenous ligand for PBR. A preliminary report suggests the existence of an endogenous ligand for PBR in plasma and urine of uremic patients (Beaumont et al., 1983). In a previous study we reported that freezing and thawing of rat kidney increase the affinity of Ro 5-4864 binding to PBR (Gavish and Fares, 1985). This finding may suggest that an endogenous ligand is present in the kidney and that it is released during the freezing and thawing process.

Of all brain tissues, two regions are prominent in their PBR density: the choroid plexus and the ependyma (Gehlert et al., 1985). These regions are closely related to CSF. The choroid plexus is the tissue generating the CSF, while the ependyma is the cell layer enveloping the ventricles, adjacent to the CSF. Therefore, we concentrated on the CSF as the putative biological source for the endogenous ligand to PBR. The method for collection of human CSF is illustrated in Figure 1 .

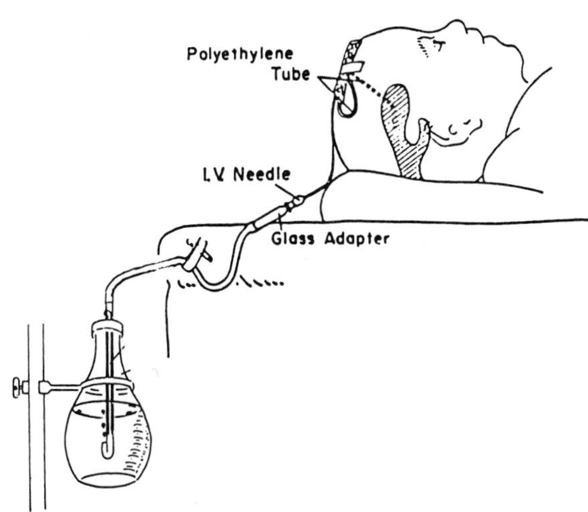

Figure 1. Collection of CSF, using intraventricular continuous drainage. During a head operation, a polyethylene catheter is placed in one of the brain ventricles. CSF is continuously drained and collected under sterile conditions, utilizing the above setting, and used for the purification of the endogenous ligand for PBR.

We found that human CSF inhibits [^{3}H]Ro 5-4864 and [^{3}H]PK 11195 binding to PBR. This inhibitory substance within the CSF exhibits the following characteristics: 1) it is precipitated by trichloroacetic acid; 2) it is precipitated by 50% ammonium sulfate; 3) it is retained in a dialysis tube with 12-kDa cut-off; and 4) during ultrafiltration, it passes through Amicon XM-IOO membrane (cut-off 100 kDa), but not through PM-IO membrane (cut-off 10 kDa).

Next, we tested the specificity of this substance. The results are summarized in Table 1. After 50-fold concentration by Amicon PM-IO membrane, CSF failed to inhibit [^{3}H]flunitrazepam binding to CBR in rat brain cortex. It also failed to affect [^{3}H]QNB (quinuclidinyl benzilate) binding to the muscarinic cholinergic receptors in rat brain cortex. However, it succeeded in fully inhibiting [^{3}H]Ro 5-4864 and [^{3}H]PK 11195 binding to kidney membranes. These findings suggest that the endogenous ligand for PBR is a protein (or a ligand bound to a protein) with a M.W. higher than 10 and lower than 100 kDa.

TABLE 1. Determination of the binding specificity of the CSF

[^{3}H] ligand	receptor	effect of CSF
PK 11195	PBR	inhibition
RO 5-4864	PBR	inhibition
Flunitrazepam	CBR	no effect
QNB	Muscarinic	no effect

PHYSICAL PROPERTIES OF THE INHIBITORY ACTIVITY

Further analysis of the stability of the inhibitory activity of this protein under various temperature conditions reveal that it is preserved after one week at 4°C and after one day at room temperature and that freezing and thawing do not harm its inhibitory activity. In addition, the inhibitory activity was found to be heat stable (60°C for 30 min). These findings have practical importance, since purification procedures are time-consuming and stability of the protein is crucial for isolation.

CHROMATOGRAFIC SEPARATION OF THE INHIBITORY ACTIVITY

After overnight dialysis against water at $4^{o}C$, CSF was loaded on a DEAE-cellulose column equilibrated with 10 mM Tris-HCl buffer, pH 7.4. The inhibitory activity was retained on the column and was eluted by NaCl with a peak of elution at 25 mM (Fig. 2). This procedure was used to obtain both concentration and purification (10-fold) of the endogenous ligand. After 10-fold purification by DEAE-cellulose, CSF was loaded on a G-75 column. The inhibitory activity was observed in the void volume.

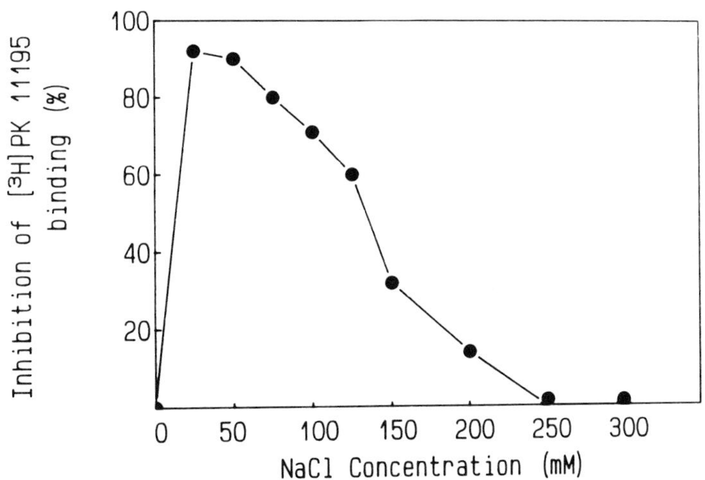

Figure 2. DEAE-cellulose chromatography of human CSF. The column was equilibrated with 10mM Tris-HCl (pH 7.4) after which 250 ml of human CSF (dialyzed against the same buffer) were applied. The column was washed with 100 ml of 10mM Tris-HCl buffer after which the elution was started by adding 20 ml NaCl at various concentrations (25-300 mM). 0.2 ml aliquates were tested for their ability to inhibit $[^3H]$PK 11195 binding to kidney membranes. Column size was 4.2x2.5 cm and fractions of 20 ml were collected.

We also used an Ultrogel AcA34 column. This column can separate proteins of M.W.'s ranging from 20 to 350 kDa. We found that the peak of inhibitory activity has a M.W. of about 70 kDa (Fig. 3). Loading a sample from the fraction which showed the inhibitory activity on SDS-PAGE revealed a major band with an approximate M.W. of 70 kDa.

ENZYMATIC DIGESTION

Tryptic digestion of CSF (concentrated by DEAE-cellulose) does not affect its inhibitory activity while the band of 70-kDa M.W. disappeared when the digest was loaded on SDS-PAGE. It is thus suggested that the 70-kDa M.W. protein is a precursor of a smaller protein- or a peptide-bound ligand which is the endogenous ligand. The 70-kDa protein may serve as stabilizer and carrier of the ligand.

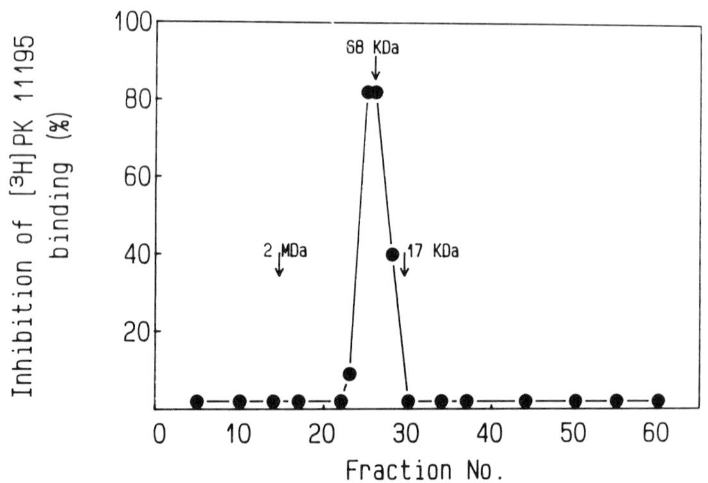

Figure 3. Gel filtration of human CSF on ultrogel AcA34. The column (1.8x110 cm) was equilibrated with 50mM NH_4HCO_3 (pH 7.8). 2.5 ml of human CSF (purified from 200 ml of CSF by DEAE cellulose) were applied on the column. 3.8 ml fractions were collected and 0.2 ml aliquots were tested for their ability to inhibit [^{3}H]PK 11195 binding to kidney membranes.

THE INHIBITORY SUBSTANCE: A BLOOD PIGMENT?

Recently, Verma et al. (1987) suggested that porphyrins are the putative endogenous ligand for PBR.

Porphyrins are tetrapyrrolic pigment, formed in the biosynthesis pathway of heme, which is the prosthetic group for mitochondrial cytochromes, hemoglobin and other heme-proteins. The naturally occurring porphyrins protoporphyrin IX, mesoporphyrin IX, deuteroporphyrin IX and hemin exhibit high affinity for PBR (Snyder et al. 1987). The exact physiological function of prophyrins is unknown, but their specificity to PBR together with their high affinity for these sites suggests that they may be the endogenous ligands for PBR.

While performing the above experiments, we noticed that the inhibitory activity correlated not only with the protein purification in a dose dependent manner, but also with the degree of a reddish-yellow pigment color that was co-eluted with the protein inhibitory activity during the various purification steps. Therefore, we investigated the inhibitory activity on PBR specific ligand binding of several pigments that might appear in the CSF. First we examined conjugated bilirubin, a product of hemoglobin, which is the abundant source of pigment in blood. Bilirubin (prepared in a diluted bicarbonate solution) produced no inhibitory effect even at concentrations as high as 10-μM. Then, we tried hemoglobin, a product of lysed red blood cells, which apparently contaminated the CSF. Hemoglobin inhibited 50% of the specific binding of [^{3}H]Ro 5-4864 and [^{3}H]PK 11195 to kidney PBR at a concentration of 10 μM.

The present study, which was aimed at identifying such a ligand, indicate the possibility that the partialy purified CSF contains large amounts of protein-bound substances, probably porphyrines (Verma et al. 1987), which inhibit peripheraly BZs ligands binding to their receptor. It is not yet clear whether this protein is a specific carrier for porphyrins in the CSF or a general carrier for porphyrins in the blood. Nevertheless, the chemical bond between the ligand and the protein is very strong, and it can not be broken by a simple physical and/or chemical manipulations, as was demonstrated in the present study. It is not inevitable that this 70 kDa protein may serve to stabilize and carry the endogenous ligand.

THE EFFECT OF HUMAN CSF AND PBR-SPECIFIC LIGANDS ON THE RAT HEART

The myocardium have been chosen as a target organ for evaluation of the physiologicl effect of the human CSF on the living tissue. As previously mentioned, the heart may serve as primary target for endogenous peripheral benzodiazepine ligands. The high density of the PBR in the heart together with the profound effects of the PBR specific ligands on the coronary blood flow (Grupp et al., 1987) may imply some of its biological properties.

A new technique published recently (Abraham et al., 1987) was adopted in this study for the evaluation of CSF (crude or partialy purified) effects in the heart. The experiments were performed on pentobarbital (25 mg/Kg, IP) anesthetized rats (450-500 gr). The right carotid artery was cannulated and the cannula (PE-50, 1000 U/ml of heparin) pushed forward to the region of the coronary ostium This cannula served for the intracarotid administration of drugs and the subsequent evaluation of their effect on the ECG pattern. A second cannula was introduced into the femoral artery for the measurement of arterial blood pressure. Lead I of the ECG was recorded using ECG Biotach (Gould). The blood pressure was recorded simultaneously using Physiological Pressure Transducer (Gould).

Dose dependent effect of the CSF injected into the carotid is demonstrated in Fig 4. Within few seconds of the CSF injection, ST segment elevation followed by continuous degraded decrease in the R wave was observed. This reduction of the R wave lasted 0.5 to 2 seconds (dose dependent) with partial recovery within 3 to 4 sec., and further recovery of the R wave within 0.5 to 1 min. Similar effect was obtained when RO 5-4864 was injected into the carotid artery. BAY K 8644 on the other hand, had considerable effect on the ST segment but almost no effect on the R wave. Although the injection of BAY K 8644 by itself, at relevant doses, never caused AV block, administration of Bay K 8644 after the CSF, even after full recovery of the ECG changes, ended with AV block and/or worsening of the iscemic status.

PK 11195 by itself, when injected intracarotidly (Abraham et al., 1987), had no effect on the ECG in this preparation but inhibited the ECG alterations induced by BAY K 8644. In this study when the PK 11195 was administered after what seemed a full recovery from the CSF injection, an ischemic effect was noted. Similar effect was recorded from both the crude and the partialy purified CSF but at different concentrations. On a contrary, in isolated retrograde perfused Langendorff rat heart preparation, at nanomolar concentrations, PK 11195 and RO 5-4864 had similar vasodilatory effect and increased the coronary flow (Grupp et al. 1987). The reason for this contradictory effect remains unknown.

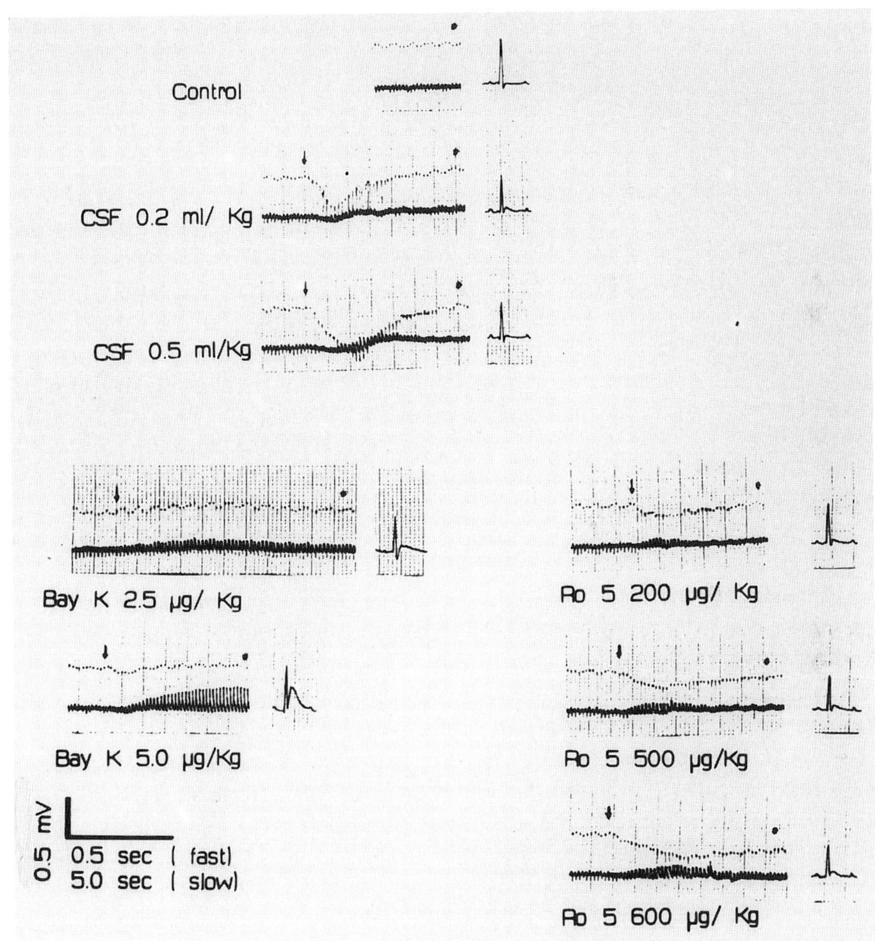

Figure 4. ECG tracings before and after the administration of CSF, Bay K 8644 (Bay K) and Ro 5-4864 (Ro 5) at various doses into the carotid artery. Arrow indicate injection of tested drugs and (*) the time of the fast ECG sweep on the right.

Human CSF has high specificity to PBR (see table 1). Maximum inhibition of the binding of [^{3}H]PK 11195 to kidney homogenate was obtained using 100 µL of crude CSF. When tested on in-vivo preparation the crude CSF injected intracarotidly demonstrated similar response as 500 µgr/Kg of RO 5-4864 which gives us rough estimate of its activity. But whereas the effect of RO 5-4864 lasted not more then 2-3 min., the CSF sensitized the myocardium for at least 10-15 min. This sensitization was characterized by the increase in the ischemic effect of BAY K 8644 and inversion of the response to PK 11195.

CONCLUSION

Human CSF contains a 70 kDa fraction which inhibit PBR specific ligands binding to their receptor. The inhibitory activity of this fraction is preserved after tryptic digestion, suggesting the presence of a low molecular weight substance, probably porphyrin like ligand. Intracarotid injection of human CSF resulted in long lasting and dose-dependent ST segment alterations. Sensitization of the animal's myocardium lasted at least 30 min. The effect could be mimicked by Ro 5-4864. It is speculated that the active compound may serve as a long lasting regulator of vascular tissue contractility.

Acknowledgement: We thank Ruth Singer for typing the manuscript. This research was supported by a grant to Dr. M. Gavish from the National Council for Research and Development, Israel and South Africa Medical Research Council.
Correspondence: Dr. Gavish, Rappaport Family Institute for Research in the Medical Sciences and Department of Pharmacology, Faculty of Medicine, Technion - Israel Institute of Technology, P.O.B. 9649, Haifa 31096, Israel.

REFERENCES

Abraham S, Amitai G, Oz N, Weissman BA (1987). Bay K 8644-induced changes in the ECG pattern of the rat and their inhibition by antianginal drugs. Br J Pharmac 92:603-608.

Anholt RRH, Pedersen PL, De Souza EB, Snyder SH (1986). The

peripheral-type benzodiazepine receptor: localization to the mitochondrial outer membrane. J Biol Chem 201:576-583.

Awad M, Gavish M (1987). Binding of [^{3}H]Ro 5-4864 and [^{3}H]PK 11195 to cerebral cortex and peripheral tissues of various species: species differences and heterogeneity in peripheral benzodiazepine binding sites. J Neurochem 49:1407-1414.

Awad M, Gavish M [1988). Differential effects of detergents on [^{3}H]Ro 5-4864 and [^{3}H]PK 11195 binding to peripheral- type benzodiazepine binding sites. Life Sci 43:167-175.

Beaumont K, Cheung AK, Geller ML, Fanestil DD (1983). Inhibitors of peripheral-type benzodiazepine receptors present in human urine and plasma ultrafiltrates. Life Sci 33:1375-1387.

Benavides J, Bagassat F., Phan T, Tur C, Uzan A, Renault C, Dubroeco Mc., Le Fur G (1984). Histidine modification with diethylpyrocarbamate induces a decrease in the binding of an antagonist, PK 11195, but not of an agonist, Ro 5-4864, of the peripheral benzodiazepine receptors. Life Sci 35:1249-1256.

Bender AS, Hertz L (1985). Pharmacological evidence that the non-neuronal diazepam binding site in primary cultures of glial cells is associated with calcium channels. Eur J Pharmacol 110:287-288.

Bergey JL, Wendt RL, Nocella K, MacCallum JD (1984). Acute coronary occlusion-reperfusion arrhythmias in pigs: antiarrhythmic and antifibrillatory evaluation of verapamil, nifedipine, prenylamine, and propranolol. Eur J Pharmacol 97:95-98.

Bosmann HB, Penny DP, Case KR, DiStefano P, Averill K (1978). Diazepam-receptor: specific binding of [^{3}H]Diazepam and [^{3}H]Flunitrazepam to rat brain. FEBS Lett 87:199-202.

Cantor EH, Kenessey A, Semenuk G, Spector S (1984). Interaction of calcium channel blockers with non-neuronal benzodiazepine binding sites. Proc Natl Acad Sci USA 81:1549-1552.

Clanchan AS, Marshall RJ (1979). Therapeutic concentrations of diazepam potentiate the effects of adenosine on isolated cardiac and smooth muscle [Abstract]. Proceedings of the biannual meeting of the British Pharmacological Society, September 1979, p 148.

Clanchan AS, Marshall RJ (1980). Diazepam potentiates the coronary vasodilator actions of adenosine in anaesthetized dogs [Abstract]. Proceedings of the biannual meeting of the British Pharmacological Society, April 1980, p 66.

Costa E, Rodbard D, Pert CB (1979). Is the benzodiazepine receptor coupled to a chloride anion channels? Nature 277:315-317.
Davis LP, Huston V (1981). Peripheral benzodiazepine binding sites in heart and their interaction with dipyridamole. Eur J Pharmacol 73:209-211.
De Souza EB, Anholt RRH, Murphy KMM, Snyder SH, Kuhar MJ (1985). Peripheral-type benzodiazepine receptors in endocrine organs: autoradiographic localization in rat pituitary, adrenal, and testis. Endocrinology 116:567-573.
Doble A, Benavides J, Ferris O, Bertrand P, Menager J, Vaucher N, Burgevin M-C, Uzan A, Gueremy C, Le Fur C (1985). Dihydropyridine and peripheral-type benzodiazepine binding sites: subcellular distribution and molecular size determination. Eur J Pharmacol 119:153-167.
Doble A, Ferris O, Burgevin MC, Menager J, Uzan A, Dubroeucq MC, Renault C, Gueremy C, Le Fur C (1987). Photoaffinity labeling of peripheral-type benzodiazepine binding sites. Mol Pharmacol 31:42-49.
French JF, Matlib MA (1988). The inhibitory effects of peripheral-type benzodiazepine specific ligands on vascular smooth muscle contractions are unrelated to their high-affinity binding site. FASEB J 2:A786.
Gavish M, Fares F (1985). The effect of freezing and thawing or of detergent treatment on peripheral benzodiazepine binding: the possible existence of an endogenous ligand. Eur J Pharmacol 107:283-284.
Gehlert DR, Yamamura HI, Wamsley JK (1985). Autoradiographic localization of "peripheral-type" benzodiazepine binding sites in the rat brain, heart, and kidney. Arch Pharmacol 328:454-460.
Grupp IL, French JF, Matlib MA (1987). Benzodiazepine Ro 5-4864 increases coronary flow. Eur J Pharmacol 143:143-147.
Harrison NL, Simmonds MA (1984). Modulation of the GABA receptor complex by steroid anesthetic. Brain Res 323:287-292.
Johonston GAR, Willow M (1982). GABA and barbiturate receptors. Trends Pharmacol Sci 6:328-330.
Le Fur G, Perrier ML, Vaucher N, Imbault F, Flamier A, Benavides J, Uzan A, Renault C, Dubroeucq MC, Gueremy C (1983a). Peripheral benzodiazepine binding sites: effect of PK 11195, 1-(2-chlorophenyl)-N-methyl-N (1-methylpropy1)-3-isoquinoline carboxamide. 1. In vitro studies. Life Sci 32:1839-1847.
Le Fur G, Vaucher N, Perrier ML, Flamier A, Benavides J,

Renault C, Dubroeucq MC, Gueremy C, Uzan A (1983b). Differentiation between two ligands for peripheral benzodiazepine binding sites, [^{3}H]Ro5-4864 and [^{3}H]PK 11195, by thermodynamic studies. Life Sci 33:449-457.

Levy RI (1982). Prevalence and epidemiology of cardiovascular disease. In Wyngaarden JB, Smith LH (eds): "Cecil's Textbook of Medicine," Philadelphia: WB Saunders, p 98-101.

Mackerer CR., Kochman R. (1978). Effects of cations and anions on the binding of ^{3}H-Diazepam to rat brain. Proc Soc Exp Biol Med 158:393-397.

Mestre M., Boutard G., Uzan A., Gueremy C., Renault C, Dubroeucq MC, Le Fur G (1985a). PK 11195, an antagonist of peripheral benzodiazepine receptors, reduces ventricular arrhythmias during myocardial ischemia and reperfusion in the dog. Eur J Pharmacol 112:257-260.

Mestre M, Carriot T, Belin C, Uzan A, Renault C, Dubroeucq MC, Gueremy C, Doble A, Le Fur G (1985b). Electrophysiological and pharmacological evidence that peripheral-type benzodiazepine receptors are coupled to calcium channels in the heart. Life Sci 36:391-400.

Mestre M, Carriot T, Belin C, Uzan A, Renault C, Dubroeucq MC, Gueremy C, Le Fur G (1984). Electrophysiological and pharmacological characterization of peripheral benzodiazepine receptors in a guinea pig heart preparation. Life Sci 35:953-962.

Mohler H, Okada T (1977). Benzodiazepine receptor: demonstration in the central nervous system. Science 198:849-851.

Moritoki H, Fukuda H, Kotani M, Ueyama T, Ishida Y, Takei M (1985). Possible mechanism of action of diazepam as an adenosine potentiator. Eur J Pharmacol 113:89-98.

Regan JW, Yamamura HI, Yamada S, Roeske WR (1981). High-affinity renal [^{3}H]flunitrazepam binding: characterization, localization, and alterations in hypertension. Life Sci 28:991-998.

Schoemaker H, Bliss M, Yamamura HI (1981). Specific high-affinity saturable binding of [^{3}H]Ro 5-4864 to benzodiazepine binding sites in the rat cerebral cortex. Eur J Pharmacol. 71:173-175.

Skowronski R, Fanestil DD, Beaumont K (1988). Photoaffinity labeling of peripheral-type benzodiazepine receptors in rat kidney mitochondria with [^{3}H]PK 14105. Eur J Pharmacol 148:187-193.

Snyder SH, Verma A, Trifiletti RR (1987). The peripheral-type benzodiazepine receptor: a protein of mitochndrial outer membranes utilizing porphyrins as endogenous ligands. FASEB J 1:282-288.

Squires R, Braestrup C (1977). Benzodiazepine receptors in rat brain. Nature 266:732-734.

Tallman JF, Thomas JW, Gallager DW (1978). GABAergic modulation of benzodiazepine binding site sensitivity. Nature 274:383-385.

Taniguchi T, Wang JKT, Spector S (1982). [^{3}H]Diazepam binding sites on rat heart and kidney. Biochem Pharmacol 31:589-590.

Van Zwieten PA, Van Meel JC, Timmermans PBMWM (1983). Functional interaction between calcium antagonists and vasoconstriction induced by stimulation of postsynaptic α2-adrenoreceptors. Circ Res 52:77-82.

Verma A, Nye JS, Snyder SH (1987). Porphyrins are endogenous ligands for the mitochondrial (peripheral-type) benzodiazepine receptor. Proc Natl Acad USA 84:2256-2260.

Wang JKT, Taniguchi T, Spector S (1980). Properties of [^{3}H]Diazepam binding on rat blood platelets. Life Sci 27:1881-1888.

Calcium Channel Modulators in Heart and Smooth Muscle: Basic Mechanisms and Pharmacological Aspects
S. Abraham and G. Amitai, editors
© 1990, VCH, Weinheim/Deerfield Beach, FL and Balaban, Rehovot/Philadelphia

The effects of nitrovasodilators on peripheral benzodiazepine receptors in the guinea pig heart

BEN A. WEISSMAN[1], PETER K. CHIANG[1], AND GORDON T. BOLGER[2]

[1]*Department of Applied Biochemistry, Division of Biochemistry, Walter Reed Army Institute of Research, Washington, DC, USA 20307-5100 and*
[2]*Department of Pharmacology, BioMega Inc., Laval, Quebec, Canada*

INTRODUCTION

High affinity, saturable and stereospecific receptors for peripheral-type benzodiazepine ligands (e.g., PK 11195, Ro5-4864, diazepam) have been identified in a wide variety of mammalian tissues (Schoemaker et al., 1983; Bolger et al., 1985; Awad and Gavish, 1987). The pharmacological profile of the binding of these ligands to peripheral benzodiazepine receptors (PBR) was studied extensively. One of the outstanding characteristics of PBR is that clonazepam that binds with a sub-nanomolar affinity to central benzodiazepine receptors (CBR) does not displace [^{3}H]Ro5-4864, the prototype ligand for PBR. Similarly, Ro5-4864 and PK 11195 are inactive as ligands for CBR (Weissman et al., 1984).

Recently, several studies provided evidence that PBR are associated with the mitochondial outer membrane (e.g. Anholt et al., 1986). It has been suggested that these sites function as modulators of respiratory control (cf Hirsch et al., 1989). In addition, PBR were shown to affect calcium-dependent events in the guinea pig heart (Mestre et al., 1985b; Mestre et al., 1986), to be associated with calcium channels in glial cells (Bender & Hertz, 1985) and to regulate $^{45}Ca^{2+}$ uptake into synaptosomes (Rampe & Triggle, 1987). Moreover, it was demonstrated that PK 11195, an antagonist of PBR

(Le Fur et al., 1983), reduced ventricular arrhythmias induced by myocardial ischemia and reperfusion in the dog (Mestre et al., 1985a) and inhibited BAY K 8644-elicited myocardial ischemia in the rat (Abraham et al., 1987).

Nitric oxide containing vasodilators such as nitroglycerine and sodium nitroprusside (SNP) are proposed to share parts of a cascade of biochemical events which culminate in the activation of guanylate cyclase, elevation of cGMP levels, reduction of intracellular calcium concentration and smooth muscle relaxation (Ignaro et al., 1981, Henry et al., 1989). In fact, recent studies have shown that SNP is a potent inhibitor of norepinephrine elicited $^{45}Ca^{2+}$ uptake to aortic strips (Karaki et al., 1988).

In the present study we investigated the possible interaction between PBR and NO-containing (or NO-producing) compounds in the rat spontaneously beating atrium and guinea pig heart membrane preparations.

METHODS

I. Intact Tissue Studies

Male Sprague-Dawley rats weighing 175-200g (Charles River, St. Constant, Quebec) were killed by decapitation. For prepariation of the atrium the hearts were rapidly removed and placed into a physiologic salt solution (PSS) which had previously been equilibrated with 95% O_2/5% CO_2 pH 7.2 at 37°C. The PSS had the following composition (mM): NaCl (137), KCl (2.7), $CaCl_2$ (1.8), $MgCl_2$ (0.49), NaH_2PO_4 (0.35), dextrose (5.6) and $NaHCO_3$ (11.9). The right and left atria were dissected free from surrounding tissue and mounted with 0.5g tension to a high compliance strain guage transducer (Grass Model FT03C, Grass Instruments) in a 20 ml organ bath containing PSS at 37°C equilibrated with 95% O_2/5% CO_2. Atrial tissues were equilibrated for 60 min at 37°C with one change of bath medium every 15 min) before any drug addition(s).

II. Radioligand Binding

Male Hartley guinea pigs (250-300g) were killed by decapitation. The brain and kidney were rapidly removed and placed into ice-cold 10 mM HEPES/10% Sucrose (pH 7.4) buffer. The cerebral cortex and the kidney were homogenized in 10 volumes HEPES buffere using a 15 second burst from a Brinkman polytron (setting 6). The homogenates were centrifuged at 50,000 X g for 60 min, the supernatant discarded and the membrane pellet resuspended and centrifuged once more. Final pellets were resuspended into 100 or 300 vols 50 mM of Tris-HCl, pH 7.4 for the cortex or kidney, respectively. The heart was rapidly removed and placed into ice cold HEPES sucrose; tissue was minced and homogenized by polytron (2 X 15 sec, setting 7) and subsequently centrifuged at 50,000 X g for 60 min. This step was repeated and the final pellet was resuspended in 250 volumes of Tris-HCl buffer.

[^{3}H]Ro5-4864 (or [^{3}H]PK 11195) binding was performed in a total assay volume of 1 ml consisting of 0.25 ml of membrane suspension (0.3-0.5 mg protein for brain membranes, 0.05-0.1 mg of protein for cardiac membrane and 0.025-0.05 mg of protein for kidney membranes), 0.1 ml radioligand and 0.55 ml of Tris-HCl. Nonspecific binding for [^{3}H]Ro5-4864 and [^{3}H]PK 11195 was determined in the presence of 5 μM PK 11195 or Ro5-4864, respectively. In most experiments tissues and drugs were incubated for 10 min at 30°C, placed on ice, ligand added and the incubation continued for 60 min. In competition assays the final concentration of [^{3}H]Ro5-4864 or [^{3}H]PK 11195 was 1 nM. Binding was terminated by rapid filtration through Whatman GF/B filters, followed by 2 X 5 ml washes with ice-cold Tris-HCl. The filters were placed in 8 ml of scintillation fluid and the radioactivity determined by liquid scintillation spectroscopy.

III. Drugs

(-)-S-BAY K 8644 was a gift from Dr. Scriabine, Miles Institute for Preclinical Pharmacology, West Haven CT, USA. The following drugs were obtained from the companies named: Ro5-4864, Hoffman-LaRoche, Nutley, NJ, USA; PK 11195, Pharmuka Laboratories, Geneviliers,

France; all other chemicals were obtained from standard commercial sources. (-)-S-Bay K 8644, Ro5-4864, 4-nitroquinoline N-oxide, PK 11195 and isosorbide dinatrate were dissolved in ethanol and diluted in water or buffer for organ bath and binding experiments, respectively.

IV. Data Processing and Statistical Analysis

IC_{50} values were calculated using the GRAFCAL program for the IBM PC (Dr. D. Kaplan, IIBR, Israel). In this program, parameters obtained from the Hill equation are substituted in an algorithm describing a sigmodial curve. Differences between groups of data were analyzed by parametric (student's t-test) using the BMDP statistical package made adaptable to the IBM PC (BMDP Statistical Software, CA).

RESULTS

SNP at a concentration range of 1μm to 1 mM had only a marginal effect on the binding of $[^3H]$Ro5-4864 (Fig. 1). Preincubation of this drug for 10 min at 30°C followed by the addition of ligand to the assay tubes at 0°C and continuing the incubation for 60 min resulted in a concentration-dependent inhibition of $[^3H]$Ro5-4864 binding to guinea pig heart membranes. The IC_{50} of SNP in this tissue is $5.6 \pm 1.7 \times 10^{-5}$ M and the Hill coefficient is 0.55 ± 0.06 (n=8). Under the same conditions the IC_{50} values for this vasodilator in kidney and brain membranes were one or two orders of magnified higher, respectively (Fig. 2).

Results presented in Table 1 demonstrate that the clinically useful vasodilator isosorbide dinitrate has a IC_{50} value of 1 mM. Higher concentrations of this drug could not be tested due to excessive ethanol amounts needed to dissolve it.

Potassium ferricyanide and sodium nitrite, an inhibitor and an activator, respectively, of guanylate cyclase, inhibited the binding of $[^3H]$Ro5-4864 to guinea pig heart membranes with IC_{50} values above 1 mM (Table 1). Sodium cyanide (results not shown), and sodium nitrate (25.4 % inhibition at a concentration of 10 mM) were essentially inactive. The NO-containing agent 4-

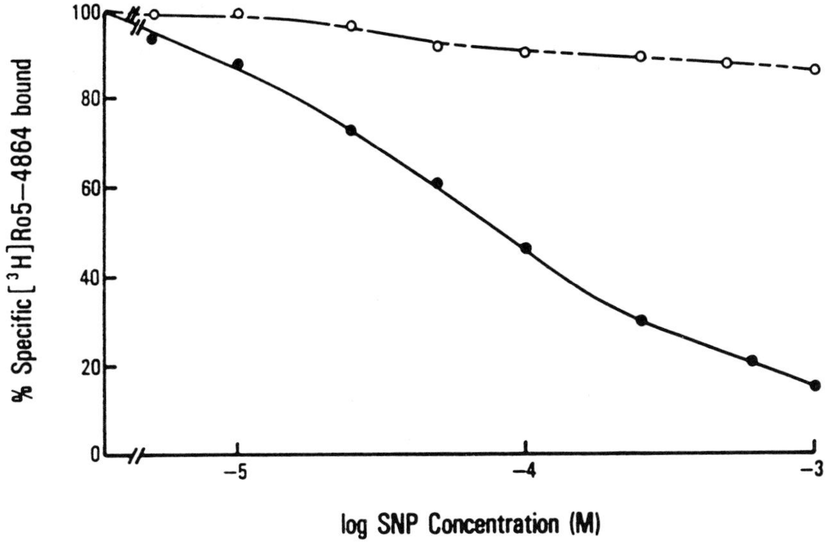

Figure 1: Effects of SNP on [^{3}H]Ro5-4864 binding to guinea pig heart membranes. Various concentrations of SNP were preincubated at 30°C for 10 min and then incubated for 60 min at 0°C (●) or at 0°C for 60 min (o). Data represent % specific [^{3}H]Ro5-4864 bound and are from one experiment repeated 3 times with identical results.

nitroquinoline-N-oxide displaced [^{3}H]Ro5-4864 with an IC_{50} of 2.0 ± 0.2 mM and a Hill coefficient of 0.68 ± 0.08 (n=4). Sodium azide, an activator of guanylate cyclase exhibited activity similar to that of sodium nitrite (Table 1). SNP exibited similar potency in inhibiting the binding of [^{3}H]PK 11195 and [^{3}H]Ro5-4864. Figure 3 depicts a representative concentration-response curve of the effect SNP and hydroxylamine on [^{3}H]PK 11195 binding to guinea pig heart membrane preparation. These data as well as results summarized in Table 2 show that there is a significant difference between the potency of SNP to displace PBR ligands. In addition to the lower potency of SNP in displacing [^{3}H]PK 11195, the Hill coefficient in this case is not different from 1.0, as compared to a value close to 0.5 for [^{3}H]Ro5-4864. Hydroxylamine competed with [^{3}H]PK 11195 for PBR on these membranes in a parallel manner to

Table 1. Effects of various agents on the binding of $[^3H]$Ro5-4864 to guinea pig heart membranes.

Agent	Concentration (M)			
	10^{-5}	10^{-4}	10^{-3}	10^{-2}
	(% Inhibition)			
$NaNO_2$		3.4 ± 2.7	21.7 ± 1.5	43.2 ± 2.8
$NaNO_3$			8.0 ± 4.2	25.4 ± 11.7
iso-sorb[a]	5.8 ± 2.1	12.7 ± 5.3	53.7 ± 2.4	
$K_3Fe(CN)_6$[b]	6.3 ± 1.7	12.7 ± 4.5	44.5 ± 6.2	52.9 ± 4.6
NaN_3	7.5 ± 4.3	14.9 ± 1.7	28.3 ± 1.8	

Data are presented as mean ± S.E.M. (n=2-4) of the percent inhibition of $[^3H]$Ro5-4864 binding to guinea pig heart membranes. Final ligand concentration was 1 nM. 4-Nitroquinoline-N-oxide had a IC_{50} of 2.0 ± 0.2 X 10^{-3} M and a Hill coefficient of 0.68 ± 0.08 (n=4).
[a]iso-sorb: isosorbide dinitrate. At 5 X 10^{-4} M it inhibited the binding by 30.4 ± 3.2%.
[b]In one experiment the IC_{50} was 5.8 X 10^{-3} M (nH=0.548).

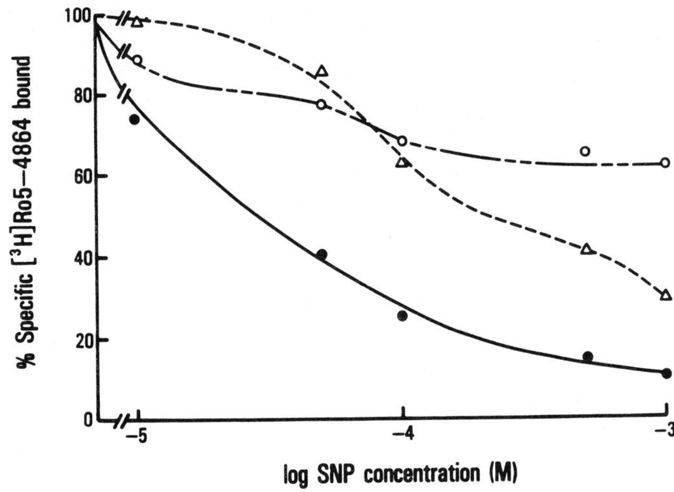

FIGURE 2: Effects of SNP on the binding of [^{3}H]Ro5-4864 to guinea pig heart (●), kidney (△) and cerebral cortex (o) membranes. Tissues were prepared and incubated as described in "METHODS". Data represent the % specific [^{3}H]Ro5-4864 bound and are from one experiment repeated 3-8 times with similar results. The IC_{50}'s and Hill coefficients are summarized below.

	Heart	Kidney	Cortex
IC_{50}	$5.6 \pm 1.7 \times 10^{-5}$	$3.0 \pm 0.2 \times 10^{-4}$	$3.7 \pm 0.3 \times 10^{-3}$
nH	0.55 ± 0.06	0.67 ± 0.05	0.44 ± 0.14

its competition with [^{3}H]Ro5-4864. The differential interactions of SNP with PBR ligands may indicate that Ro5-4864 and PK 1195 bind to two distinct proteins (McCabe et al., 1989).

Scatchard analyses of the binding of [^{3}H]Ro5-4864 to membranes prepared from guinea pig hearts revealed that in the presence or absence of SNP only a single, high affinity site can be detected (Fig. 4). However, SNP at a concentration of 40 μM causes a significant decrease in the affinity of [^{3}HRo5-4864 binding (Kd values of 1.75 ± 0.14 nM and 2.81 ± 0.19 nM for control and SNP treated, respectively) without a significant alteration of the maximal binding site density.

In another set of experiments it was observed that high concentrations of SNP (>1 mM) failed to affect tritiated N-methyl scopolamine ([^{3}H]NMS) binding to guinea pig heart membranes (results not shown).

Using the paradigm reported by Bolger et al. (1989), spontaneously beating rat atrium were incubated in organ bath in the presence of the calcium channel activator BAY K 8644. The latter increased the inotropy of these tissues by 135±32% (n=7). Ro5-4864 and SNP had no effect on basal activity (Table 3). Nevertheless, the addition of 10 μM Ro5-4864 to a bath containing 1 μM (-)-BAY K 8644 elicited a 102% elevation of the effect of the activator, whereas SNP (100 μM) was inactive under the same conditions. The combined addition SNP (100 μM) and Ro5-4864 to atrium incubated in the presence of 1 μM (-)-BAK K 8644 resulted in a complete inhibition of Ro5-4864-induced inotropy.

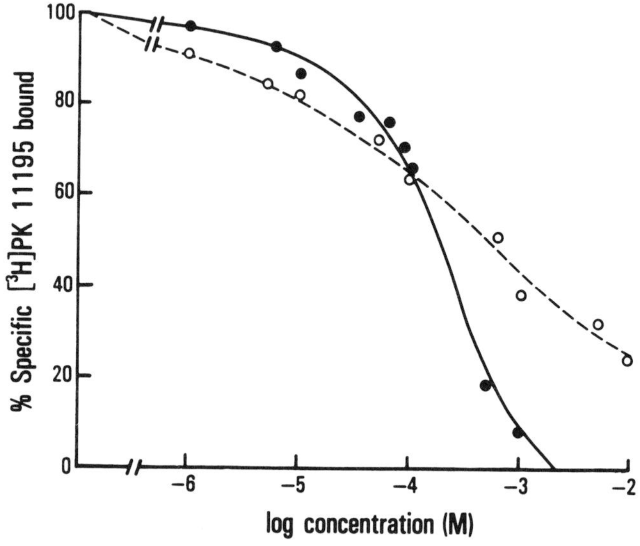

Figure 3. Concentration-response curves for SNP (●) and hydroxylamine (o) on [^{3}H]PK 11195 binding to guinea pig heart membranes. Incubations were performed as described in "METHODS". Data represent % specific [^{3}H]PK 11195 bound from one experiment repeated 3-4 times with similar results.

Table 2. Effects of SNP and hydroxylamine on the binding of PBR ligands

Compound	$[^3H]$Ro5-4864		$[^3H]$PK 11195	
	IC_{50} (M)	nH	IC_{50}(M)	nH
SNP	$5.61 \pm 1.72 \times 10^{-5}$	0.5 ± 0.06	$1.21 \pm 0.26 \times 10^{-4}$	0.95 ± 0.06
Hydroxylamine	$5.12 \pm 1.86 \times 10^{-4}$	0.56 ± 0.05	$2.42 \pm 0.57 \times 10^{-4}$	0.56 ± 0.07

Data represent the mean ± S.E.M. of 4-8 experiments. Binding assays were carried out as described in "METHODS" with a final ligand concentration of 1 nM.

(-)-BAY K 8644 produced a concommitent elevation of chronotropy which was not significantly affected by either SNP, Ro5-4864 or their combination.

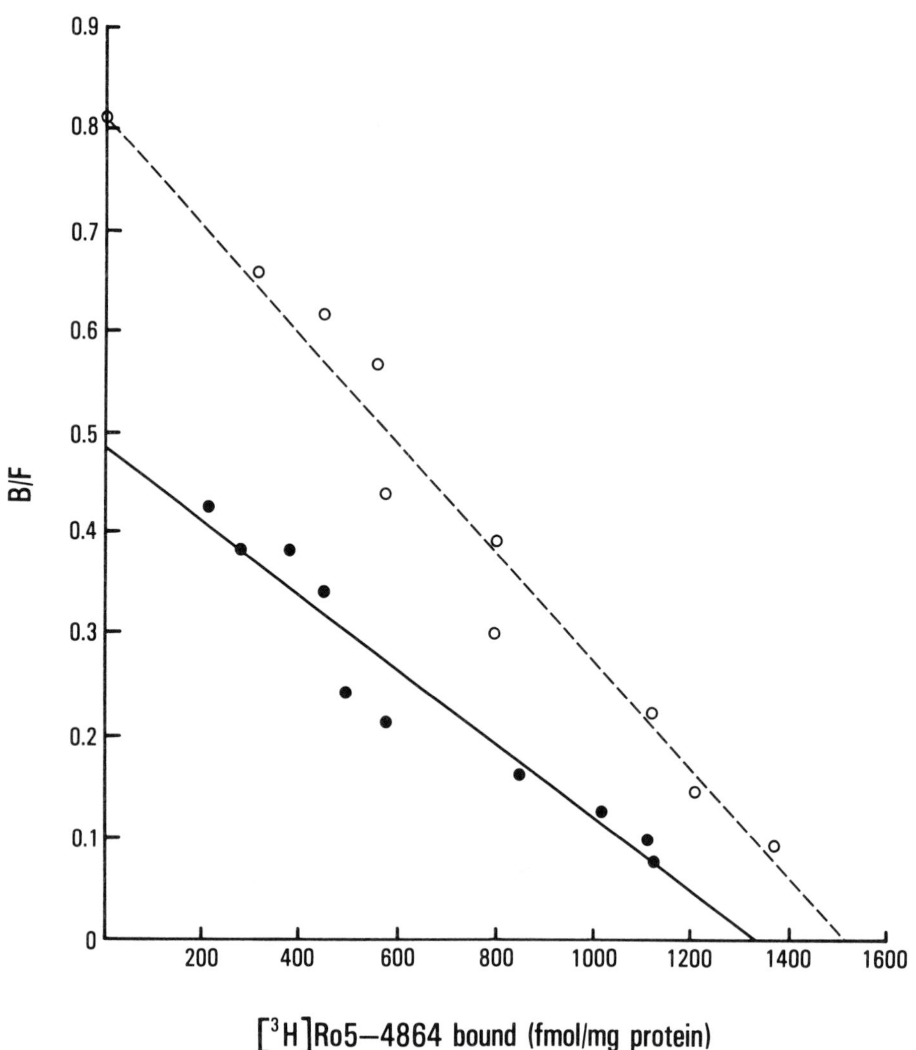

Figure 4. Scatchard plots of the binding of [^{3}H]Ro5-4864 to guinea pig heart membranes. Tissues were prepared as described under "METHODS" and incubated for 10 min at 30°C in the absence or presence of 4 X 10^{-5}M SNP. Incubations were then continued for 60 min at 0°C. Data represent the results of a single experiment repeated 5 times with similar results.

Table 3. Effects of SNP on Ro5-4864 modulation of (-)-BAY K 8644 stimulated activity of spontaneously beating rat atrium.

Agent(s) (μM)	Chronotropy[a] (beats/min)	Inotropy[a,b] (% increase of control)	(N)
Control	240±7		10
(-)-BAY-K 8644 (1)	300±18*	135±32*	7
+ SNP (100)	298±40*	128±13*	3
+Ro5-4864 (10)	297±8*	237±33*,**	7
+Ro5-4864 (10) + SNP (100)	270±12*	141±22*	7

[a]Results are presented as the mean ± S.E.M. of three sec measurements. SNP and Ro5-4864 were incubated for 10 min before the introduction of (-)-BAY K 8644 (5 min).
[b]Inotropy is presented as the % increase compared to control force of contraction.

SNP and Ro5-4864 had no effect on inotropy or chronotropy in the absence of (-)-BAY K 8644.

Significantly different from * control, $p < 0.05$; from ** (-)-BAY K 8644, $p < 0.05$, Student's t-test.

REFERENCES

Abraham S, Amitai G, Oz N, Weissman BA (1987). (±)-BAY K 8644 induced changes in the ECG pattern of the rat and their blockade by antianginal drugs. Brit J Pharmacol 92:603-608.

Anholt RRH, Pederson PL, DeSouza EB, Snyder SH (1986). The peripheral benzodiazepine receptor: Localization to the mitochondrial outer membrane. J Biol Chem 261:576-583.

Awad M, Gavish M (1987). Binding of [^{3}H]Ro5-4864 and [^{3}H]PK 11195 to cerebral cortex and peripheral tissues of various species: species differences and heterogeneity in peripheral benzodiazepine binding sites. J Neurochem 49:1407-1414.

Bender A, Hertz L (1985). Pharmacological evidence that the non-neuronal diazepam binding in primary cultures of glial cells is associated with a calcium channel. Eur J Pharmacol 110:287-288.

Bolger GT, Newman AH, Rice KC, Lueddens HWM, Basile AS, Skolnick P. (1989). Characterization of the effects of AHN 086, an irreversible ligand of peripheral benzodiazepine receptors on contraction in guinea-pig atria and ileal longitudinal smooth muscle. Can J Physiol Pharmacol 67:126-134.

Bolger GT, Weissman BA, Lueddens HWM, Basile AS, Mantione CR, Barrett JE, Witkin JM, Paul SM, Skolnick P (1985). Late evolutionary appearance of "peripheral-type" binding sites for benzodiazepines. Brain Res 338:366-370.

Grupp IL, French JF, Matlib MA (1987). Benzodiazepine Ro5-4864 increases coronary flow. Eur J Pharmacol 143:143-147.

Henry PJ, Horowitz JD, Louis WJ (1989). Nitroglycerine-induced tolerance effects multiple sites in the organic nitrate bioconversion cascade. J Pharmacol Exp Therap 248:762-768.

Hirsch JD, Beyer CE, Malkowitz L, Beer B, Blume AJ (1989). Mitochondrial benzodiazepine receptors modrate inhibition of mitochondrial respiratory control. Mol Pharmacol 34:157-163.

Ignaro LJ, Lippton H, Edwards JC, Baricos WH, Hyman AL, Kadowitz PJ, Gruetter CA (1981). Mechanism of vascular smooth muscle relaxation by organic nitrates, nitrites, nitroprusside and nitric oxide: Evidence for the involvement S-nitrosothiols as active intermediates. J Pharmacol Exp Ther 218:739-749.

Karaki H, Sato K, Ozak H, Murakami K (1988). Effects of sodium nitroprusside on cytosolic calcium level in vascular smooth muscle. Eur J Pharmacol 158:259-266.

Le Fur G, Vacher N, Perrier MI, Flamier A, Benevides J, Renault C, Dubroeucq MC, Gueremy C, Uzan A (1983). Differentiation between two ligands for peripheral benzodiazepine binding sites, [^{3}H]Ro 5-4864 and [^{3}H]PK 11195, by thermodynamic studies. Life Sci 33:449-457.

McCabe RT, Schoenheimer JA, Skolnick P, Newman AH, Rice KC, Reig JA, Kline DC (1989). [^{3}H]AHN 086 acylates peripheral benzodiazepine receptors in the rat pineal gland. FEBS Lett 244:263-267.

Mestre M, Boutard G, Uzan A, Gueremy C, Renault C, Dubroeucq MC, Le Fur G (1985a). PK 11195, an antagonist of peripheral benzodiazepine receptors, reduces ventricular arrhythmias during myocardial ischemia and reperfusion in the dog. Eur J Pharmacol 112:257-260.

Mestre M, Carriot T, Belin C, Uzan A, Renault C, Dubroeucq MC, Guermy C, Le Fur G (1985b). Electrophysiological and pharmacological characterization of peripheral benzodiazepine receptors in a guinea-pig heart preparation. Life Sci 36:953-962.

Mestre M, Carriot T, Neliat G, Uzan A, Renault C, Dubroeucq MC, Gueremy C, Doble A, Le Fur G (1986). PK 11195, an antagonist of peripheral benzodiazepine receptors, modulate Bay K 8644 sensitive but not β- or H_2-receptor sensitive voltage operated calcium channels in the guinea pig heart. Life Sci 39:329-339.

Rampe D, Triggle DJ (1986). Benzodiazepines and calcium channel function. Trends Pharm Sci 11:461-464.

Schoemaker H, Boles RG, Horst D, Yamamura HI (1983). Specific high affinity binding sites for [^{3}H]Ro5-4864 in rat brain and kidney. J Pharmacol Exp Ther 225:61-69.

Weissman BA, Bolger GT, Isaas L, Paul SM, Skolnick P (1984). Characterization of the binding of [^{3}H]Ro5-4864, a convulsant benzodiazepine, to guinea pig brain. J Neurochem 42:969-975.

Calcium Channel Modulators in Heart and Smooth Muscle: Basic Mechanisms and Pharmacological Aspects
S. Abraham and G. Amitai, editors
© 1990, VCH, Weinheim/Deerfield Beach, FL and Balaban, Rehovot/Philadelphia

Cardiovascular effects of mammalian endothelins and snake venom sarafotoxins

Z. WOLLBERG, A. BDOLAH AND E. KOCHVA

Department of Zoology, George S. Wise Faculty of Life Sciences, Tel Aviv University, Tel Aviv 69978, Israel

INTRODUCTION

It is now well established that certain substances that are produced by the vascular endothelium act as relaxing factors, while others show vasoconstrictor effects (Taylor and Weston, 1988; Vanhoutte and Katusic, 1988). One of these factors is the recently discovered group of endothelins (ETs), which act as potent vasoconstrictors in several organ systems (e.g. Yanagisawa et al., 1988a; Goetz et al., 1988; Saito et al., 1989; Badr et al., 1989; Chabrier et al., 1989a). Additional compounds, albeit from a totally different source, belong functionally to the group of vasoactive substances. These are the sarafotoxins (SRTXs) which were isolated from the venom of the Israeli Burrowing Asp (*Atractaspis engaddensis*, Atractaspididae, Ophidia, Fig.1). The endothelins and sarafotoxins are 21-residue peptides that contain two disulfide bridges connected in positions 1-15 and 3-11 (Kochva et al., 1982; Takasaki et al., 1988; Kumagaye et al. 1988; Yanagisawa et al., 1988b). The two groups bind with high affinity to the same population of hitherto undetected receptors (Ambar et al., 1989; Gu et al., 1989) and induce phosphoinositide (PI) hydrolysis which is to a large extent independent of extracellular Ca^{2+} concentration (Kloog et al., 1988, 1989; Ambar et al., 1988; Ambar et al., 1989). The vasoconstrictor potency of these highly homologous peptides is attributed to the elevation of intracellular free calcium ions in the vascular smooth muscles. This may be achieved by two different mechanisms: 1) mobilization of Ca ions from internal sources by the activation of the phosphoinositide cycle (Miasiro et al., 1988; Marsden et al., 1989); 2) activation of a non-selective membranal cation channel which depolarizes the membrane to a level that activates L-type Ca^{2+} channels, resulting in a massive influx of extracellular calcium ions (Van Renterghem et al., 1988; Chabrier et al., 1989b). In this report we demonstrate some pharmacological effects of endothelin and sarafotoxin on the cardiovascular system. We also discuss some possible structure-function relationships of these compounds.

Fig. 1. The Israel Burrowing Asp *Atractaspis engaddensis*.

TOXICITY

The sarafotoxins are three highly homologous isotoxins designated as SRTX-a, SRTX-b and SRTX-c (Fig. 2 and Takasaki et al. 1988). Of these three isotoxins, SRTX-b is the most toxic and is among the most lethal toxins of snake venom origin (i.v. LD_{50} in mice: ≃15 ug/kg bw). SRTX-a is about as toxic, while SRTX-c is of a considerably lower lethality (LD_{50}: 300 ug/kg bw). The endothelins (ET-1, ET-2 and ET-3; Hiley, 1989) are very similar in structure and toxicity to the sarafotoxins (Fig. 2 and Bdolah et al. 1989a). The toxicity of ET-1 is essentially the same as that of SRTX-b (LD_{50}: 15 ug/kg bw; Bdolah et al. 1989b), while ET-3 is somewhat less toxic (unpublished results); ET-2 has not yet been tested. The action of both ET-1 and SRTX-b is very rapid, and an i.v. administration of a lethal dose induces death in mice within minutes, apparently as a result of coronary vasospasm and disturbances in cardiac activity (Weiser et al., 1984; Wollberg et al., 1988; Lee et al., 1988; Bdolah et al., 1989b). Under natural conditions, the endothelins presumably operate as normal modulators of blood pressure at very small doses (Lippton et al., 1988; Payne and Whittle, 1988; Wright and Fozard, 1988; Braquet et al., 1989).

SARAFOTOXINS

SRTX-a	C	S	C	K	D	M	T	D	K	E	C	L	N	F	C	H	Q	D	V	I	W
SRTX-b	C	S	C	K	D	M	T	D	K	E	C	L	Y	F	C	H	Q	D	V	I	W
SRTX-c	C	T	C	N	D	M	T	D	E	E	C	L	N	F	C	H	Q	D	V	I	W

ENDOTHELINS

ET-1	C	S	C	S	S	L	M	D	K	E	C	V	Y	F	C	H	L	D	I	I	W
ET-2	C	S	C	S	S	W	L	D	K	E	C	V	Y	F	C	H	L	D	I	I	W
ET-3	C	T	C	F	T	Y	K	D	K	E	C	V	Y	Y	C	H	L	D	I	I	W
	1				5					10					15					20	

Fig. 2. Amino acid sequences of the sarafotoxins and endothelins. The standard one-letter notation is used for the amino acids (IUPAC-IUB, 1968). ET-1, ET-2 and ET-3 are the products of three human endothelin genes. ET-1 is the form originally isolated from porcine endothelial cells and ET-3 is the sequence found in the rat genome.

EFFECTS ON ECG AND ISOLATED HEART PREPARATIONS

Intravenous administration of lethal doses (>25 ug/kg bw) of either ET-1 or SRTX-b to anesthetized mice induces changes in most ECG components within a few seconds. These changes are essentially the same for both compounds (Fig. 3). The first noticeable alterations consist of a transient but very marked S-T elevation and an increase in the amplitude of the R wave. After 15 to 30 seconds from the onset of drug administration, the S-T elevation turns gradually into a conspicuous S-T depression which may be transitory or may last for several minutes until the death of the animal. Concomitantly, an A-V block, increasing in severity, is developed. It starts with a gradual but

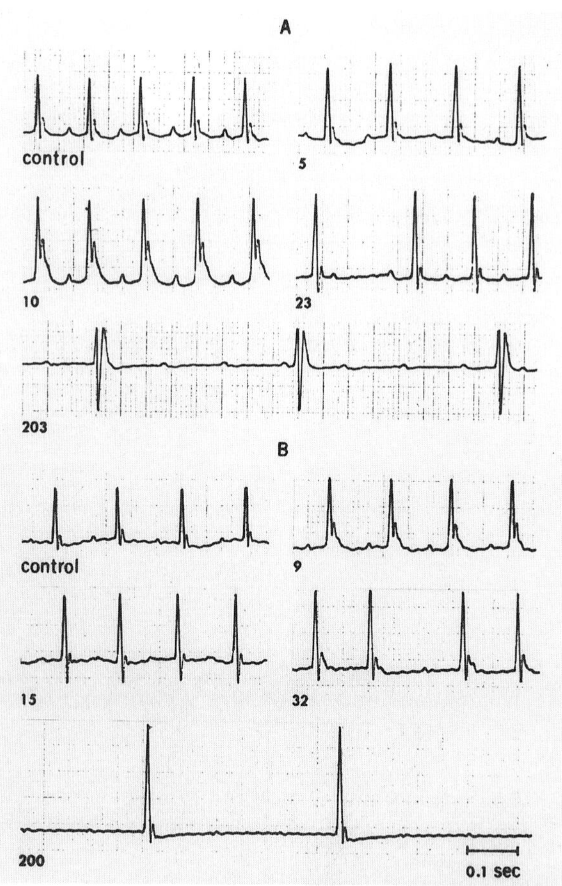

Fig. 3. Effects of lethal doses (25 ug/kg bw) of ET-1 (A) and SRTXb (B) on the ECG of anesthetized mice. Selected samples from continuous paper recordings. Recording was carried out with bipolar limb leads comparable to lead II. The number under each trace indicates time lapse (in seconds) from drug administration. Note the transient S-T elevation, the increased amplitude of the R wave , the S-T depression and the severe A-V block which leads to cardiac arrest.

consistent increase in the P-R interval (1st degree), followed by the appearance of "dropped beats" (2nd degree) and leading to a complete atrioventricular dissociation (3rd degree) and cardiac arrest. It should be stressed that not always are all three typical stages of the A-V block discerned. This is especially true when higher doses are injected. In these circumstances, the A-V block develops very fast; the 1st degree does not usually appear at all and the duration of the 2nd degree is shortened significantly. Intravenous administration of SRTX-b to anesthetized rats and dogs results in essentially the same effects (not shown). In the intact perfused rat heart ("Langendorff's preparation"), bolus injections of SRTX-b into the perfusion system, close to the heart, result in an almost immediate and marked increase in the amplitude of the R wave and an A-V block of increasing severity, usually going through all three stages (Fig. 4). A very prominent coronary vasoconstriction can be detected by a drop counter which shows a change in the flow rate of the perfusate.

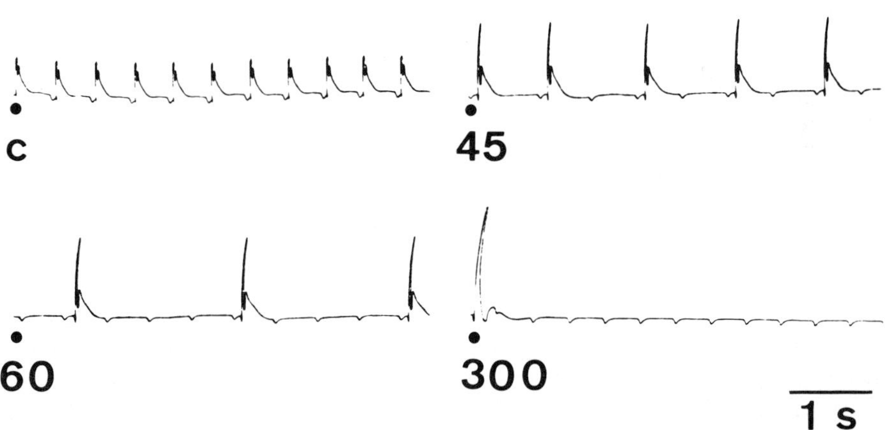

Fig. 4. Effect of SRTXb on ECG in a Langendorff perfused rat heart preparation. Selected samples from continuous recordings. Note the gradual increase in severity of the A-V block. Time lapse, in seconds, from onset of toxin administration is indicated below each sample. C: control.

The electrocardiographic events observed in the intact animals and the effects of the toxin on the isolated heart preparation, indicate a state of coronary insufficiency resulting from acute coronary vasospasm and disturbances in the atrioventricular conduction system. Several lines of evidence suggest that these two pathological events develop independently. Nonetheless, a contribution of the ischemic state to the progress of the A-V block is not unlikely (Wollberg et al., 1988). Interestingly, similar syndromes were reported for human victims bitten by the snake, but were wrongly diagnosed as secondary (Alkan and Sukenik, 1974).

The similarity in the patterns of ECG disturbances caused by ET-1 and SRTX-b, suggests that these two peptides affect the heart of intact mice by a similar mechanism(s). This notion is corroborated by the finding that both these compounds induce PI hydrolysis and compete for the same atrial receptors (Ambar et al., 1989; Bdolah et al., 1989b; Kloog and Sokolovsky, 1989). The high toxicity of endothelin and its experimentally induced pathogenic effects on the heart, raises the question of possible damage that might be caused by this endogenous substance under failure or malfunction of its regulatory mechanism.

The direct effects of SRTX-b on cardiac contractility were studied with rat and human atria as well as with human papillary muscles. The toxin produced a marked positive inotropic effect in both spontaneously contracting right rat atria (under S-A node control) and in right and left rat atria and human right atrial and papillary muscle strips, electrically driven by external stimulation at a rate of 0.5 or 0.75 Hz (Fig. 5). The human tissue, discarded during cardiac surgery, was obtained with the approval of the Hospital Ethics Committee. In the rat atrial preparations, addition of 15 nM to 1.5 uM of the toxin to the organ bath resulted in a slow dose-dependent increase in the maximal contraction force (about 200% over control), becoming noticeable after about one min and reaching its maximum within about 15 min. The rate of spontaneous contractions was not affected by the toxin, suggesting that it does not have a direct effect on the S-A node. The inotropic effects of the toxin were similar to those induced by alpha-adrenoceptors (Shavit et al. 1986). However, pretreatment with either alpha or beta-adrenergic blockers did not abolish the effect of the toxin. Recent studies showed that synthetic ET-1 has essentially the same inotropic effects on isolated guinea pig atrial preparations (Hu et al., 1988; Ishikawa et al., 1988).

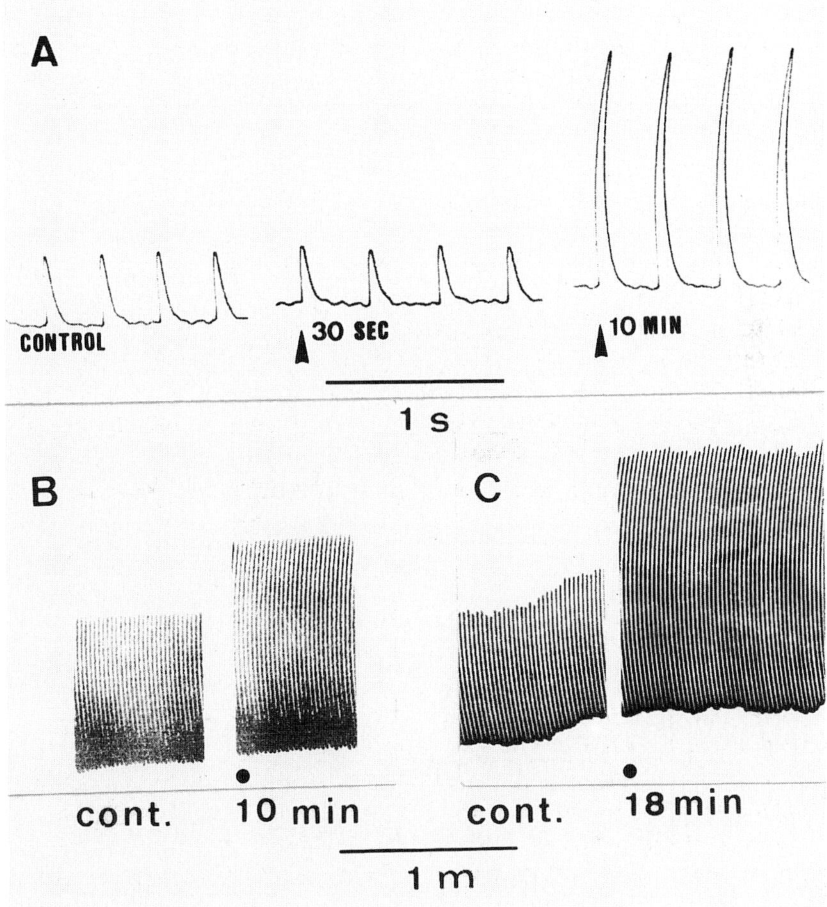

Fig. 5. Inotropic effects of SRTXb on cardiac muscle strips. Selected samples from continuous recordings illustrating force of contraction. Time lapse after addition of the toxin is indicated below each sample. A: rat right atrium beating spontaneously. B: adult human papillary muscle. C: rat left atrium. In B and C the preparations were driven by external stimulation (rates: 0.75 and 0.5 Hz, respectively).

VASOCONSTRICTOR EFFECTS ON RABBIT AORTIC STRIPS

Figure 6 illustrates typical cumulative constrictions of rabbit aortic strips induced by the three SRTXs and by

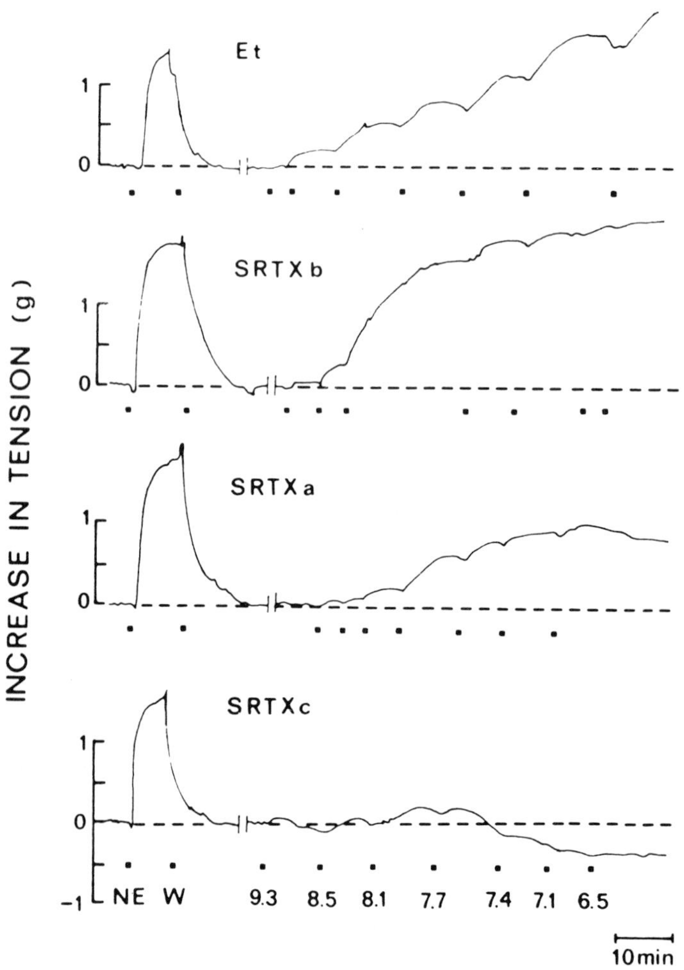

Fig. 6. Isometric contractile responses of rabbit aortic strips to norepinephrine (NE; 6 uM) and to cumulatively applied synthetic ET-1, SRTX-a, SRTX-b and SRTX-c. Filled squares at the bottom of each recording represent timing of drug application or washout (W). Concentrations (expressed as negative log molar) are designated by numbers under the lower tracing and refer also to all other tracings.

synthetic ET-1. Averaged cumulative dose response relationships for the three isotoxins are summarized in Figure 7. It can be seen that under these experimental conditions, ET-1 induces a concentration dependent

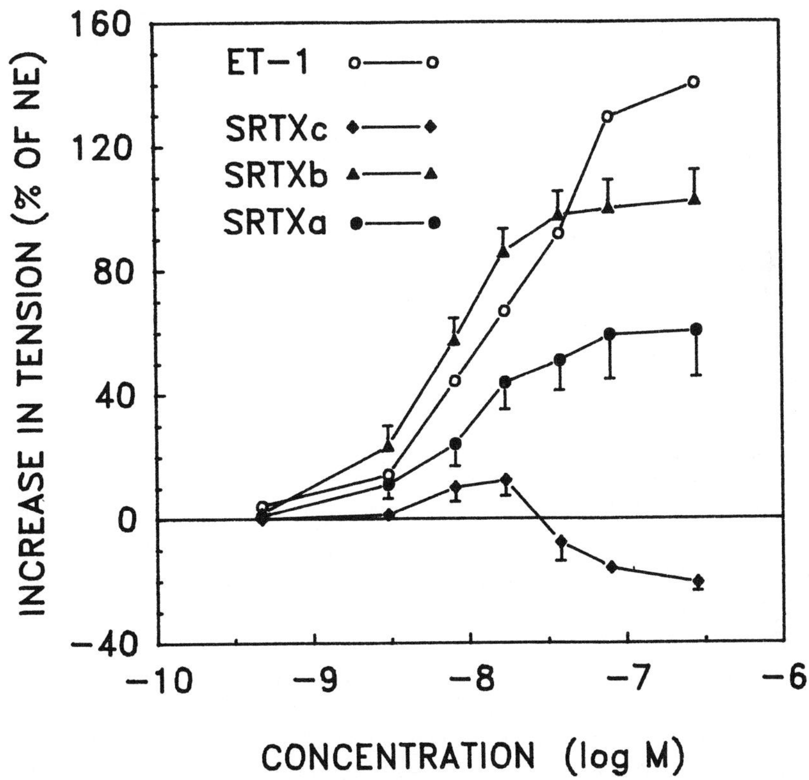

Fig. 7. Mean (± SEM) contractile dose response relationships of rabbit aortic strips to cumulatively applied SRTX-a, SRTX-b and SRTX-c (n = 5 in each case). The dose response of ET-1 is based on a single experiment and is comparable to previously described dose responses of this peptide. Responses are described as % of maximal response to NE.

constriction of the rabbit aorta similar in shape to that shown previously for the porcine coronary artery (Yanagisawa et al., 1988a; Kimura et al., 1988) and other mammalian tissues (Maggi et al., 1989a,b). SRTX-a and SRTX-b also induce a dose dependent contraction that builds up slowly. Typically, a contraction was followed by a long lasting decay, which was very difficult to wash out. The ED_{50} values of SRTX-a and SRTX-b (cumulative effective doses that induce half maximum response) were about 3.5 nM and 6.4 nM, respectively and were not statistically different. The maximum increase in tension induced by SRTX-b, however, was significantly higher than that produced by SRTX-a (102% and 60% of the maximal contraction induced by NE, respectively; see also Wollberg et al., 1989). Of special interest was the extremely weak vasoconstriction potency of SRTX-c (about 12% of maximal NE contraction, at a concentration of about 18 nM). This contraction was followed by a spontaneous prolonged relaxation that continued even beyond the resting equilibrated tension. It is noteworthy that the lethality and cardiotoxicity of SRTX-c are also significantly lower than those of the other two isotoxins (Takasaki et al., 1988; Bdolah et al., 1989a,b).

In several experiments intended to examine possible additive effects of the three sarafotoxins, a different isotoxin was added to the bath after the aortic strip reached a maximal response to one of the toxins and ceased to respond to supplemental doses of this isotoxin. Fig 8 (upper tracing) illustrates a case in which maximal doses of SRTX-a and SRTX-b were added sequentially on top of an SRTX-c cumulative response. SRTX-c induced the typical dual response; SRTX-a induced a somewhat weaker contraction than that usually induced by this toxin when applied separately, and the response to SRTX-b reached its regular values. The response to NE at the end of the experiment was essentially the same as the control. Fig 8 (lower tracing) demonstrates another case in which a maximal dose of SRTX-b was added on top of cumulative doses of SRTX-a. It can be seen that the response to SRTX-b on top of a typical cumulative dose response of SRTX-a, remained unaffected. These results may point to the possibility that each of the three isotoxins activates preferentially a different receptor subtype, a notion that is supported by a previous suggestion which was based on binding studies (Ambar et al., 1988).

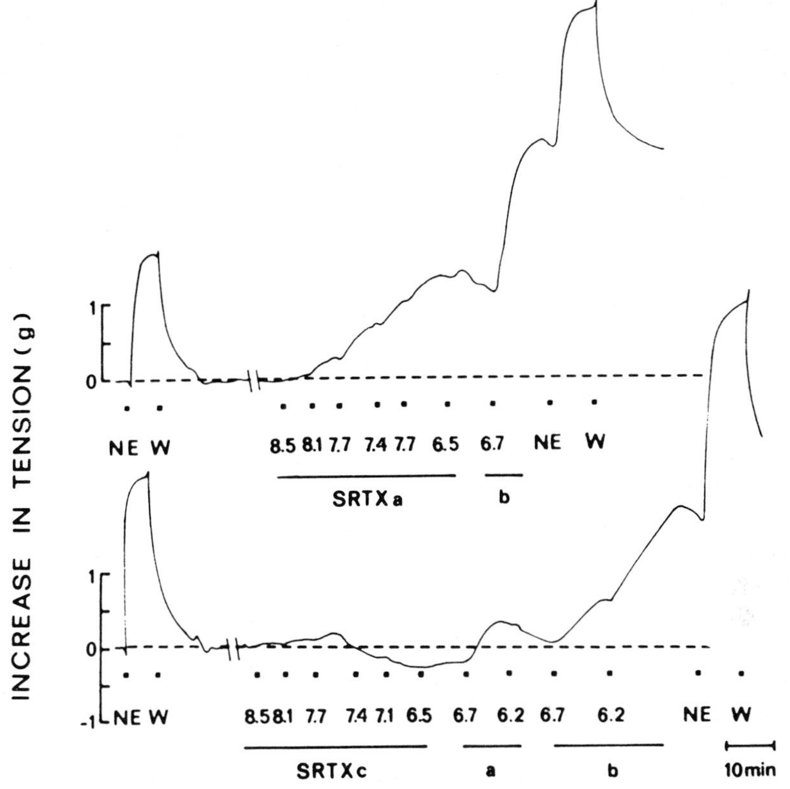

Fig. 8. Cumulative combined responses of rabbit aortic strips to consecutive application of the three *Atractaspis* isotoxins and NE. The sequence in the lower tracing: SRTX-c, SRTX-a, SRTX-b, NE. In the upper tracing: SRTX-a SRTX-b, NE. All other details as in Fig. 6.

CONCLUSIONS

The high structural homology between the snake venom sarafotoxins and the mammalian endothelins points to a common evolutionary ancestry of these two groups of peptides (see also Graur et al., 1988/9). Since the most distant peptides within the SRTX/ET family, SRTX-a and SRTX-c on the one hand and ET-3 on the other, share in common 11 out of 21 amino acids (52%), the probability that these two lineages represent a case of parallel evolution is very unlikely.

The sarafotoxins and endothelins also share several functional properties. SRTX-b and ET-1 compete for the same receptors in the heart and brain and induce phosphoinositide (PI) hydrolysis in these tissues (Ambar et al., 1989). Both these peptides are highly lethal and cause similar cardiotoxic effects in mice (Bdolah et al., 1989b); they are powerful vasoconstrictors of the aorta and coronary blood vessels and show a positive inotropic effect in heart muscles (Yanagisawa et al., 1988a,b; Ishikawa et al., 1988; Wollberg et al., 1988, 1989; Han et al., 1989; Kneupfer et al., 1989).

The differences in vasoconstriction of the three SRTXs provide information on the structure-function relationships of the ET/SRTX group. The ETs and the SRTXs all have a hydrophobic C-terminal "tail" which is essential for their activities (Kimura et al., 1988). However, this is certainly not sufficient in itself, as SRTX-c which has a C-terminal tail identical to the other two SRTXs, is a very poor vasoconstrictor. It is thus evident that other domains are involved in the function of these peptides. SRTX-c is the most acidic peptide of this group. It has five acidic residues without any Lys (or Arg), whereas all the other peptides have at least one Lys residue in the main intramolecular loop. The maximal vasoconstrictor potency of SRTX-b is about twice as high as that of SRTX-a. The only difference between the two is in position 13, where Tyr is replaced by Asn in SRTX-a (also in SRTX-c). All three ETs also have a Tyr residue in position 13, which indicates its importance inside the loop between Cys^{11} to Cys^{15}. The different pattern of activity of SRTX-c can thus be explained by several changes in the N-terminal half of the molecule. When the three SRTXs are applied sequentially to rabbit aortic strips, they cause a summated constrictor response. These results point to the possibility that the *Atractaspis* isotoxins and the mammalian endothelins may have more than one receptor subtype.

REFERENCES

Alkan ML, Sukenik S (1975) Atrioventricular block in a case of snakebite inflicted by *Atractaspis engaddensis*. Trans R Soc Trop Med Hyg 69:166.

Ambar I, Kloog Y, Kochva E, Wollberg Z, Bdolah A, Oron U, Sokolovsky M (1988) Characterization and localization of a novel neuroreceptor for the peptide sarafotoxin. Biochem Biophys Res Commun 157:1104-1110.

Ambar I, Kloog Y, Schvartz I, Hazum E, Sokolovsky, M (1989) Competitive inhibition between endothelin and sarafotoxin: Binding and phosphoinositide hydrolysis in rat atria and brain. Biochem Biophys Res Commun 158: 195-201.

Badr FK, Murray JJ, Breyer MD, Takahashi K, Inagami T, Harris RC (1989) Mesangial cell, glomerular and renal vascular response to endothelin in rat kidney - elucidation of signal transduction pathway. J Clin Invest 83: 336-342.

Bdolah A, Wollberg Z, Kochva E (1989a) Sarafotoxins: A new group of cardiotoxic peptides from the venom of *Atractaspis*. In: Snake Venoms (Harvey AL, ed) Pergamon Press (in press).

Bdolah A, Wollberg Z, Ambar I, Kloog Y, Sokolovsky M, Kochva E (1989b) Disturbances in the cardiovascular system caused by endothelin and sarafotoxins. Biochem Pharmacol (in press).

Braquet P, Touvay C, Lagente V, Vilain B, Pons F, Hosford D, Chabrier PE Mencia-Huerta JM (1989) Effect of endothelin-1 on blood pressure and bronchopulmonary system of the guinea pig. J Cardiovascular Pharmacol 13 (suppl 5) (in press).

Chabrier PE, Mencia-Huerta JM, Braquet P (1989a) Pharmacological modulation of the bronchopulmonary action of the vasoactive peptide, endothelin, administration by aerosol in the guinea pig. Biochem Biophys Res Commun 158: 625-632.

Chabrier PE, Auguet M, Roybert P, Lonchampt M O, Gillard V, Guillon JM, Delaflotte S, Braquet P (1989b) Vascular mechanism of action of endothelin-1: Effect of Ca^{2+} antagonists. J Cardiovascular Pharmacol 13 (suppl 5) (in press).

Goetz KL, Wang BC, Madwed JB, Zhu JL, Leadley Jr. RJ (1988) Cardiovascular, renal and endocrine response to intravenous endothelin in conscious dogs. Amer J Physiol 255: R1064-R1068.

Graur D, Bdolah A, Wollberg Z, Kochva E (1988/9) Homology between snake venom sarafotoxins and mammalian endothelins. Israel J Zool 35: 171-175.

Gu XH, Casley DJ, Nayler WG (1989) Sarafotoxin S6b displaces specifically bound ^{125}I-endothelin. Europ J Pharmacol 162: 509.

Han SP, Kneupfer MM, Trapani AJ, Fok KF, Westfall TC (1989) Cardiovascular effects of endothelin (ET) and sarafotoxin S6b (SRT): *In vivo* and *in vitro* studies. Fed Am Soc Exp Biol. Abst No 54.

Hiley CR (1989) Functional studies on endothelin catch up with molecular biology. Trends Pharmacol Sci 10: 47-49.

Hu JR, Von Harsdorf R, Lang RE (1988) Endothelin has potent inotropic effects in rat atria. Europ J Pharmacol 158: 275-278.

Ishikawa T, Yanagisawa M, Kimura S, Goto K, Masaki S (1988) Positive inotropic action of novel vasoconstrictor peptide endothelin on guinea pig atria. Am J Physiol 255: H970-973.

IUPAC-IUB Commission on Biochemical Nomenclature. (1968) A one-letter notation for amino acid sequences: Tentative rules. Europ J Biochem 5: 151-153.

Kimura S, Kasuya Y, Sawamura T, Shinmi O, Sugita Y, Yanagisawa, M, Goto K, Masaki T (1988) Structure-activity relationship of endothelin: Importance of the C-terminal moiety. Biochem Biophys Res Commun 156: 1182-1186.

Kloog Y, Ambar I, Sokolovsky M, Kochva E, Wollberg Z, Bdolah, A (1988) Sarafotoxin, a novel vasoconstrictor peptide: Phosphoinositide hydrolysis in rat heart and brain. Science 242: 268-270.

Kloog Y, Ambar I, Kochva E, Wollberg Z, Bdolah A, Sokolovsky M (1989) Sarafotoxin receptors mediate phosphoinositide hydrolysis in various rat brain regions. FEBS Lett 242: 387-390.

Kloog Y, Sokolovsky M (1989) Similarity in mode and sites of sarafotoxins and endothelins. Trends Pharmacol Sci 10: 212-214.

Kochva E, Viljoen CC, Botes DP (1982) A new type of toxin in the venom of snakes of the genus *Atractaspis* (Atractaspidinae). Toxicon 20: 581-592.

Kneupfer MM, Han SP, Trapani AJ, Fok KF, Westfall TC (1989) Cardiac and vascular actions of endothelin and sarafotoxins in rats. Fed Am Soc Exp Biol. Abst No 3784.

Kumagaye SI, Kurodo H, Nakagima K, Watanabe TX, Kimura T, Masaki T, Sakakibura S (1988) Synthesis and disulfide structure determination of porcine endothelin: An endothelium-derived vasoconstricting peptide. Int J Protein Peptide Res 32: 519-526.

Lee SY, Lee CY, Chen YM, Kochva E (1986) Coronary vasospasm as the primary cause of death due to the venom of the burrowing asp, *Atractaspis engaddensis*. Toxicon 24: 285-291.

Lippton H, Goff J, Hyman H (1988) Effect of endothelin in the systemic and renal vascular beds in vivo. Europ J Pharmacol 155: 197-198.

Maggi CA, Patacchini R, Giuliani S, Meli A (1989a) Potent contractile effect of endothelin in isolated guinea-pig airways. Europ J Pharmacol 160: 179-1182.

Maggi CA, Giuliani S, Patacchini R, Santicioli P, Turini D, Barbanti G, Meli A (1989b) A potent contractile activity of endothelin on the human isolated urinary bladder. Brit J Pharmacol 96: 755-758.

Marsden PA, Danthuluri NR, Brenner BM, Ballermann BJ, Brock TA (1989) Endothelin action on vascular smooth muscle involves inositol triphosphate and calcium mobilization. Biochem Biophys Res Commun 158: 86-93.

Miasiro N, Yamamoto H, Kanaide H, Nakamura M (1988) Does endothelin mobilize calcium from intracellular store sites in rat aortic vascular smooth muscle cells in primary culture? Biochem Biophys Res Commun 154: 312-317.

Payne AN, Whittle BJR (1988) Potent cyclo-oxygenase-mediated bronchoconstrictor effects of endothelin in the guinea-pig in vivo. Europ J Pharmacol 158:303-304.

Saito A, Shiba R, Kimura S, Yanagisawa M, Goto K, Masaki T (1989) Vasoconstrictor response of large cerebral arteries of cat to endothelin, an endothelium-derived vasoactive peptide. Europ J Pharmacol 162: 353-358.

Shavit G, Gitter S, Barak Y, Vidne BA, Oron Y (1986) Positive inotropic response to alpha-adrenergic stimulation in electrically-driven rat left atrium: The role of external calcium. J Cardiovas Pharmac 8: 324-331.

Takasaki C, Tamiya N, Bdolah A, Wollberg Z, Kochva E (1988) Sarafotoxins S6: Several isotoxins from *Atractaspis engaddensis* (Burrowing Asp) venom that affect the heart. Toxicon 26: 543-548.

Taylor SG, Weston AH (1988) Endothelium-derived hyperpolarizing factor: a new endogenous inhibitor from vascular endothelium. Trends Pharmacol Sci 9: 272-274.

Vanhoutte PM, Katusic ZS (1988) Endothelium-derived contracting factor: Endothelin and/or superoxide anion? Trends Pharmacol Sci 9:229-230.

Van Renterghem C, Vigne P, Barhanin J, Schmid-Allina A, Ferlin C, Lazdunsky M (1988) Molecular mechanism of action of the vasoconstrictor peptide endothelin. Biochem Biophys Res Commun 157: 977-985.

Weiser E, Wollberg Z, Kochva E, Lee SY (1984) Cardiotoxic effects of the venom of the Burrowing Asp, *Atractaspis engaddensis* (Atractaspididae, Ophidia). Toxicon 22: 764-774.

Wollberg Z, Shabo-Shina R, Intrator N, Bdolah A, Kochva E, Shavit G, Oron Y, Vidne BA, Gitter S (1988) A novel cardiotoxic polypeptide from the venom of *Atractaspis engaddensis* (Burrowing Asp): Cardiac effects in mice and isolated rat and human heart preparations. Toxicon 26: 525-534.

Wollberg Z, Bdolah A, Kochva E (1989) Vasoconstriction effects of sarafotoxins in rabbit aorta: Structure-function relationships. BBRC (in press).

Wright CE, Fozard JR (1988) Regional vasodilation is a prominent feature of the haemodynamic response to endothelin in anaesthetized, spontaneously hypersensitive rats. Europ J Pharmacol 155: 201-203.

Yanagisawa M, Kurihara H, Kimura S, Tomobe Y, Kobayashi M, Mitsui Y, Yazaki Y, Goto K, Masaki T (1988a) A novel potent vasoconstrictor peptide produced by vascular endothelial cells. Nature 332: 411-415.

Yanagisawa M, Inoue A, Ishikawa T, Kasuya Y, Kimura S, Kumagaye SI, Nagajima K, Watanabe TX, Sakakibara S, Goto K, Masaki T (1988b) Primary structure, synthesis, and biological activity of rat endothelin, an endothelium-derived vosoconstrictor peptide. Proc Natl Acad Sci USA 85: 6964-6967.

Calcium Channel Modulators in Heart and Smooth Muscle: Basic Mechanisms and Pharmacological Aspects
S. Abraham and G. Amitai, editors
© 1990, VCH, Weinheim/Deerfield Beach, FL and Balaban, Rehovot/Philadelphia

Gingival hyperplasia induced by calcium channel blockers

ABRAHAM ZLOTOGORSKI, TREVOR WANER AND ABRAHAM NYSKA

Department of Dermatology, Hadassah University Hospital, Jerusalem and Life Science Research Israel, Ness Ziona, Israel

The gingiva is that part of the oral mucous membrane that covers the alveolar processes of the jaws and surrounds the necks of the teeth. In the healthy state it is salmon pink in colour, with no exudate or plaque accumulations. Histologically, the epithelium and the connective tissue contain few inflammatory cells and can be described as a layer of keratinized stratified squamous epithelium separated from the connective tissue by a basal lamina membrane (Goldman and Taynor, 1987).

Gingival hyperplasia can occur in a variety of different diseases. The hyperplasia can occur in Vitamin C deficiency, leukemia, endocrine imbalances induced at puberty or pregnancy and in Crohn's disease. Iatrogenic drug-induced gingival hyperplasia is another important cause of this phenomenon.

Since the introduction of sodium diphenylhydantoin (Dilantin) for the treatment of epilepsy (Kimball, 1939), gingival hyperplasia has been described for a number of other drugs. These include valproic acid (Syrjanen and Syrjanen, 1979), cyclosporine A (Rateitschak-Pluss et al., 1983) and nifedipine (Ramon et al., 1984). The latter drug is a calcium channel blocker which has been in wide use since 1977 for the treatment and prophylaxis of ischemic heart disease and hypertension.

In a search for an appropriate animal model and to study the pathogenesis of gingival hyperplasia, rats (do Nascimento et al., 1985), cats (Latimer et al., 1986), ferrets (Hall and Squier, 1982), guinea pigs (Carrel et al., 1983), and only recently dogs (Waner et al., 1988) have been used.

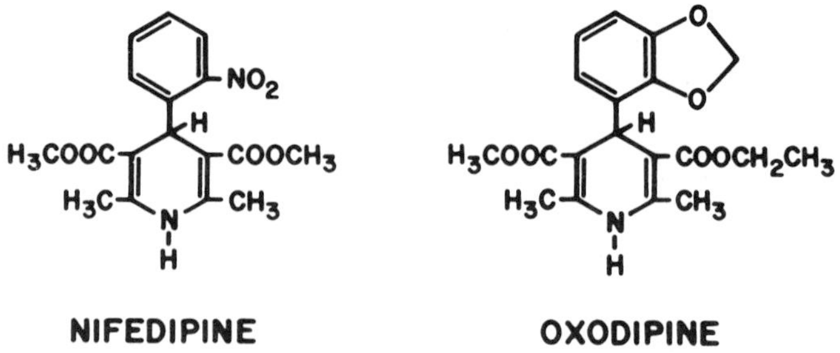

NIFEDIPINE

OXODIPINE

Fig. 1. Structure of calcium channel blockers, nifedipine and oxodipine.

We describe here our experiences with oxodipine (Galiano, 1987), a new light stable calcium channel blocking agent, in dogs and rats treated subchronically and chronically. A comparison of the nature of the hyperplastic changes is made between these species and humans treated with nifedipine.

It is interesting to note that whereas both the calcium-channel blockers and the hydantoin-derived anti-convulsants act with similar pharmacophysiological effects on a cellular level, on different therapeutic target organs, their toxic side-effects on the gingiva are similar. The result of action of the calcium channel blockers is to prevent the breakdown of ATP by calcium dependent ATPase and, in so doing decrease the consumption of high energy phosphate. In the case of hydantoin-derived anti-convulsants, there is a suppression of the sodium-potassium ATPase pump, thereby lowering the excitatory threshold of the affected neurons in the motor cortex.

Gingival hyperplasia in man, induced by nifedipine and hydantoin were found to be similar in terms of histopathologic morphological studies (Lucas et al., 1985). The clinical picture of the gingival hyperplasia mediated by nifedipine appears mainly in the labial gingiva of the lower anterior teeth, around the maxillary molars and/or interdental gingiva. The edentulous gums are unaffected (Ramon et al., 1984; Zlotogorski et al., 1989). The hyperplasia occurs after one to nine months at dosages of 30 to 100 mg. No correlation has been found between drug dosage and the degree of the hyperplasia (Bencini et al., 1986), while a more severe hyperplasia with higher dosages was observed by Barak et al., 1987.

Cessation of nifedipine therapy leads to clinical improvement within days to months (Shaftic et al., 1986; Ramon et al., 1984). Rigorous oral hygiene has been suggested (van der Wall et al., 1985; Bencini et al., 1985) to retard the progress and the severity of the hyperplasia. Although the idea seems reasonable, only a minor improvement has so far been observed. Furthermore, nifedipine induced gingival hyperplasia has been reported in patients with good oral hygiene. Cessation of the drug causes a regression of the hyperplastic changes, even in patients with poor oral hygiene. Gingivectomy is sometimes required and in severe cases dental extraction and reconstitution with a prosthesis is warranted (Zlotogorski et al., 1989).

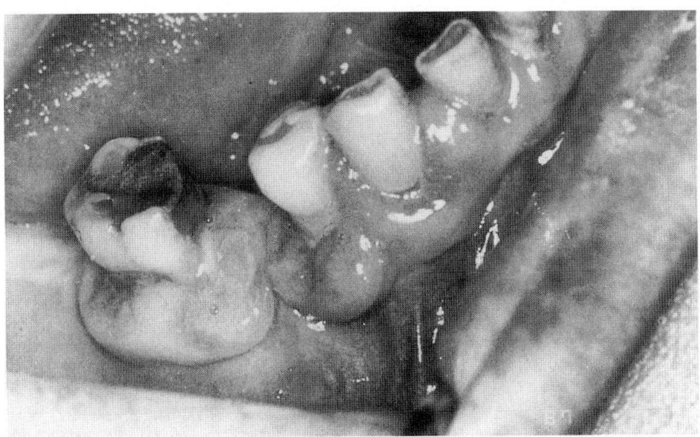

Fig. 2. A 64-year-old male with severe gingival hyperplasia of the lower molar and premolar teeth, following treatment with nifedipine. From Zlotogorski et al., *Isr. J. Med. Sci.* 1989, by permission.

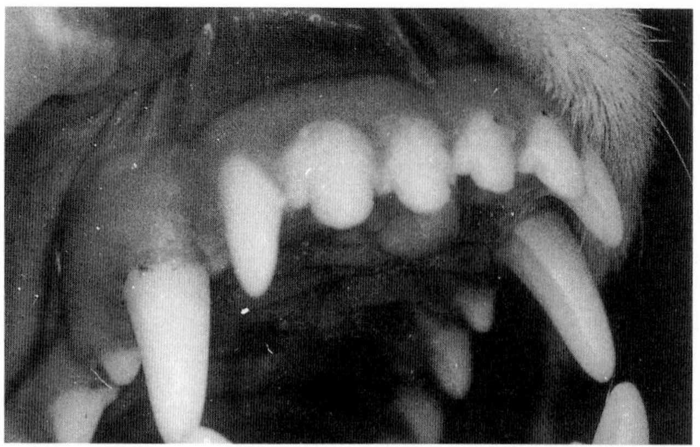

Fig. 3. Gingiva of a control dog.

Gingival hyperplasia in the rat induced by calcium channel blockers has not been documented in the literature. As our preliminary findings show, it is remarkable the fact that no induction of inflammation is necessary to achieve the gingival hyperplastic changes (Nyska, unpublished data).

The role of the fibroblasts in the genesis of gingival hyperplasia has been emphasized by a number of authors. In the gingival hyperplastic changes induced by both nifedipine and hydantoin drugs, an increased amount of extracellular ground substance, as well as an increase in the number of fibroblasts, containing strongly sulfated acid mucopolysaccharides, has been demonstrated (Hassel et al., 1976). An important factor in the abnormal proliferation of fibroblasts appears to be the selection by the drug of a population of fibroblasts which synthesize an inactive collagenase. This then results in a disturbance in the balance between the production and synthesis of collagen (Hassel et al., 1976). Furthermore, it has been postulated that the activity of the inactive form of collagenase may be modulated by calcium flux, and that the alterations in the movement of this ion may be responsible for the hyperplastic process.

Recently, gingival hyperplasia with a similar clinicopathological picture has been reported, in humans, using another calcium channel blocker - diltiazem (Colvard et al., 1986). Oxodipine has not yet been demonstrated in clinical trials to induce gingival hyperplasia in man (Galiano, personal communication).

A picture similar to that of nifedipine and diltiazem, was observed in dogs treated with oxodipine for a period of three months (Waner et al., 1988). Dogs developed signs of gingival hyperplasia after seven weeks of treatment. The gingivae were observed to swell in the region of the incisor and canine teeth. By the ninth week of treatment all the teeth were involved; the necks were obscured by the gingiva and only the tips could be seen (Waner et al., 1988). After a recovery period of six weeks, during which time no oxodipine was administered, a decline in the mass of gingiva was detected. It is considered that with a longer recovery period complete reversion to normal could be achieved (Laor et al., 1988). Salient features of the pathological changes in dogs were: Acanthosis and parakeratosis of the gingival epithelium, proliferation and elongation of the rete pegs, fibroblastic and capillary proliferation and aggregation of mononuclear inflammatory cells. This histopathological picture is similar to that of nifedipine (reviewed by Zlotogorski et al., 1989).

Preliminary studies in rats, treated with oxodipine, have demonstrated similar histopathological changes to those seen in man and dogs. The most prominent feature of the gingival hyperplastic change was the lack of inflammatory infiltration seen in the corium (Nyska, unpublished data).

The role of chronic gingival inflammation to gingival overgrowth has been observed in rats, ferrets, monkeys and cats. This factor has also been incriminated in the development of gingival hyperplasia in man. Studies in rats have concentrated on this mechanism, in the attempt to induce gingival inflammation by tying wire or silk about the base of the teeth, and then subsequently dosing the rats with phenytoin-like drugs. In general the rat has been found to be refractory to gingival hyperplastic changes (Swenson, 1954; Conrad et al., 1975). Isolated cases of successful induction of gingival hyperplasia in rats using phenytoin have been described (Yamada et al., 1977).

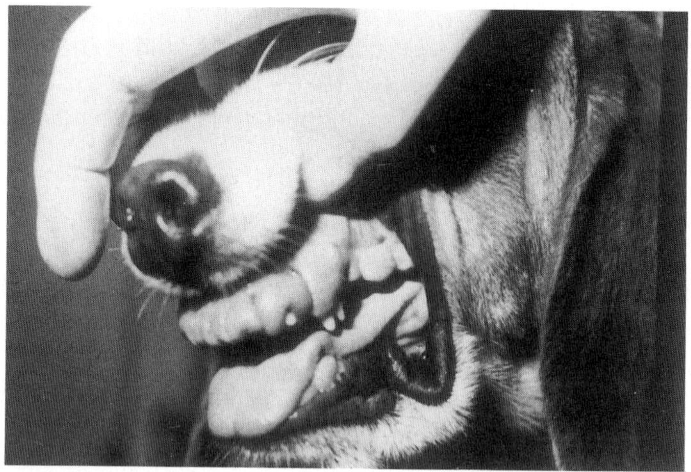

Fig. 4. Maxilla of an oxodipine treated dog. Note the marked generalized gingival hyperplasia. From Waner et al., *Toxicol. Pathol.* 1988, by permission.

Tissue culture studies using phenytoin sodium have also pointed to the role of the fibroblasts in the generation of gingival hyperplasia. Here, an increase in mitotic activity as measured by autoradiography was demonstrated (Al-Ubaidy et al., 1981).

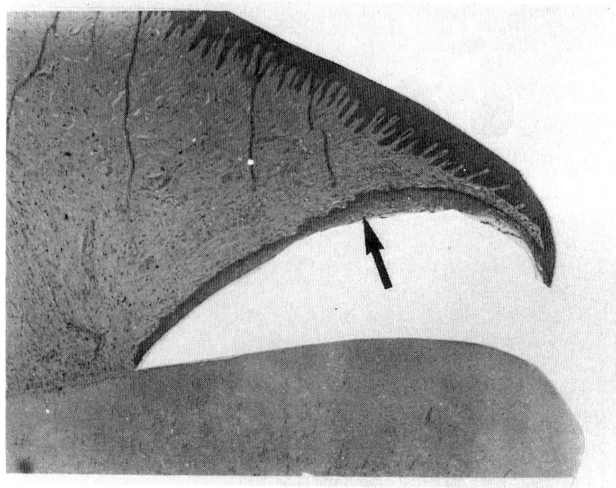

Fig. 5. Gingiva of a control dog. Note (arrow) the absence of rete pegs in the epithelium adjacent to the tooth (H&E x50). From Waner et al., *Toxicol. Pathol.* 1988, by permission.

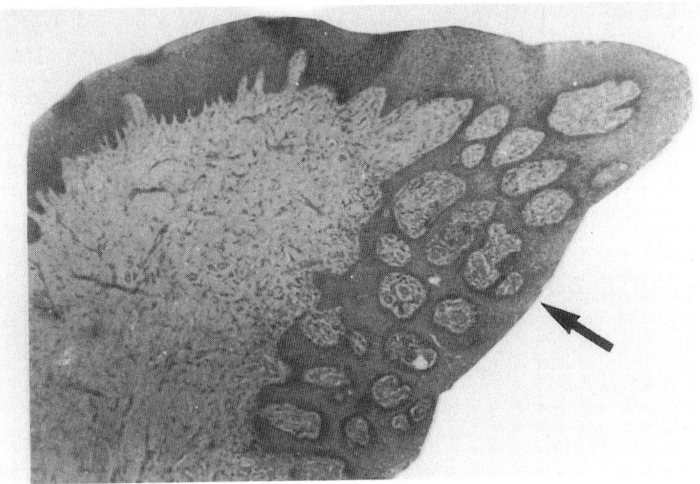

Fig. 6. Gingiva of oxodipine treated dog. The stratified squamous epithelium adjacent to the tooth (arrow) is markedly hyperplastic with long anastomosing rete pegs (H&E x50). From Waner et al., *Toxicol. Pathol.* 1988, by permission.

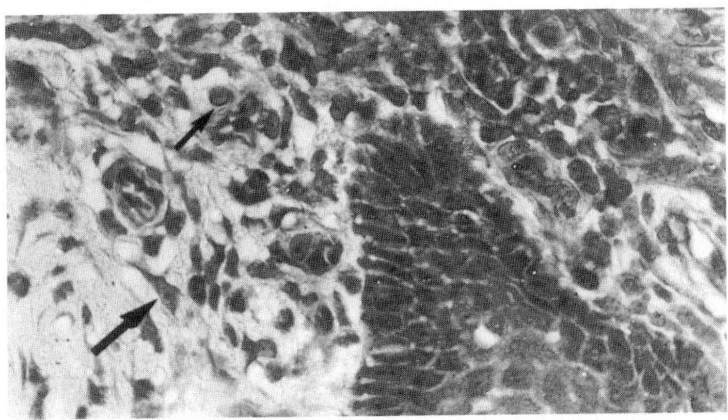

Fig. 7. Gingiva of an oxodipine treated dog. The lamina propria lying adjacent to the hyperplastic rete pegs shows fibroblastic (large arrow) and capillary proliferations, with mononuclear cell infiltrations (small arrow) (H&E x500).

In conclusion, both our studies in dogs and rats appear to point to the primary induction of fibroblastic elements as the initiation of hyperplastic changes in the gingiva. It is interesting to note that the gingival enlargement observed in the cyclosporine treated patients was not due to increased tissue collagen but was probably a result of the increased epithelium combined with the accumulation of non-collagenous extracellular matrix material (Pisanty et al., 1988).

Acknowledgement

The authors thank Dr. Galiano for his cooperation and support.

References

Al-Ubaidy SS, Al-Janabi NY, Al-Tai SA (1981). Effect of Phenytoin on mitotic activity of gingival tissue and cultured fibroblasts. J Periodontol 52:747-749.

Barak S, Engelberg IS, Hiss J (1987). Gingival hyperplasia caused by nifedipine. Histopathologic findings. J periodontol 58:639-642.

Bencini PL, Crosti C, Sala F, Ambroso GC, Marini D (1986). Gingival hyperplasia by nifedipine. G Ital Dermatol Venereol 121:29-31.

Carrel R, Ba TT, Chapman MK (1983). Gingival aberrations in Dilantin-treated guinea pigs. J Pedod 7:229-240.

Colvard MD, Bishop J, Weissman D, Gargiulo AV (1986). Cardizem induced gingival hyperplasia: a report of two cases. Periodont Case Rep 8:67-68.

Conrad GJ, Osborn JC, Pekary AD, Scholle RH (1975). Relationship of drug metabolism and inflammation to the gingival response of rats treated with diphenylhydantoin. J Dent Res 54:B68-74.

do Nascimento A, Barreto R de C, Bozzo L, de Almeida OP (1985). Interaction of phenytoin and inflammation induces gingival overgrowth in rats. J Periodont Res 20:386-391.

Galiano A (1987). Oxodipine. Drugs of the future 12:633-635.

Goldman RS, Taynor LZ (1987). Odontologic and Periodontal diseases. In NL Novick (ed.) "Clinics in Dermatology" Vol. 5, Philadelphia, J.B. Lippincott, pp 59-65.

Hall BK, Squier CA (1982). Ultrastructural quantitation of connective tissue changes in phenytoin-induced gingival overgrowth in the ferret. J Dent Res 61:942-952.

Hassell TM, Roebuck S, Page RC, Wray SH (1982). Quantitative histopathologic assessment of developing phenytoin-induced gingival overgrowth in cat. J Clin Periodontol 9:365-372.

Kimball OP (1939). The treatment of epilepsy with sodium-diphenylhydantoinate. J Am Med Assoc 112:1244-1245.

Laor O, Waner T. Galiano A, Pirak M, Nyska A (1988). Recovery from gingival hyperplasia induced by oxodipine, a calcium channel blocker. 29th Congress of the European Society of Toxicology, Munich.

Latimer KS, Rakich PM, Purswell BJ, Kircher IM (1986). Effects of Cyclosporin-A administration in cats. Vet Immunol Immunopathol 11:161-173.

Lucas RM, Howell LP, Wall BA (1985). Nifedipine-induced gingival hyperplasia. A histochemical and ultrastructural study. J Periodontol 56:211-215.

Pisanty S, Shoshan S, Chajek T, Maftsir G, Sacks B, BenEzra D (1988). The effect of cyclosporin A (CsA) treatment on gingival tissue of patients with Behcet's disease. J Periodontol 59:599-603.

Ramon Y, Behar S, Kishon Y, Engelberg IS (1984). Gingival hyperplasia caused by nifedipine- a preliminary report. Int J Cardiol 5:195-206.

Rateitschak-Pluss EM, Hefti A, Lortscher R, Theil G (1983). Initial observation that cyclosporin-A induces gingival enlargement in man. J Clin Periodontol 10:237-246.

Shaftic AA, Widdup LL, Abate MA, Jacknowitz AI (1986). Nifedipine-induced gingival hyperplasia. Drug Intell Clin Pharm 20:602-605.

Swenson HM (1954). A failure to produce enlargement of gingiva in rats by using dilantin sodium. J Dent Res 33:468.

Syrjanen SM, Syrjanen KJ (1979). hyperplastic gingivitis in a child receiving sodium valproate treatment. Proc Finn Dent Soc 75:95-98.

van der Wall EE, Tuinzing DB, Hes J (1985). Gingival hyperplasia induced by nifedipine, an arterial vasodilating drug. Oral Surg Oral Med Oral Pathol 60:38-40.

Waner T, Nyska A, Nyska M, Sela M, Pirak M, Galiano A (1988). Gingival hyperplasia in dogs induced by oxodipine, a calcium channel blocking agent. Toxicol Pathol 16:327-332.

Yamada S, Sato T, Miake K (1977). Experimental pathological studies on dilantin gingival hyperplasia in rat. (I) Macroscopical and light-microscopical observations. Bull Tokyo Dent Coll 18:181-193.

Zlotogorski A, Nyska M, Sela M, Nyska A, Gotsman M (1989). Nifedipine-induced gingival hyperplasia: Case reports and literature review. Isr J Med Sci. in press

Model systems for myocardial ischemia and calcium channels

Calcium Channel Modulators in Heart and Smooth Muscle: Basic Mechanisms and Pharmacological Aspects
S. Abraham and G. Amitai, editors
© 1990, VCH, Weinheim/Deerfield Beach, FL and Balaban, Rehovot/Philadelphia

Modulation of cardiac calcium channels expressed in Xenopus oocytes

ELI GERSHON AND NATHAN DASCAL

Department of Physiology and Pharmacology, Sackler School of Medicine, Tel Aviv University, Ramat Aviv 69978, Israel

ABSTRACT

This paper summarizes our attempts to establish the Xenopus oocyte as a model system for the study of modulation of voltage-dependent Ca channels (VDCC). Voltage-dependent calcium channels currents have been expressed in Xenopus oocytes following injection of heart RNA from 7 day-old rats. The currents were resolved with Ba^{2+} as charge carrier using the two electrode voltage clamp. The current consisted of two components: a slow (L-type), dihydropyridine-sensitive current, and a fast transient dihydropyridine-insensitive one. The fast current was much more sensitive to inhibition by Ni^{2+} than the slow one. The slow current was selectively inhibited by benzodiazepines with micromolar affinity, with potency order of Ro5-4864 > diazepam > clonazepam. Beta adrenergic agonists enhanced the slow current; this effect was abolished by propranolol. Cyclic AMP and forskolin also potentiated the slow current. The alpha adrenergic ligands methoxamine and norepinephrine, in the presence of propranolol, reduced the slow current; this effect required intracellular Ca^{2+} and was abolished by intracellularly injected EGTA. Activation of protein kinase C by beta phorbol esters produced a biphasic effect: enhancement of the slow current for several minutes, followed by a decrease below the basal level. The results of our work establish the oocytes as a suitable and supportive model system for studying heart calcium channels, in particular their various modulations by neurotransmitters and enzyme activators known to act in the original tissue. The modulations studied so far permit further studies which are difficult or complex to perform in the original tissue and help to better specify the interaction of various modulators on the calcium channel currents.

INTRODUCTION

The use of Xenopus oocytes in the study of ion channels and neurotransmitter receptors has increased tremendously in recent years (Dascal, 1987 ; Lester, 1988). Xenopus oocytes are injected with RNA (or DNA) from the tissue of interest resulting in the "transplantation" of the various proteins encoded by the injected nucleic acids. In most cases expression of the various proteins is achieved within one to several days and is highly reproducible (Soreq, 1985). In regard to ion channels various methods can be applied to study both their biophysical properties and the pharmacological and interaction with various modulators. Since the oocyte is a huge cell, it is amenable to modern electrophysiological techniques such as voltage clamp and patch clamp. Furthermore, one can exploit the endogenous signal transduction pathways including their enzyme cascade which are related to the cell's own needs, to study the biochemical control of the expressed channel. The latter includes, in particular, phosphorylation and dephosphorylation of the channel protein by the various enzyme systems (cAMP dependent protein kinase A, kinase C, etc...).

Heart RNA was shown to cause the appearance of calcium channel currents (Dascal et al., 1986). The currents were resolved using Ba ions and were separated into two components: a slow current which displayed high sensitivity towards the dihydropyridine compounds and a fast current unaffected by dihydropyridines. In most cases the slow current was the dominant one; both currents activated at potentials > -30mV. The similarity of the slow current to the L-type current of the heart was verified and demonstrated by both voltage-dependency of activation, single-channel conductance, and modulation of the current by dihydropyridines and cAMP-elevating agents (Dascal et al., 1986; Moorman et al., 1987; Lotan et al., 1989; Gershon & Dascal, 1989). Thus it was established that the expressed channel may be subject to various modulations that are known in the heart and are yet to be better clarified.

The present study aims towards developing the oocyte as a convenient model system for studying in particular various modulatory effect of enzyme activation and neurotransmitter application on the expressed calcium ion channels.

METHODS

The oocytes are usually defolliculated by collagenase treatment, injected with total or poly(A) RNA, incubated in sterile physiological solution for 2 to 5 days, and the currents tested using the two-electrode voltage clamp technique (for details, see Dascal et al., 1986, Boton et al., 1989).

RESULTS

1) Characterization of the calcium-channel evoked currents: electrophysiological and pharmacological studies.

Injection of RNA into Xenopus oocytes results in the expression of calcium channel currents within 48-72 hours (Dascal et al., 1986). The currents were evoked using the two electrode voltage clamp technique in a high Ba^{2+}, Ca-free solution (Gershon & Dascal, 1989). Figure 1 depicts the Ba^{2+} current evoked by a depolarization step, in a protocol that we usually employed in most of our pharmacological studies. The inset at the top shows the protocol used to elicit the current. The membrane potential was first stepped from the holding level for 5 sec to -100 mV in order to remove the inactivation of the fast current, and then to -50 mV for 2.5 sec. This step did not elicit any time-dependent currents. The cell was then kept again at the holding potential for 10 sec. This was followed by a 5 sec step to -100 mV and then a 2.5 sec step to 0 mV. The last step evoked the Ba^{2+} current through the Ca^{2+} channel(s). The leak current evoked by the step from -100 to -50 mV was multiplied by two and digitally subtracted from the current obtained during the step from -100 to 0 mV, to give the net Ba^{2+} current (the total Ba^{2+} current).

A slow time-dependent outward current developed in the oocytes at potentials higher than -10 mV (Lotan et al., 1989). This current was not suppressed by Ca^{2+} channel blockers (not shown). At 0 mV, it reached 3 to 10 nA 2.5 seconds after the beginning of the test pulse and could therefore lead to underestimation of the Ba^{2+} current amplitude at this time point. Since _Xenopus_ oocytes were shown to have endogenous calcium-dependent chloride current $I_{Cl(Ca)}$ (Barish 1983), the oocytes were usually injected with EGTA (typically 200-500 pmole) several hours before the experiments. Though the current through Ca^{2+} channels was resolved with Ba ions, large barium currents could activate $I_{Cl(Ca)}$ by an unknown mechanism, resulting in a contamination of the Ba^{2+} current (not shown). The injection of EGTA resulted in a further decrease of the leak current and elimination of $I_{Cl(Ca)}$. As shown in Figure 1, the total Ba^{2+} current evoked from -100 mV to 0 mV can be separated into a slow component and a fast one. To study the effect of various ligands on the slow, L-type, channel separately from their effect on the fast current, a simple separation procedure was used. The protocol used is based on the deferences in inactivation properties of the fast and slow Ba^{2+} current (Lotan et al, 1989 ; Gershon & Dascal, 1989).

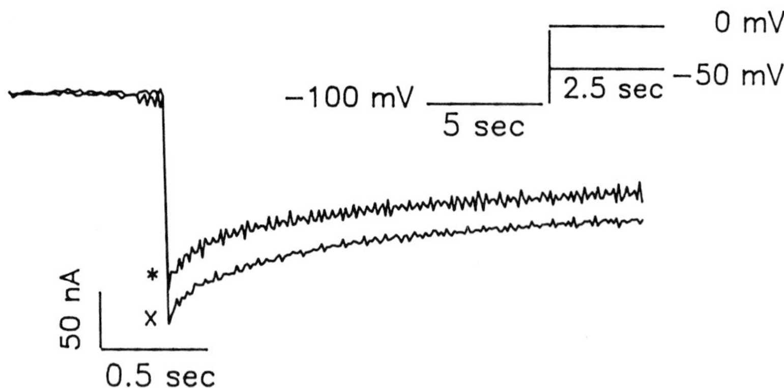

Fig 1: Effect of 25 uM phenylephrine on the total Ba^{2+} current. The inset at the top depicts the protocol used to evoke the current (for details see text). Control current is x labeled and the current in the presence of phenylephrine is * labeled. Solution contains (in mM) 40 $Ba(OH)_2$, 2 KOH, 60 NaOH and 5 Hepes. The pH of the solution was set to 7.5 by titrating with methanesulfonic acid. Phenylephrine was applied in the presence of 50 uM propranolol (beta blocker).

Figure 2 describes the protocol used to eliminate the fast component of the total current. Total Ba^{2+} current was obtained by depolarization from -100 mV (a 5 second long prepulse) to 0 mV for 2.5 seconds. After a 10 sec interval, a 5 sec step to -20 mV was applied (to inactivate the fast current) followed by a 2.5 sec step to 0 mV. The latter step evoked mainly the slow Ba^{2+} current through the L-type Ca^{2+} channels. Note that at -20 mV the slow Ba^{2+} was marginally inactivated, and subtraction of the current evoked by the step from -20 to 0 mV (the slow current) from the total current revealed mainly the fast current (evoked by stepping the voltage from -100 mV to 0 mV), but also a small portion of the slow Ba^{2+} current.

The slow current displays the pharmacological and electrophysiological characteristics of the cardiac, L-type, channel as was found for the classical Ca-channel antagonists (e.g. DHP, phenylalkylbenzens and benzothiazepines; Reuter, 1984). Thus application of 10 uM nifedipine resulted in an almost complete abolishment of the slow current (Dascal et al., 1986). The current was also inhibited by verapamil and diltiazem (not shown). The Ca-channel activator Bay K 8644 (0.05 uM) enhanced the current by about 3 fold (Moorman et al 1987 ; Gershon & Dascal, 1989).

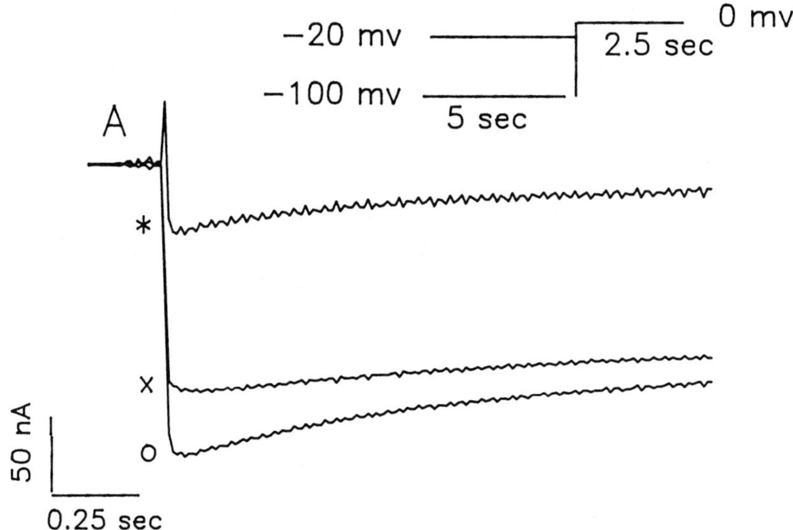

Fig 2: Electrophysiological separation of the fast and slow components of the total current. The inset at the top depicts the protocol used to evoke the currents. Traces shown are total (fast +slow) current (o labeled) and the slow current (x labeled). The fast current (* labeled) was revealed by subtracting the slow current from the total current (for details see text). The oocyte was injected with 200 pmole EGTA 24 hours prior to the experiment. Solution composition as in Figure 1.

The fast current was found to activate at about -30 mV with a peak at about 15 mV (Dascal et al., 1986; Umbach & Gundersen, 1987). The steady state inactivation shows 50% inactivation at about -30 mV. This current is DHP-insensitive and can be blocked by polyvalent cations such as Cd^{2+}, Co^{2+}, Gd^{3+} (Gershon and Dascal ,unpublished results). In most injected oocytes this current is completely abolished by 100 uM Cd^{2+}. Unfortunately, the sensitivity of the fast current to organic calcium channel blockers is poorly characterized (this also applies to the fast transient current found in the heart (Nowicky et al, 1986 ; Bean 1985 ; Nilius et al, 1985). Therefore we can not fully attribute this current to originate only from the heart and in part it may contain the endogenous fast Ca-channel current of the oocytes (Umbach & Gundersen, 1987).

2) Modulation of the slow, L-type, Ca-channel currents by adrenergic agonists

To further resolve the nature of the slow current, specifically its modulation by second messengers and enzyme activators, several experiments were done using various techniques (see also section 4). Intracellular injection of cAMP (5-20 pmole) into voltage clamped cells resulted in a considerable enhancement of the current. Similar effect was found by extracellular application of forskolin (40 uM). The use of cAMP elevating agents like beta-adrenergic agonists caused the same effect. Thus application of adrenalin (10 uM) or isoproterenol (10 uM) caused a significant increase of the Ba^{2+} current. The beta-adrenergic effect was completely abolished by previously applied propranolol (10 uM) (Dascal et al., 1986; Gershon et al., 1988). In addition it was shown that once the current is enhanced by beta-agonist, application of acetylcholine (ACh) decreased the current in a few minutes. While the last observation is widely recognized in the heart (Reuter, 1983), some debate exists concerning whether or not ACh modulates the calcium channel current without previous enhancement by beta-adrenergic agonists. In our experiments we found that in most oocytes ACh decreased the basal Ba^{2+} current, too. The latter effect was not observed when the oocytes were previously injected with 200 pmole EGTA (24 hr. before the experiment). However, it is still unclear whether this effect occurs in the heart or is peculiar to the oocyte.

To study the effect of alpha adrenergic ligands, the currents were evoked by voltage steps as seen in Figure 1. The alpha agonists decreased the slow current without an appreciable effect on the fast one (Gershon et al, 1988). Application of phenylephrine (alpha and beta agonist) together with 10 uM propranolol (beta blocker) resulted in a 20% decrease of the peak amplitude of the Ba^{2+} current. This effect was completely abolished by phentolamine (alpha blocker). When phenylephrine was applied without propranolol the current was enhanced, as expected from its beta agonist activity. Similar results were obtained with epinephrine and methoxamine. The decrease of the current was not observed when the oocyte were previously injected with EGTA (200 pmole). It should be noted that the alpha adrenergic action was not observed in all the oocytes injected with different sources of heart RNA, suggesting variable expression of the alpha adrenergic receptors.

3) Effect of benzodiazepines on the calcium channel currents

Benzodiazepines (BZ) were found to exert negative inotropic effect, decrease action potential duration and inhibit nitrendipine binding in the heart (Mestre et al, 1985; Holck et al, 1985). These effects were suspected to result from an interaction with calcium channel currents either directly or via a BZ-peripheral binding site in the heart (see Mestre et al,1985; Holck et al, 1985; Rampe et al, 1986). In our study (Gershon & Dascal, 1989) we have shown that the BZ ligands including the central and peripheral high affinity ligands decrease the L-type current in the micromolar range with potency order Ro5-4864 > diazepam > clonazepam. The inhibitory effect could not be detected at concentration less than 5 uM. The effect of diazepam on the slow and fast components of the total Ba^{2+} current (not shown) revealed that the drug mainly inhibit the slow current with a marginal effect on the fast one. The central antagonist Ro15-1788 did not interfere with the inhibitory effect of the drug tested, suggesting their effect could not be exerted through a BZ receptor of the central type (Tallman et al, 1980).

4) Effect of protein-kinase C activators on the Ca-channel currents

The well known protein kinase C activators phorbol esters have been shown to modulate Ca-channel currents in several tissues including the heart (Lacerda et al, 1988; Doerner et al, 1988). Recently it was shown that the potent phorbol ester TPA caused an increase of the calcium channel current (within seconds) followed by a decrease of the current in heart cells (Lacerda et al, 1988). Further it was shown that the diacylglycerol derivative DOG only enhances the current. In order to further study the modulatory pathways of the calcium channel currents expressed in the injected oocytes, we applied phorbol esters ligands to the oocytes (Gershon and Dascal, manuscript in preparation). Figure 3 depicts the effect of 0.05 uM PDBu on the

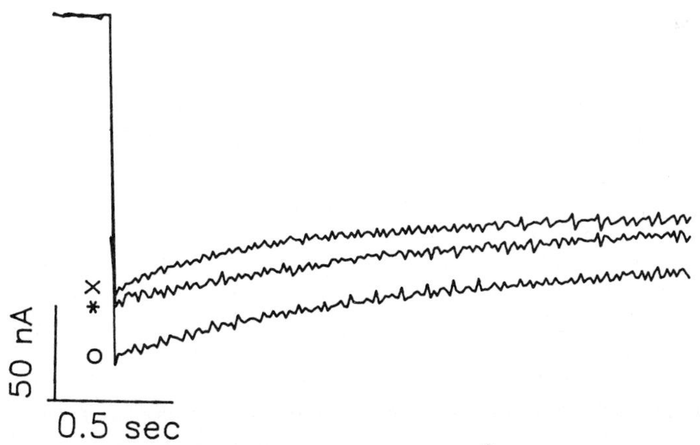

Fig 3: Effect of 0.05 uM PDBu on the total Ba^{2+} current evoked as in Figure 1. Control current is x labeled and the currents in the presence of PDBu are o labeled (after 12 minutes) and * labeled (after 25 minutes). Solution composition as in Figure 1. The oocyte was injected with 200 pmole EGTA 24 hours prior to the experiment.

evoked current as explained in Figure 1. The current was enhanced within a few minutes and later decreased to the basal level within 25-30 min. A biphasic effect was detected with PMA (0.05 uM) with a faster time course; the decrease after 20 min. was below basal level (not shown).

DISCUSSION

The use of Xenopus oocytes to study heart calcium-channels has several important advantages. Since the oocyte is a large cell, it is convenient to apply modern electrophysiological techniques such as voltage clamp and patch clamp techniques. Secondly, the oocyte can be used to study modulatory pathways of the expressed channels by either affecting endogenous modulatory pathways of the oocyte (which are common to most excitable cells) or by intracellularly introducing molecular components such as second messengers and enzymes. Thus, activation of protein kinases by cAMP or by phorbol esters can be easily handled. Furthermore once the transplanted channel is fully characterized, one can apply molecular biology methods to study the role of the subunits composing the channel, as indeed was performed in the case of heart calcium channels (Lotan et al, 1989). In particular it is possible to study a desired clone or a modified sequence (by site-directed mutagenesis) of the channel under investigation in the oocyte model system by applying the various methods used here.

The results presented here shade no doubt about the identity of the slow component of the Ba current. This current was firmly established as the slow, L-type, Ca-channel current by studying its electrophysiological properties and interaction with classical calcium channel ligands. However, in regard to the inactivation of the current, one major difference exist between the expressed current and the native L-current in the heart, since the expressed current does not inactivate (up to -10 mV, Gershon and Dascal, 1989), whereas the native current has a distinct inactivation pattern (e.g. Sanguinetti et al., 1986). It is possible that the channel subunits are partially expressed or even differently assembled in the oocyte, without losing the ability to interact with classical calcium ligands.

The fast component of the total Ba^{2+} current is yet to be further characterized as it partially resembles the endogenous current of the oocyte. Therefore we can not strongly conclude as to the effect of the various ligands on the fast current.

The effect of the BZ ligands on the slow current further demonstrate the advantage of the oocyte model, since these drugs probably interact with multiple currents in high concentrations in the original tissue (Holck et al, 1985) which makes them difficult to resolve in the heart. In the oocyte these drugs were shown to uniquely interact with the L-type current (probably with the open and resting states of the channel) with a marginal interaction with the fast current. These results may support the existence of a low affinity benzodiazepine receptor in the heart (Taft & Delorenzo, 1984 ; Johansen et al, 1985)

REFERENCES

Barish ME (1983) A transient calcium-dependent chloride current in the immature _Xenopus_ oocytes. J Physiol 342:309-325

Bean BP (1985) Two kinds of calcium channels in canine atrial cells: Differences in kinetics, selectivity, and pharmacology. J Gen Physiol 86:1-30

Boton R, Dascal N, Gillo B, Lass Y (1989) Two calcium-activated conductances in Xenopus laevis oocytes permeabilized with the ionophore A23187. J Physiol 408:511-534

Dascal N (1987) The use of _Xenopus_ oocytes for the study of ion channels. CRC Crit Rev Biochem 22:317-387

Dascal N, Landau EM, Lass Y (1984) _Xenopus_ oocyte resting potential, muscarinic responses and the role of calcium and cyclic GMP. J Physiol 352:551-574

Dascal N, Snutch TP, Lubbert H, Davidson N, Lester HA (1986) Expression and modulation of voltage gated calcium channels after RNA injection in _Xenopus_ oocytes. Science 231:1147-115

Gershon E, Lass Y, Dascal N (1988) Ca currents in Xenopus oocytes injected with RNA from rat heart: Regulation By alpha-adrenergic agonists. Biophys J 53:538a

Gershon E, Dascal N (1989) Effect of benzodiazepine agonists on Ca^{2+} channel currents in Xenopus oocytes injected with rat heart RNA. Submitted to Pflugers Arch

Holck M, Osterrieder W (1985) The peripheral high affinity benzodiazepine binding site is not coupled to the cardiac Ca^{2+} channel Eur J Pharmacol 118:293-301

Johansen J, Taft WC, Yang J, Kleinhaus AL, Delorenzo RJ (1985) Inhibition of Ca^{2+} conductance in identified leech neurons by benzodiazepines. Proc Natl Acad Sci USA 82:3935-3939

Lester HA (1988) Heterologous expression of excitability proteins: Route to more specific drugs? Science 241:1057-1063

Lotan I, Goelet P, Gigi A, Dascal N (1989) Specific block of calcium channel expressed by a fragment of dihydropyridine receptor cDNA. Science 243:666-669

Mestre M, Belin C, Carriot T, Doble A, Dubroeucq MC, Gueremy C, Le Fur G, Renault C, Uzan A (1985) Electrophysiological and pharmacological evidence that peripheral type benzodiazepine receptors are coupled to calcium channels in the heart. Life Sci 36:391-400.

Moorman JR, Zhou Z, Kirsch GE, Lacerda AE, Cafferey JM, Lam DMK, Joho RH, Brown AM (1987) Expression of single calcium channels in Xenopus oocytes after injection of mRNA from rat heart. Am J Physiol 253:H985-H991

Nilius B, Hess P, Lansman JP, Tsien RW (1985) A novel type of cardiac calcium channel in ventricular cells. Nature 316:443-446

Nowicky MC, Fox AP, Tsien RW (1985) Long opening mode of gating of neuronal calcium channels and its promotion by the dihydropyridine calcium agonist Bay K 8644. Proc Natl Acad Sci 82:2178-2182

Rampe D, Triggle DJ (1986) Benzodiazepines and calcium channel function. Trends Pharmacol Sci 7:461-463

Reuter H (1983) Calcium channel modulation by neurotransmitters, enzymes and drugs. Nature 301:569-574

Reuter H (1984) Ion channels in cardiac cell membranes. Ann Rev Physiol 46:473-484

Sanguinetti MC, Krafte DS, Kass RS (1986) Voltage dependent modulation of Ca^{2+} channel current in heart cells by Bay K8644. J Gen Physiol 88:369-392

Soreq H (1985) The biosynthesis of biologically active proteins in mRNA microinjected _Xenopus_ oocytes. CRC Crit Rev Biochem 18:199-238

Taft WC, Delorenzo RJ (1984) Micromolar affinity benzodiazepine receptors regulate voltage-dependent calcium channels in nerve terminal preparations. Proc Natl Acad Sci USA 81:3118-3122

Tallman JF, Paul SM, Skolnick P, Gallager DW (1980) Receptors for the age of anxiety : Pharmacology of the benzodiazepines. Science 207:274-281

Umbach JA, Gundersen CB (1987) Expression of an w-conotoxin sensitive calcium channel in _Xenopus_ oocytes injected with mRNA from _Torpedo_ electric lobe. Proc Natl Acad Sci 84:5464-5468

Calcium Channel Modulators in Heart and Smooth Muscle: Basic Mechanisms and Pharmacological Aspects
S. Abraham and G. Amitai, editors
© 1990, VCH, Weinheim/Deerfield Beach, FL and Balaban, Rehovot/Philadelphia

BAY K 8644-induced myocardial ischemia in the rat: An animal model for the evaluation of antianginal drugs

S. ABRAHAM, G. AMITAI, N. OZ, AND B.A. WEISSMAN
Israel Institute for Biological Research, P.O. Box 19, Ness-Ziona 70450, Israel

INTRODUCTION

Myocardial ischemia reflects a condition of inadequate blood supply to the myocardium. This condition occurs when the coronary blood supply does not meet the myocardial requirements, most often due to diminished perfusion or increased demands. The extent of diminished perfusion varies from transient spasms of the coronary arteries to a total occlusion. The recognition of coronary vasospasm as a clinical entity can be traced as early as 1867 when Brunton expressed the first hypothesis to link between coronary vasospasm and angina (Brunton, 1867). Almost a century afterwards, Prinzmetal et al,(1959; 1960) have described several cases of ST segment elevation and rest angina. The syndrome have been termed variant angina and was ascribed to an increased coronary artery tone at the site of atherosclerotic stenosis. Subsequent studies have established that transient reversible coronary artery occlusion results in myocardial ischemia and angina and are associated with alterations of the ST segment amplitude (Lambert et al, 1986).

Three major approaches have been undertaken in the past three decades for the induction of myocardial ischemia : a. Induction of atherosclerosis by high cholesterol diet; b. Mechanical obstruction of the coronary blood flow (mostly by ligation of the coronary artery); c. Chemical, drug-induced, obstruction of the coronary blood flow. Close examination of the literature reveal that in contrast to a large number of animal preparations available for the induction of myocardial ischemia by way of mechanical obstruction (see Botting et al, 1985; Curtis et al, 1987), little attention have been paid to the chemical approach. Moreover, in spite of the clinical importance of coronary vasospasm, there is a shortage of animal preparations for this purpose. The chemical approach to obstruct coronary blood flow is based on chemicals known to cause vasoconstriction (Ginsburg et al 1983), thereby mimicking the clinical situation of coronary vasospasm. Among these, two chemicals merit special attention. The first compound is ergonovine, an ergot alkaloid derivative which is utilized clinically during coronary arteriography as a diagnostic test to induce coronary artery vasospasm and identify patients with variant angina (Lambert et al, 1986). The second compound is methacholine, a muscarinic agonist that have been used by Sakai et al (1981) to produce myocardial ischemia in the rats. However, in agreement with Vergona et al (1987) the use of this preparation should be taken with caution as methacholine, being a muscarinic agonist, may identify potent antimuscarinic compounds without any clinical relevance as♥ antianginal drugs. Thus, due to the shortage of animal preparations for the screening of antianginal drugs, Bay K 8644 is proposed as an additional compound to provoke myocardial ischemia through direct action on the contractile machinary.

Bay K 8644 is a calcium channel agonist, with unique pharmacological properties which render it as a suitable candidate for this purpose. Studies pioneered by Schramm et al (1983) and followed by others (Gross et al,1985; Wada et al,1985) have shown that in intact animals, Bay K 8644 increased blood pressure, LVP and dV/dt. Subsequently, in isolated tissues Bay K 8644 possessed positive inotropic effect (Finet et al, 1985), caused vasoconstriction of both arteries and veins (Gopalarikhan et al, 1985; Mikelsen et al, 1985; Su et al, 1984) and decreased coronary blood flow (Ishii et a, 1986). Based on the above reported data, we have reasoned that in intact animals Bay K 8644 would inevitably cause myocardial ischemia due to both coronary vasoconstriction and positive inotropism.

In the present study we aimed to overview results pertaining to an animal preparation based on Bay k 8644-induced myocardial ischemia in the rat and examine its suitability for the evaluation of antianginal drugs.

MATERIALS AND METHODS

The present study was performed on pentobarbital (25 mg/Kg, i.p.) anesthetized Sprague-Dawley rats (male, 450-500 g) as described earlier (Abraham et al, 1987). Briefly, three cannulae (PE-50) were introduced. The first two cannulae (heparin,100U/ml) were placed in the right femoral vein and artery and these served for the administration of drugs and measurment of systemic arterial blood pressure, respectively. The third cannula (heparin, 1000U/ml) was introduced in the right carotid artery and pushed forward to the region of the coronary ostium (see Figure 1). This cannula was used for the administration of Bay K 8644 and the induction of myocardial ischemia. Parameters recorded were arterial blood pressure and ECG (Lead I, using human electrocardiogram). The heart rate was derived from the ECG tracing.

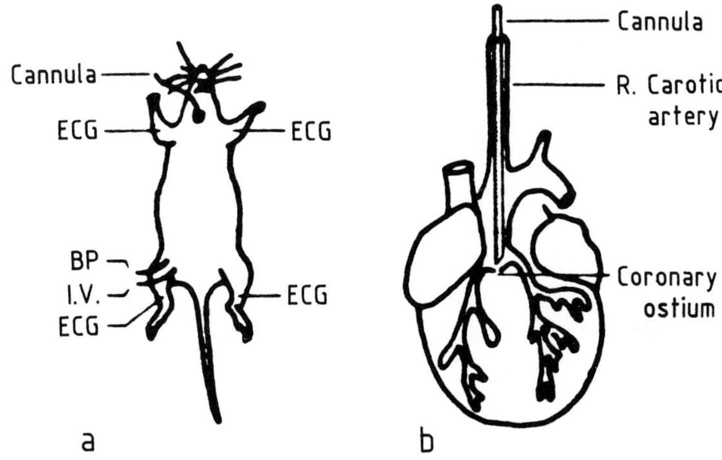

Figure 1

Schematic illustration of the experimental animal preparation. (a) Position of cannulae and ECG leads; (b) position of cannula in the proximity of the coronary ostium.

Drugs used were Atropine sulfate (Sigma), Diltiazem hydrochloride (Sigma), Nifedipine (Sigma), Nitroglycerin (Teva Pharmaceuticals, Israel), Phentolamine (Ciba-Geigy Corporation,USA), (±) Bay K 8644 (A gift from Dr M. Schramm, Bayer, West Germany), PK 11195 (a gift from Dr G. Le Fur (Pharmuka Laboratories, France). All chemicals except for Bay K 8644, nifedipine and PK 11195 were dissolved in saline

(0.9% NaCl solution). PK 11195 was dissolved in a mixure of Emulphor EL-620 (GAF, West Germany) and ethanol (1:1) and diluted with saline. Bay K 8644 and nifedipine were dissolved in absolute ethanol and diluted with saline. Chemicals were injected in volumes not exceeding 100 µl and containing less than 7 µl of ethanol. 100 µl Of vehicle alone had no effect on blood pressure or ECG pattern.

Data are presented as mean ± s.e.m. Statistical analysis was performed using the Student's paired t test (two-tailed).

RESULTS

Effects of Bay K 8644 on the ECG pattern and hemodynamic variables.

Profound effects on both arterial blood pressure and the ECG pattern have been observed following intracarotid administration of Bay K 8644 in pentobarbital anesthetized rats. The following doses of Bay K 8644 were administered: 0.5, 1, 2, 5, 10, 20, and 50 ug/Kg. At a dose range of 0.5-10 ug/Kg Bay K produced a dose related increase in blood pressure. However, at 20 and 50 ug/Kg the magnitude of the pressor response was decreased. Results are depicted in Figure 2. The increase in blood pressure was discernible within 5 sec after the administration of Bay K 8644 and lasted for 30-140 sec. Blood pressure returned to control levels at 0.5-20 ug/Kg whereas at a dose of 50 ug/Kg the pressor response was followed by short hypotensive phase and death of the animals (Fig 2). All the rats (n=5) administered with 50 ug/Kg of Bay K 8644 have died within 1 min post injection due to electromechanical dissociation.

The effect of Bay K 8644 on heart rate is given in Table 1. This table demonstrates that Bay K 8644 did not affect heart rate at doses below 2 ug/Kg. However, at doses of 5 ug/Kg or above heart rate was reduced by 40-60 beats/min. Due to the arrhythmias developed at 50 ug/Kg, the changes in heart rate were not measured.

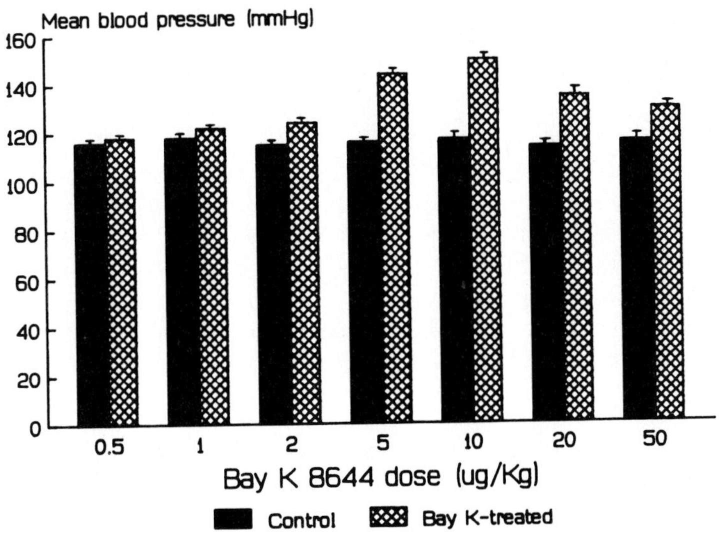

Figure 2

Dose-related effects of intracarotid injections of Bay K 8644 on the mean arterial blood pressure of anesthetized rats. Results are expressed in mmHg and vertical lines indicate s.e.m where n values are as follows: 0.5 µg/Kg- n=5; 1 µg/Kg- n=7; 2 µg/Kg- n=6; 5 µg/Kg- n=15; 10 µg/Kg- n=7; 20 µg/Kg- n=7; 50 µg/Kg- n=5.

Table 1
Effect of intracarotid administration of Bay K 8644 on heart rate in pentobarbital-nesthetized rats.

Dose (µg/Kg)	n	Heart rate (beats/min) Pre	Post
0.5	5	372.0 ± 6.6	378.8 ± 6.6
1	7	381.4 ± 8.6	372.9 ± 6.1
2	6	390.0 ±10.9	375.0 ±10.2
5	15	384.0 ± 5.2	344.3 ± 4.0
10	7	381.4 ± 5.5	321.4 ± 5.5
20	7	390.0 ± 9.3	317.6 ± 6.1
50	5	384.0 ±11.2	NM

NM= Non-measurable
Results are expressed in terms of mean ± s.e.m.

The pressor response caused by Bay K 8644 was immediately (within 6-10 sec) followed by changes in the ECG pattern. These changes reached maximal values within 12-17 sec and lasted for 30-240 sec. Typical ECG tracings are shown in figure 3. Included in this figure are tracing obtained fron different animals administered with the indicated dose of Bay K 8644. Shown here are representative ECG tracings at the time and doses indicated. At the doses shown, the effect of Bay K 8644 on the ECG pattern were confined mostly to the ST segment where elevation or depression have been noticed. Except for occasional increase in R wave amplitude no effects were noticed on other variables of the ECG such as the P-R or the QT intervals.

Whereas ECG pattern restored its control shape within 4 minutes at doses below 20 ug/Kg, the highest dose of Bay K 8644, 50 ug/Kg, produced A-V block and resulted with death of all the animals (n=5) within 1 minute. Representative tracing at this dose level are shown in Figure 4.

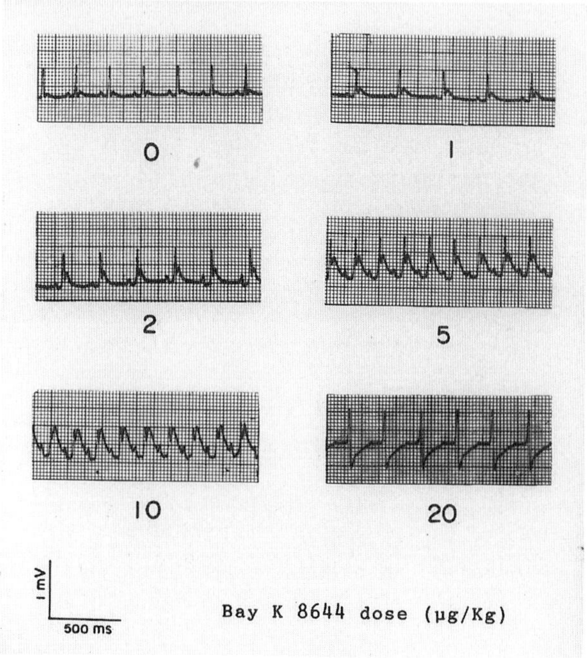

Figure 3

Representative ECG (Lead I) tracings obtained from different animals administered with the indicated dose of Bay K 8644. These tracings were sampled at 12-17 sec after the administration of Bay K 8644.

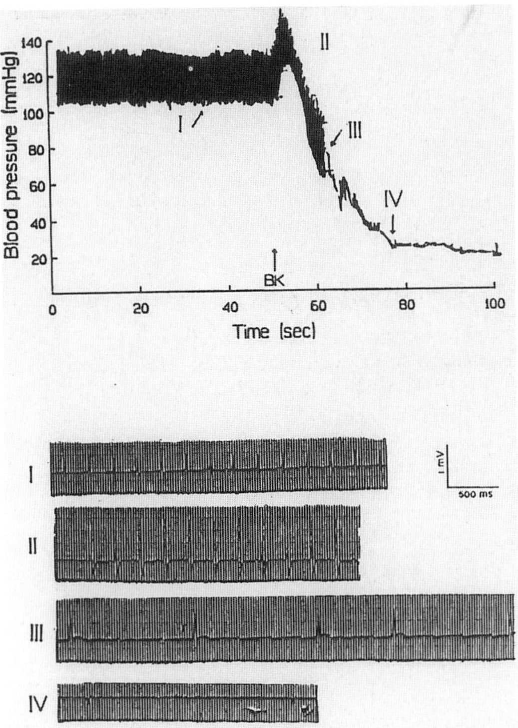

Figure 4

Representative ECG (Lead I) and arterial blood pressure tracing obtained following intracarotid injection 50 ug/Kg of Bay K 8644 into anesthetized rats. ECG tracings were sampled at various time intervals after the administration of Bay K 8644 as indicated by the arrows. Tracing No I represents predrug recording; the others were sampled at 7, 15 and 40 seconds.

In order to quantify the effects of Bay K 8644 on the ST segment amplitude, we have measured the ST segment elevation or depression in mV relative to the isoelectric potential (Abraham et al, 1987; Vergona et al, 1984). Results are depicted in Figure 5. This figure demonstrates that the effect of Bay K on the ST segment amplitude is bidirectional: At doses up to 10 ug/Kg ST segment was elevated whereas depression occured at higher doses.

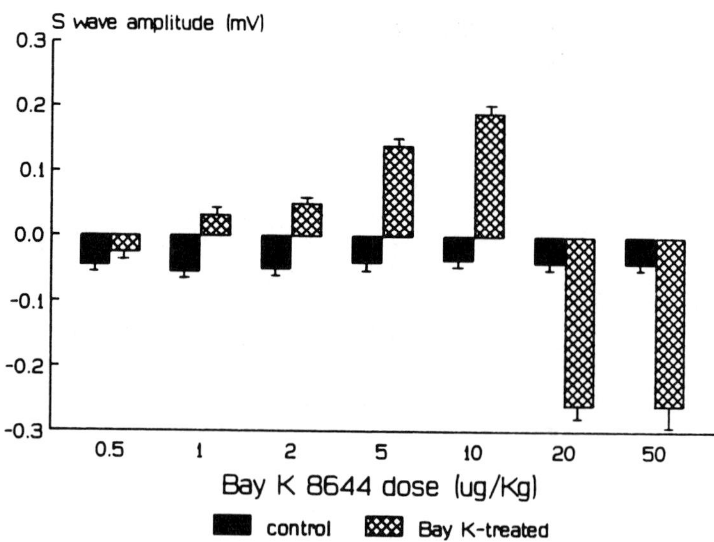

Figure 5

Dose-related effects of intracarotid injections of Bay K 8644 on the S wave amplitude of anesthetized rats. Results are expressed in mV and vertical lines indicate s.e.m where n values are as follows: 0.5 µg/Kg- n=5; 1 µg/Kg- n=7; 2 µg/Kg- n=6; 5 µg/Kg- n=15; 10 µg/Kg- n=7; 20 µg/Kg- n=7; 50 µg/Kg- n=5.

The pressure-rate index (PRI), a measure of myocardial oxygen demands, was calculated as the product of mean arterial blood pressure and heart rate according to Hock et al. (1986). Results, summarized in Figure 6, indicate that Bay K 8644 caused only a slight increase in this parameter.

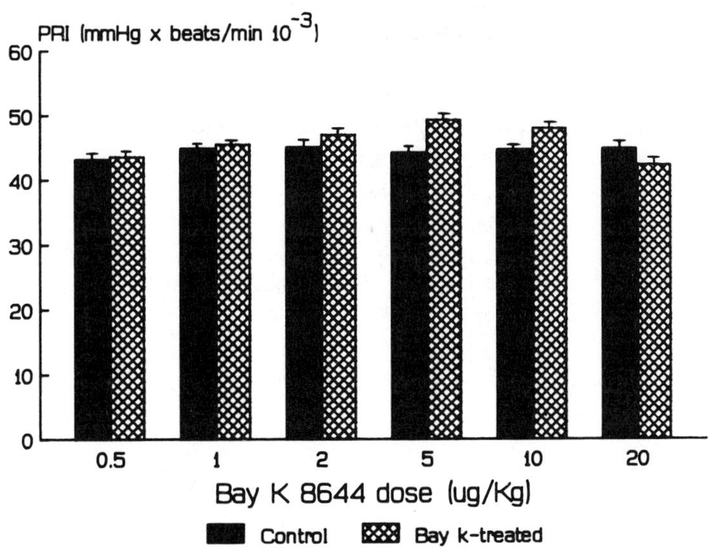

Figure 6
Dose-related effects of intracarotid injections of Bay K 8644 on the pressure rate index (PRI) of anesthetized rats. PRI values were calculated as the product of mean arterial blood pressue and heart rate obtained at the peak effect of Bay K 8644. Results are expressed as mmHg x beats/min and vertical lines indicate s.e.m where n values are as follows: 0.5 µg/Kg- n=5; 1 µg/Kg- n=7; 2 µg/Kg- n=6; 5 µg/Kg- n=15; 10 µg/Kg- n=7; 20 µg/Kg- n=7; 50 µg/Kg- n=5.

Influence of drugs on the effects of Bay K 8644.

Intracarotid administration of Bay K 8644 produced two major effects: pressor response and ST segment alterations. In order to assess the ability of drugs to reverse or attenuate the above effects, several compounds were chosen. Among these, nitroglycerin, nifedipine and diltiazem are established antianginal drugs; PK 11195, an experimental compound with reported antiarrhythmic activity in reperfusion -induced arrhythmias in dogs (Mestre et al, 1985a). Two control groups of animals were included: phentolamine served to control for the hypotensive response elicited by all antianginal drugs; Atropine , reported by Sakai et al (1981) as an extremely effective in reversing the the methacholine-induced ischemia.

Except for PK 11195, all other compounds exhibited a hypotensive response which recovered spontaneously thereafter. However, in the course of evaluating their antianginal activity, Bay K 8644 was administered on top of these drugs at a time when there was still a residual decrease in blood pressure in some of the cases. The extent of the residual decrease presented as % of control values are given in Table 2. This table shows that in those animal injected with nitroglycerin (25 ug/Kg), diltiazem and atropine blood pressure reached control levels before the administration of Bay K 8644.

Table 2
Residual decrease in blood pressure (% of control) at the time of Bay K 8644 administration.

Compound	n	Dose ug/Kg	Time (min)	Percent decrease
Atropine	3	5000	15	NONE
Diltiazem	6	250	15	NONE
	5	500	15	NONE
Nifedipine	5	50	15	9.2 ± 1.8
	6	100	15	19.8 ± 2.8
Nitroglycerin	4	25	1	NONE
	5	50	1	6.8 ± 1.2
Phentolamine	5	5000	15	37.4 ± 5.2
	5		60	32.8 ± 5.8
PK11195	3	250	15	NONE

Data are presented as mean ± s.e.m.

In order to evaluate the ability of the above compounds to reverse Bay K 8644-induced ischemia we challenged the rats to two doses of Bay K 8644, 5 and 20 ug/Kg. Whereas 5 ug/Kg was chosen as it produced submaximal effects on both ST segment amplitude and blood pressure, 20 ug/Kg represent the highest dose devoid of arrhythmias. Results are summarized in Table 3. All the compounds were injected before Bay K 8644 intracarotid administration at the specified time interval. Doses of drugs and time interval between drugs and Bay K 8644 were chosen according to preliminary experiments and published data. All compounds were given i.v. except for

PK 11195 which was given intracarotidly due to short supply. Nevertheless we have established that similar result with PK 11195 could be obtained via the i.v. route provided higher doses are used. Results are summarized in Table 3. This table demonstrates that phentolamine or atropine failed to block the ECG changes produced by Bay K 8644. However, the antianginal drugs, nitroglycerin, diltiazem and nifedipine as well as the experimental compound,PK 11195 were very effective in blocking Bay K 8644-induced ST segment alterations. Diltiazem seems to be more effctive than nifedipine. Diltiazem(500 ug/Kg) and nitroglycerin 50 ug/Kg) fully blocked the ST segment depression following 20 ug/Kg of Bay K 8644. In contrast, blockade of Bay K 8644-induced pressor response did not exceed 15%.

Table 3

Inhibitory effect of antianginal drugs and experimental compounds Bay K 8644-induced S Wave elevation.

Compound	n	Dose ug/Kg	Time (min)	Blockade of Bay K-induced S wavs alterations (% of control) 5 ug/Kg	20 ug/Kg
Diltiazem	6	250	15	95.5 ± 2.7	
	5	500	15	100	
	3	500	15		100
Nifedipine	5	50	15	60.4 ± 4.2	
	6	100	15	76.6 ± 5.9	
Nitroglycerin	4	25	1	46.6 ± 4.6	
	5	50	1	100	
	3	50	1		100
PK11195	3	100	15	100	
Atropine	3	5000	15	0	
Phentolamine	5	5000	15	0	
	5		60	0	

Drugs were administered intravenously except for PK 11195 which was injected via the carotid artery. Data are expressed as mean ± s.e.m.

DISCUSSION

The present study has demonstrated that intracarotid injections of Bay K 8644 in anesthetized rats produce two major effects: ST segment alteration and pressor response. ST segment alterations are generally considered as a sensitive measure for cardiac ischemia (Brezinski et al, 1986). Therefore, it may be suggested that the observed in the present study reflects a condition of myocardial ischemia, subsequent to the use of the calcium channel agonist.

Bay K 8644 has been shown to produce coronary vasoconstriction (Gopalakrishnan et al,1985; Mikkelsen et al,1985; Sue et al, 1984) and to decrease coronary blood flow (Ishii et al, 1986; Schramm et al, 1983a; Wada et al, 1985). In addition, positive inotropic and chronotropic effects in various cardiac preparations have been also reported (Finet et al, 1985; Satoh et al, 1984; Schramm et al, 1983b; Wada et al, 1985). Thus, it is suggested that two machanisms may be involved in the Bay K 8644-induced myocardial ischemia, namely decreased myocardial supply due to coronary vasospasm and increased metabolic demands due to positive inotropic effect. For the following reasons we believe that coronary vasospasm is more likely to play a major role: a. Wada et al (1985) have analysed the cardiovascular profile of Bay K 8644 in the dog and found that coronary vasoconstriction was more pronounced than its inotropic, chronotropic and dromotropic effects. b. The calculated PRI in the present study have only slightly increased and this preclude increased myocardial oxygen demands under the influence of Bay K 8644.

The Bay K 8644-induced myocardial ischemia could be effectively blocked by the antianginal drugs, nitroglycerin, nifedipine and diltiazem as well as by the experimental compound, PK 11195. Whereas nitroglycerin is known as a vasodilator (Abrams 1980), the calcium channel blockers exert their action via inhibition of voltage dependent calciun channels in both cardiac and coronary tissue (Braunwald, 1982). With respect to PK 11195, this peripheral benzodiazepine receptor antagonist exert its action by indirect inhibition of myocardial voltage-dependent calcium channels (Mestre et al, 1985; Mestre et al, 1986). In addition, its effect on coronary blood flow was found to be biphasic. In the nanomolar concentration, PK 11195 produces coronary dilation whereas in the micromolar range which is more relevant to the present study, PK 11195 produces no change or coronary vasocontriction. Therefore it can be stated that blockade of Bay K 8644- induced cardiac ischemia can be achieved by either coronary vasodilation (nitroglycerin), negative inotropism (PK 11195) or both (nifedipine, diltiazem). We would like to speculate that the beneficial effect of the calcium channel blockers may derive from the prevention of ischemic injury.

PK 11195, in contrast to nitroglycerin, nifedipine and diltiazem, had no effect on blood pressure but still completely reversed the ST segment alterations. This observation indicates that blockade of Bay K 8644- induced pressor response is not essential for the inhibition of myocardial ischemia. It was rather expected to note that none of the compounds tested effectively blocked the pressor response induced by Bay K 8644. Similarly, Lefer et al (1986) have demonstrated incomplete inhibition of the pressor response to Bay K 8644 by two calcium channel blockers, nitrendipine and nisoldipine. Both in the present study and that reported by Lefer et al (1986), the α-adrenergic blocker, phentolamine was found ineffective in blocking the pressor response to Bay K 8644. Moreland et al (1988) have analysed the pressor resopnse to Bay K 8644 and concluded that it may result from two separate mechanisms: activation of vascular and cardiac calcium channels as well as release of adrenal catecholamines.

In contrast to increased rate in isolated sino-atrial preparations (Wada et al, 1985), Bay K 8644 has been reported to produce baradycardia in whole animal studies (Gross et al 1985), including the present one. It is thus concluded that the Bay K 8644-induced bradycardia is reflexly mediated.

In conclusion, in the present study we have demonstrated that intracarotid injection of Bay K 8644 in the rat produce ST segment alteration, indicative of myocardial ischemia. It is of importance to note that Bay K 8644 exerts its effect through direct activation of the contractile machinary of the vascular tissue and this renders the present preparation greater selectivity. This animal preparation has been found suitable for the evaluation of antianginal drugs as clinically used antianginal compounds were found effective in correlation with their clinical efficacy. Therefore we may recommend Bay K 8644-induced myocardial ischemia in the rat as an excellent tool for the screening of antianginal drugs. However, we are aware that further studies may be required to clarify the proposed mechanism of action of Bay K 8644 in inducing myocardial ischemia and that which underly the protective effect of antianginal drugs.

REFERENCES

Abraham S, Amitai G, Oz N, Weissman, BA (1987). Bay K 8644-induced changes in the ECG pattern of the rat and their inhibition by antianginal drugs. Br. J. Pharmac. 92:603-608.

Abrams J (1980). Nitroglycerin and long acting nitrates. New Engl. J. Med. 302:1234-1237.

Botting JH, Curtis MJ and Walker MJA (1985). Arrhythmias associated with myocardial ischaemia and infarction. Molec. Aspects Med. 8:307-422.

Braunwald E (1982). Mechanism of action of calcium channel blocking agents. New Engl. J. Med. 307:1618-1627.

Brezinski ME, Darius H, Lefer AM (1986). Cardioprotective actions of new calcium channel blocker in acute myocardial ischemia. Arzneimittel-Forschung, 36:464-466.

Brunton TL (1867). On the use of nitrite of amyl in angina pectoris. Lancet 2:97-98.

Curtis MJ, Macleod BA and Walker MJA (1987). Models for the study of arrhythmias in myocardia ischemia and infarction: the use of the rat. J. Mol. Cell Cardiol. 19:399-419.

Finet M, Godfraind T, Khoury G (1985). The positive inotropic action of the nifedipine analogue, Bay K 8644, in guinea pig and rat isolated cardiac preparations. Br. J. Pharmacol. 86:27-32.

Gensini GG, Digiorgi S, Murad-Netto S, Black A (1962). Arteriographic demonstration of coronary artery spasm and its release after the use of a vasodilator in a case of angina pectoris and in the experimental animal. Angiology 13: 550-553.

Gettes LS (1983). The cause of the ST-segment changes in acute ischemia. In " What is angina?", Julian DG, Lie KI, Wilhelmsen L (eds): Molndal, Sweden, AB Hassle, pp 171-184.

Ginsburg R, Baim DS (1983). Regulation of tone in the human coronary artery. In "What is angina?", Julian DG, Lie KI, Wilhelmsen L (eds): Molndal, Sweden, AB Hassle, pp 50-60.

Gopalakrishnan V, Park LE, Triggle CR (1985). The effect of the calcium channel agonist, Bay K 8644 on human vascular smooth muscle. Eur. J. Pharmacol. 113:447-451.

Gross R, Kayser M, Schramm M, Taniel R, Thomas G (1985). Cardiovascular effects of the calcium-agonistic dihydropyridine Bay K 8644 in conscious dogs. Arch. Int. Pharmacodyn. 277:203-216.

Hock CE, Brezinski ME, Lefer, AM (1986). Anti-ischemic actions of new thromboxane receptor antagonist, SQ-29,548, in acute myocardial ischemia. Eur.J.Pharmacol. 122:213-219.

Ishii K, Sato Y, Taira N (1986). Similarity and dissimilarity of the vasoconstrictor effects of Bay K 8644 on coronary, femoral, mesenteric and renal circulations of dogs. Br. J. Pharmacol. 88:369-377.

Lambert CR, Pepine CJ (1986). Coronay artery spasm and parmacology of coronary vasodilators. In "The pathophysiology and pharmacotherapy of myocardial infarction. El-Sherif N, Reddy CVR (eds). Academic press Inc., Orlando, Florida pp 117-154.

Lefer AM, Whitney III CC, Hock CE (1986). Mechanism of the pressor effect of the calcium agonist, Bay K 8644, in the intact rat. Pharmacology 32:181-189.

Mestre M, Carriot T, Belin C, Uzan A, Renault C, Dubroueucq MC,Gueremy C, Doble A, LeFur G (1985). Electrophysiological and pharmacological evidence that peripheral type benzodiazepine receptors are coupled to calcium channels in the heart. Life Sci. 36:391-400.

Mestre M, Carriot T, Neliat G, Uzan A, Renault C, Dubroueucq MC,Gueremy C, Doble A, LeFur G (1985). PK 11195, an agonist of peripheral type benzodiazepine receptors, modulate Bay K 8644 sensitive but not β- or H_2-receptor sensitive voltage operated calcium channels in the guinea pig heart. Life Sci. 39:329-339.

Mikkelsen E, Nyborg NCB, Kazda S (1985). A novel 1,4-dihydropyridine, Bay K 8644, with contractile effects on vascular smooth muscle. Acta Pharmacol. Toxicol. 56:44-49.

Moreland S, Ushay MP, Kimball SD, Powell JR, Moreland RS (1988). Pressor responses induced by Bay K 8644 involve both release of adrenal catecholamines and calcium channel activation. Br.J.Pharmacol. 93:994-1004.

Prinzmetal M, Kennamer R, Merliss R, Wada T, Bor N (1959). Angina pectoris I. Variant form of angina pectoris. Am. J. Med. 27:375-388.

Prinzmetal M, Ekemecki A, Kennamer R, Kwoczynski JK, Shubin H, Toyoshima H (1960). Variant form of angina pectoris. Previously undelineated syndrome. J.A.M.A. 174: 1794-1800.

Sakai K, Akima M, Aono J (1981). Evaluation of drug effects in a new experimental model of angina pectoris in intact anesthetized rats. J. Pharmacol. Meth. 5: 325-336.

Satoh K, Wada Y, Taira N (1984). Differential effects of Bay K 8644, a presumed calcium channel activator, on sinoatrial nodal and ventricular automaticity of the dog heart. Naunyn-Schmiedebergs Arch Pharmacol. 326:190-192.

Schramm M, Thomas G, Towart R, Franckowiak G (1983a). Novel dihydropyridines with positive inotropic action through activation of Ca^{2+} channels. Nature 303:535-537.

Schramm M, Thomas G, Towart R, Franckowiak G (1983b). Activation of calcium channels by novel 1,4-dihydropyridines. A new mechanism for positive inotropics or smooth muscle stimulants. Arzneimittel Forschung 33:1268-1272.

Su CM, Swamy VC, Triggle DJ (1984). Calcium channel activation in vascular smooth muscle by Bay K 8644. Can. J. Physiol. Pharmacol. 62:1401-1410.

Vergona RA, Herrott C, Garippa R, Hirkaler G (1984). Mechanisms of methacholine-induced coronary vasospasm in an experimental model of variant angina in the anesthetized rat. Life Sci. 35:1877-1884.

Wada Y, Satoh K, Taira N (1985). Cardiovascular profile of Bay K 8644, a presumed calcium channel activator in the dog. Naynyn-Schmiedebergs Arch, Pharmacol. 328:382-387.

Calcium Channel Modulators in Heart and Smooth Muscle: Basic Mechanisms and Pharmacological Aspects
S. Abraham and G. Amitai, editors
© 1990, VCH, Weinheim/Deerfield Beach, FL and Balaban, Rehovot/Philadelphia

Rat models for studying arrhythmic and other adverse responses to myocardial ischaemia: Beneficial actions of calcium channel blockade

B.A. MACLEOD AND M.J.A. WALKER

Department of Pharmacology and Therapeutics, Faculty of Medicine, The University of British Columbia, 2176 Health Sciences Mall, Vancouver, B.C., Canada, V6T 1W5

INTRODUCTION

In the search for drugs which ameliorate the adverse sequelae of myocardial ischaemia and infarction use has been made of a variety of models (Fozzard, 1975; Kelliher et al., 1983). In many cases the approach underlying the use of such models has been based on an infectious disease paradigm. That is to say, models attempt to mimic the exact disease condition which is to be influenced by the drug. However, such an approach may not be very rewarding when searching for drugs of value in ameliorating adverse sequelae to human conditions of ischaemia and infarction since such conditions are too varied (Botting et al., 1987). Myocardial ischaemia and infarction in humans involves a varied spectrum including the variable pre-existence of collaterals, partial or complete ischaemia, vasospasm, pre-existing infarction and concomitant athero-sclerosis. Attempts to model such a complex situation of necessity ignores a number of possibly important factors. In addition, the infectious disease paradigm may not be appropriate. In developing H_2 antagonists to treat peptic ulcer, Black (Black et al; 1972) relied less on models of the disease and more on appropriate screens to detect the pharmacological property (H_2 receptor block-ade) associated with therapeutic efficacy. By analogy, it may be more important in ischaemia and infarction to con-centrate on screens which reliably detect pharmacological actions of interest rather than on partial attempts to model a complex disease entity.

Screens for pharmacological activity require precision and accuracy, i.e., they should accurately detect drugs with the required activity and precisely measure the extent of such activity. With such screens the degree of clinical relevance is of little direct importance.

As an example, it is useful to consider the use of dog models in the search for drugs of value in treating ischaemia and infarction. For reasons that were never scientifically tested, it was assumed that the dog heart was similar to the human heart. This reason, as well as expediency in using a large readily available and inexpensive species, resulted in the dog being extensively used. This is despite variations in coronary collateral supply between dogs resulting in large "between-animal variance" which masks all but the most profound drug effects. This is especially so with infarct size determinations, or ischaemia-induced arrhythmias. It has been calculated (Trolese-Mongheal et al., 1985) for dogs that group sizes of 50 are needed to unambiguously detect an antiarrhythmic effect of a single dose of a drug.

The overall aim of our studies is to identify pharmacological properties which confer antiarrhythmic activity against arrhythmias due to myocardial ischaemia and infarction. This goal requires precise and accurate data. It is also important to have dose-response data describing the relationship between doses and responses obtained. Limitations to efficacy, i.e., side effects and toxicity, should be recognized and quantified. These considerations led to careful consideration of the most suitable screens for detecting antiarrhythmic activity during myocardial ischaemia and infarction. After initial experiments with dogs, pigs and primates (Au et al., 1979; Harvie et al., 1978), and unpublished observations) we decided to:

1) Use a readily available and inexpensive species.
2) Avoid recent surgery or anaesthesia.
3) Use a model with consistent response to ischaemia.
4) Use an easily used species to obtain relevant pharmacological and toxicological information.

As a result we used rats, despite reservations as to the electrophysiological similarity between rat and other mammalian hearts (Langer, 1978). The following justifies the use of rats for determining which pharmacological properties confer antiarrhythmic activity against ischaemia-induced arrhythmias. We have previously reviewed the use of rats for studying myocardial ischaemia and infarction (Curtis et al., 1987).

METHODS

We use a number of rat models of ischaemia and infarction together with ancillary _in vivo_ and _in vitro_ preparations. Confirmatory studies were performed in pigs. All experiments were executed according to a blind and random design. Those involving ischaemia and infarction were performed under conditions similar to those outlined by the Lambeth Conventions (Walker et al., 1988). Exclusion criteria were applied to exclude animals on the basis of anomalous ECG changes, or occluded zone sizes. Serum potassium values were routinely measured.

Rat Models of Ischaemia and Infarction

Chronically-prepared conscious rats (Johnston et al., 1981; 1983). Our primary model involved LAD coronary artery occlusion rats in which blood pressure, heart rate and ECG were recorded before and after occlusion. Rats were surgically prepared under halothane anaesthesia with a loose LAD occluder (constructed of polyethylene and polypropylene) and permanent aortic and vena caval cannulae, as well as ECG leads. Seven days were allowed for recovery from surgery before drug treatment and occlusion. On the basis of pharmacokinetic and pharmacodynamic data, suitable dose regimens were developed to ensure pharmacological blood concentrations during the period for arrhythmias, and as infarction developed. One day after occlusion, rats were subject to recording (30 min) before their hearts were processed _in vitro_ for estimation of occluded and infarcted zones by dye exclusion and tetrazolium, respectively (Johnston et al., 1983). The model assessed drug effects (at various doses) in terms of:

1) Arrhythmias (VT, VF, VPB).

2) Blood pressure and heart rate.
3) Normal and ischaemic ECG: P-R, QRS and Q-T intervals, R-wave enlargement and "S-T" segment elevation.
4) Occluded and infarcted zone sizes at 24 h.
5) Morbidity, mortality and behavioural changes.
6) Post-mortem findings, such as lung oedema.

Such data allowed antiarrhythmic effects to be quantified and compared with cardiovascular, ECG, infarct size responses, etc. to give a full profile of drug action.

Abbreviated chronically-prepared conscious rat model (Au et al., 1987). In this model rats were prepared with an occluder, but without cannulae or ECG leads. The test drug was injected via a tail vein before and after occlusion. The number of animals dying, and the symptoms associated with death (abrupt convulsions or gradual demise), were recorded for 24 h before animals were killed and their hearts processed for occluded and infarcted zones. This model quickly identified antiarrhythmic drugs.

Acutely-prepared anaesthetised rat model (Au et al., 1979; 1983). Rats were acutely prepared under pentobarbitone anaesthesia (50 mg/kg i.p.) and artifical ventilation (10 ml/kg, 60/min). The chest was opened and an occluder implanted around the LAD for occlusion 30 min after preparation. Blood pressure, heart rate and occluded zone were recorded.

Reperfusion models. In the above models reperfusion could be induced after a period of ischaemia by releasing the occluder. After reperfusion for 30 min or 24 h, the heart was removed and occluder re-tightened so as to delineate the occluded zone. Extensive exclusion criteria, involving ECG and blood pressure changes, were used to ensure that both occlusion and reperfusion conditions were identical in all animals; these were applied both during occlusion, and after reperfusion.

Additional Models

Electrical stimulation. To test for drug effects upon the ECG and resistance to electrical stimulation, studies were performed in open-chest anaesthetised or conscious rats fitted with left ventricular electrodes (Curtis et al., 1984). Electrical measurements included threshold currents (for VPB and VF) and pulse widths, effective refractory periods and maximum following frequency.

Epicardial intracellular potentials recorded in vivo. In order to assess drug effects on electrogenesis in terms of action potential morphology, intracellular potentials were recorded from the epicardium of pentobarbitone-anaesthetised open-chest rats. Transmembrane potentials were recorded before and after drug administration, with or without occlusion, using a floating-tip technique.

Ancillary studies. The above were complemented by pharmacokinetic analyses. Plasma and tissue concentrations of drugs were assayed by HPLC techniques. Direct cardiac actions were measured in perfused rat hearts (Curtis et al., 1987).

RESULTS

The above were used in many studies (see Curtis et al., 1987) to determine which pharmacological properties are antiarrhythmic in myocardial ischaemia and infarction. We also identified pathological factors which influence arrhythmias in rat models.

Factors Influencing Ischaemia-Induced Arrhythmias

Major factors influencing the arrhythmias induced by occlusion included the size of occluded and infarct zones. The relationship between zone size and arrhythmias was expressed in various ways. When the incidence of VF and/or VT were plotted against occluded zone size, a sigmoid relationship was obtained which fitted a logistic function. If arrhythmias were recorded on a normally-distributed linearly additive scale as an arrhythmia score (Curtis and Walker, 1988) it was linear with the square root of the occluded zone size resulting in a mathematical model in which border zone area determined arrhythmias. Since occluded zone size determines arrhythmias, it is important to exclude animals which, for technical or anatomical reasons, have abnormal occluded zones.

In various experiments we assessed the possible importance of factors such as the central nervous system, the sympathetic system, eicosanoids (and related substances) in the genesis of ischaemia-induced arrhythmias. The results led us to question their importance. For example, despite its supposed importance, we failed to obtain evidence consistently supporting a role for the sympathetic system in the rat (Botting et al., 1983; Curtis et al., 1985). In keeping with this, isolated hearts subjected to LAD coronary occlusion at physiological K^+ concentrations, i.e. 3.8 rather than the usual 5.9 mM, have arrhythmias similar to those in conscious rats.

Studies into mechanisms of arrhythmogenesis alerted us to the role serum potassium concentration (K^+) plays in reducing arrhythmias in acutely-prepared anaesthetised rats, in some beta-blocked rats, and in other situations. We manipulated (K^+) in a variety of ways and found that ischaemia-induced arrhythmias were inversely related to (K^+) at the time of arrhythmias, as shown in Figure 1.

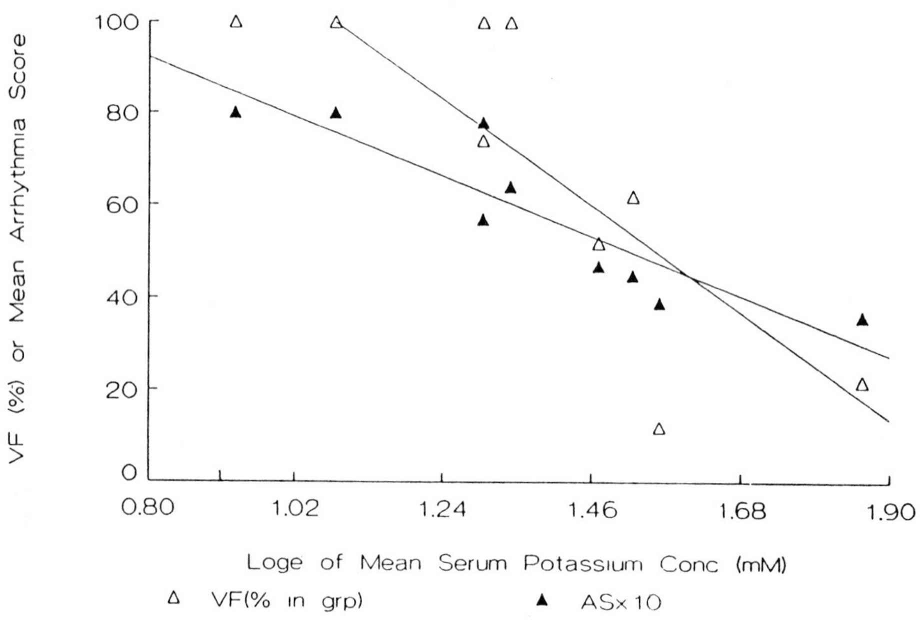

Figure 1. The relationship between $\log_e$ serum potassium and the incidence and severity of major arrhythmias. Serum potassium was manipulated in various ways in separate groups (n=9) of rats before LAD occlusion.

The relationship explains the lower incidence, as compared with conscious rats, of ischaemia-induced arrhythmias in rat models where (K^+) is elevated. To guard against this confounding effect of (K^+), we routinely measure (K^+) in all rats. The effect of (K^+) is not so marked with reperfusion arrhythmias (Table 1).

TABLE 1. Effects of Elevated Serum Potassium on Ischaemic and Reperfusion Arrhythmias in Conscious Rats

	Range of serum K+ in mM							
	3.0-3.9		4.0-4.8		4.9-5.7		5.8+	
	VT	VF	VT	VF	VT	VF	VT	VF
	Number having VT or VF (normalised for n=9)							
Ischaemia	8	6	3	5	6	7	2	4
Reperfusion	8	6	3	6	5	5	7	5

Groups of rats (n=9-27) were infused with KCl to elevate serum K^+, to levels indicated, for 10 min of occlusion and subsequent reperfusion. KCl did not change the relationship between duration of ischaemia and arrhythmias.

Studies with Calcium Channel Blockers

Amoungst various classes of drugs tested for antiarrhythmic efficacy, only ion-channel blocking antiarrhythmics had significant antiarrhythmic actions against ischaemia-induced arrhythmias in rats. Sodium, calcium and potassium channels blockers were all effective.

Extensive studies were performed with calcium channel blockers, including verapamil (and enantiomers), D-888, diltiazem, anipamil, nifedipine and felodipine (Curtis et al., 1984; 1986; 1987; Curtis and Walker, 1988; Walker and Beatch, 1989). Anipamil is a long acting verapamil analoque with a similar pharmacological spectrum. All calcium channel blockers had antiarrhythmic actions if given at a high enough dose. With nifedipine and felodipine, such doses were supra-maximal with regard to blood pressure lowering. Comparison of dose-response curves for blood pressure, heart rate and ECG effects suggested that the blockers were only antiarrhythmic at doses blocking myocardial calcium channels. This view was supplemented by mortality and morbidity data since doses preventing arrhythmias increased the proportion of animals with congestive heart failure. For example, maximally vasodilator doses of nifedipine were not antiarrhythmic and did not increase the incidence of congestive heart failure following coronary occlusion, whereas antiarrhythmic doses did.

Ancillary electrophysiological experiments showed that blockers were not acting by Class I actions (sodium channel blockade) on normal myocardium. Additional experiments in isolated hearts, together with pharmacokinetic data, suggested that calcium channel blockade in ischaemic myocardium depends upon factors such as raised extracellular potassium (Curtis and Walker, 1988).

Another consistent finding was that antiarrhythmic doses produced effects on ischaemia-induced changes in the ECG, despite not reducing occluded or infarcted zones. The blockers also delayed the rate at which R-wave enlargment and "S-T" segment elevation developed with ischaemia. Within seconds of the onset of ischaemia the R-wave enlarged to reach 2-3X its original value. The rise in "S-T" segment was slower and took 4-6 min to maximise. Hours later the R-wave diminished and a Q-wave developed. Such changes were not readily recorded in acutely-prepared rats.

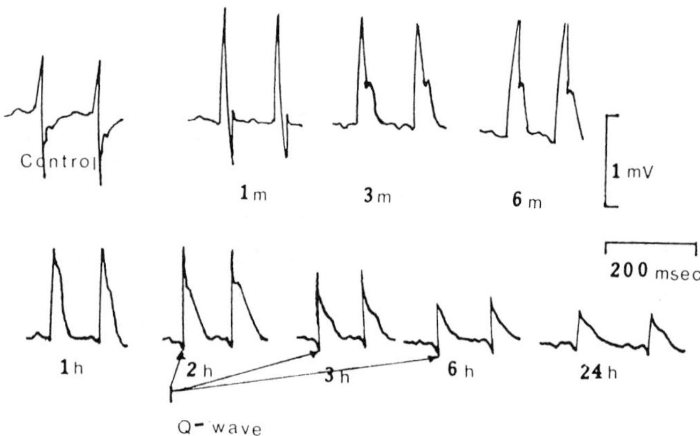

Figure 2. Typical ECG changes.

If the above ECG findings were due to anti-ischaemic effects, anipamil was chosen to evaluate how they relate to antiarrhythmic effects. To determine the relationship between anti-ischaemic and antiarrhythmic effects, data for arrhythmias were divided into appropriate time periods related to development of arrhythmias and an ischaemic ECG.

TABLE 2. Incidence of VF in Anipamil-Treated Rats

	Time periods (mins) after occlusion				
	0-3	3-5	5-7	7-10	10+
	Number of animals having VF in group of n = 18				
Control	0	9	11	1	0
A 1 mg/kg	0	2*	6	4	2
A 5 mg/kg	0	2*	4*	0	0
	Cumulative totals for VF at different times				
	3 min	5 min	7 min	10 min	30 min
Control	0	8	13	14	14
A 1 mg/kg	0	2*	7	8	9
A 5 mg/kg	0	2*	4*	4*	4*

A is anipamil given i.v. Incidences are the number of animals in a group of 18 having one or more episode of VF.

Anipamil slowed the appearence rate for VF after onset of ischaemia and, in addition, reduced the total number of arrhythmias in all periods, even when ischaemia was fully developed (Table 3). Table 3 shows that increasing doses of anipamil prolonged the time taken for the "S-T" elevation to reach a maximum. Nevertheless, the same maxima were reached, in keeping with anipamil's failure to alter the size of occluded or infarcted zones.

Table 3 Effect of Anipamil on "S-T" Segment Elevation

	Time after occlusion (min)						
	1 min	2 min	3 min	5 min	10 min	15 min	30 min
Control	17	29	36	42	40	40	43
A 1 mg/kg	10	12	18	26	33	40	40
A 2.5 mg/kg	4	9	18	24	43	43	44
A 5 mg/kg	6	5	10	14	27	32	35

Anipamil (A) was given before occlusion. Values are group means (n = 9) of "S-T" elevation as a percentage of R-wave. R-wave findings were similar (not shown).

DISCUSSION

The above is a summary of the manner in which we used rat models to evaluate potential beneficial effects of different drug classes in the setting of myocardial ischaemia and infarction. To assess drug action it is vital to have both models and screens, but excessive emphasis should not be placed on the appropriateness of models to human disease conditions. In searching for drugs to prevent the adverse sequelae of ischaemia and infarction, it may be of less importance to follow a disease paradigm and more important to identify beneficial pharmacological actions. With ischaemic arrhythmias, ion channel blockade is the pharmacological property of importance. What remains to be defined is how such blockade can be maximised to improve therapeutic ratios.

Our studies show how important are precision and accuracy and an understanding of the model used. By using rats, we were able to show, for this species, the importance of occluded zone size and (K^+) in arrhythmogenesis. Their importance has also been demonstrated in isolated rat hearts (Curtis and Hearse, 1989a,b). If ischaemic zone size and (K^+) are important in other species, we might ask how many studies have falsely reported the anti-arrhythmic efficacy of a drug when variations in ischaemic zone size or (K^+) were responsible

While it is difficult scientifically to show the importance of experimental design, it should be given due prominence as should dose-response studies. This is especially true when one considers the subjective errors that intrude into ischaemia and infarction-studies. Progress has been stultified not by the use of inappropriate models, but by the inappropriate scientific and statistical use of varyingly appropriate models. For example, studies have assessed antiarrhythmic effects as changes in VPB numbers without realizing that such data has to be log transformed. The number and duration of tachyarrhythmias should be similarly transformed (Johnston et al., 1983).

Our studies have illustrated that blockade of myocardial calcium channels is antiarrhythmic against ischaemia-induced arrhythmias. With those blockers which possess marked vascular selectivity, large doses had to be given to influence the myocardium. Thus, with the dihydropyridines, nifedipine and felodipine, antiarrhythmic doses were higher than maximally hypotensive doses. Previous studies (see review by Opie et al; 1988) have shown calcium channel blockers to be antiarrhythmic, but such studies were often fragmentary in terms of number of drugs examined, doses used, or availablity of supplemental pharmacokinetic or pharmacological data. Studies in the dog similar to those we have accomplished in the rat would be almost impossible. We used over 1,000 rats, an impossibly high figure in large animals. This figure does not include rats used in ancillary experiments.

Careful consideration of all our data led us to suggest a theoretical route to more efficacious calcium channel blockers for the treatment of ischaemia-induced arrhythmias. The examination of the relative efficacy of different drugs, toxicities, pharmacokinetics and cardiac actions in the presence of an elevated (K^+) or stimulation frequency suggested routes to more efficacious drugs.

In conclusion, careful choice of models and screens allowed the use of limited resources to show that myocardial calcium channel blockade is antiarrhythmic against ischaemia-induced arrhythmias and that the presence of ancillary properties controlled the antiarrhythmic efficacy and potency of the blockers used in the study.

ACKNOWLEDGEMENTS

In developing and utilising the models described above we are grateful for the help of M. J. Curtis, T.L.S. Au, C.J. Harvie, M. Moult and K.M. Johnston. Funds were provided by the Canadian and British Columbia Heart Foundations, Knoll and Haessle Pharmaceuticals.

REFERENCES

Au TLS, Collins GA, Harvie, CJ, Walker MJA (1979). The actions of prostaglandins I_2 and E_2 on arrhythmias produced by coronary occlusion in the rat and dog. Prostaglandins 18:707-720.

Au TLS, Collins GA, MacLeod BA, Walker MJA. (1983). Effects of prostaglandin E_2, propranolol and nitroglycerin with halothane, pethidine, or pentobarbitone anaesthesia on arrhythmas and other responses to ligation of a coronary artery in vitro. Br J Pharmacol 79:929-937.

Au TLS, Curtis MJ, Walker MJA (1987). Effects of (-), (±) and (+) on coronary occlusion-induced mortality and infarct size. J Cardiovasc Pharmacol 10:327-331.

Bernier M, Curtis MJ, Hearse DJ (1989). Ischemia-induced and reperfusion-induced arrhythmias: Importance of heart rate. Am J Physiol (Heart & Circ Physiol) 256:H21-H31.

Black JW, Duncan WAM, Durant CJ, Ganellin CR, Parsons EM (1972) Definition and antagonism of H_2-receptors. Nature 236:385-390.

Botting, JH, Curtis MJ, Walker MJA (1987). Arrhythmias associated with myocardial ischaemia and infarction. Molec Aspects Med 8:312-422.

Botting JH, Johnston KM, MacLeod BA, Walker MJA (1983). The effect of modification of sympathetic activity on responses to ligation of a coronary artery in the conscious rat. Br J Pharmacol 79:265-271.

Curtis MJ, Hearse DJ (1989a). Ischemia-induced and reperfusion-induced arrhythmias differ in sensitivity to potassium: Implications for mechanisms of initiation and maintenance of ventricular fibrillation. J Molec Cell Cardiol 21:21-40.

Curtis MJ, Hearse DJ (1989b). Reperfusion-induced arrhythmias are critically dependent upon occluded zone size: Relevance to the mechanism of arrhythmogenesis. J Molec Cell Cardiol (in press).

Curtis MJ, Johnston KM, MacLeod BA, Walker MJA (1985). The actions of felodipine on arrhythmias and other responses to myocardial ischaemia in conscious rat. Am J Pharmacol 117:169-178.

Curtis MJ, MacLeod BA, Tabrizchi R, Walker MJA (1986). An improved perfusion apparatus for small animal hearts. J Pharmacol Methods 15:87-94.

Curtis MJ, MacLeod BA, Walker MJA (1984). Antiarrhythmic actions of verapamil against ischaemic arrhythmias. Br J Pharmacol 83:373-385.

Curtis MJ, MacLeod BA, Walker MJA (1986). An improved pithed rat preparation: the actions of the optical enantiomers of verapamil. Asia Pacific J Pharmacol 1:73-79.

Curtis MJ, MacLeod BA, Walker MJA. The effects of ablation in the central nervous system on arrhythmias induced by coronary occlusion. Br J Pharmacol 86:663-670.

Curtis MJ, MacLeod BA, Walker MJA (1987). Models for the study of arrhythmias in myocardial ischaemia and infarction: The use of the rat. J Molec Cell Cardiol 19:399-419.

Curtis MJ, Walker MJA (1986). The mechanism of action of the optical enantiomers of verapamil agonists against ischaemia-induced arrhythmias in the conscious rat. Br J Pharmacol 89:137-149.

Curtis MJ, Walker MJA (1988). The mechanisms of action of calcium antagonists on arrhythmias in early myocardial ischaemia: studies with nifedipine and DHM-9. Br J Pharmacol 94:1275-1286.

Curtis MJ, Walker MJA (1989). Quantification of arrhythmias using scoring systems: An examination of seven scores in an _in vivo_ model of regional myocardial ischaemia. Cardiovasc Res 22:656-665.

Curtis MJ, Walker MJA, Yuswack T (1986). Actions of the verapamil analogues, anipamil and verapamil, against ischaemia-induced arrhythmias in conscious rats. Br J Pharmacol 88:355-361.

Fozzard HA (1975). Validity of myocardial infarction models. Circulation 52:131-146.

Harvie CJ, Collins GA, Miyagishima RT, Walker MJA (1978). The actions of prostaglandins E_2 and $F_{1\alpha}$ on myocardial ischaemia-infarction arrhythmias in the dog. Prostaglandins 17:885-899.

Johnston KM, MacLeod BA, Walker MJA (1980). ECG and other responses to ligation of a coronary artery in the conscious rat. In: Rat electrogram in Pharmacology & Toxicology. Ed. Budden R, Deitweiler DK, Zbindin G, Pergamon Press, UK, pp 243-252.

Johnston KM, MacLeod BA, Walker MJA (1983). Responses to ligation of a coronary artery in conscious rats and the actions of antiarrhythmics. Can J Physiol Pharmacol 61:1340-1353.

Kelliher GJ, Dix RA, Soifer, BE (1983). Animal studies in artificially induced myocardial infarction. In: Myocardial Infarction and Sudden Death. ed. Marguiles E, Academic Press, pp 21-59.

Langer GA (1978). Interspecies variation in myocardial physiology: The anomalous rat. Environ Health Perspect 26:175-179.

Opie LH, Coetzee WA, Dennis SC, Thandroyen FT (1988). Apotential role of calcium ions in early ischemic and reperfusion arrhythmias. Ann NY Acad Sci 522:464-477.

Trolese-Mongheal Y, Duchene-Marullaz P, Trolese JF, Leinot M, Lamar J-C, Lacroix P (1985). Sudden death and experimental acute myocardial ischaemia. Am J Cardiol 56:677-681.

Walker MJA, Beatch GW (1989). Amelioration by Ca^+ antagonists of adverse responses to myocardial ischaemia. In: Japan USA Symposium on Cardiovascular Drugs. Recent Advances on Ca^{++}-channels and Ca^{++}-antagonists. Pergamon Press, New York (in press).

Walker MJA et al., (1988). The Lambeth Convention on arrhythmia studies in ischaemia, infarction and reperfusion. Cardiovas Res 22:447-455.

Drug design and new approaches

Calcium Channel Modulators in Heart and Smooth Muscle: Basic Mechanisms and Pharmacological Aspects
S. Abraham and G. Amitai, editors
© 1990, VCH, Weinheim/Deerfield Beach, FL and Balaban, Rehovot/Philadelphia

Ca-agonists

RAINER GROSS, MARTIN BECHEM, SIEGBERT HEBISCH AND MATTHIAS SCHRAMM

Institute of Pharmacology, Bayer AG, P.O. Box 101709, 5600 Wuppertal 1, FRG

INTRODUCTION

Calcium plays a fundamental role for the function of cells. Therefore, interest of both basic scientists and of pharmacologists has been directed to the actions and movement of this ion. The discovery and the clinically proven therapeutic benefit of compounds specifically interacting with the Ca-channel like the prototypes verapamil (Haas and Härtfelder, 1962) and nifedipine (Vater et al., 1972) have further intensified electrophysiological, biochemical, pharmcological and clinical research related to the Ca-channel.

This research has been greatly enhanced by the discovery of dihydropyridine-derivatives (DHP´s) with pharmacological properties just opposite to those of all the Ca-antagonists known so far (Schramm et al., 1983 a, Schramm et al. 1983 b). After intensive pharmacological studies, the mode of action of these so-called Ca-agonists became evident. At the same time the dihydropyridine-receptor has been studied (Bellemann et al., 1981), isolated and characterized (Curtis and Catterall, 1984, Glossmann et al., 1985), leading to knowledge of the channel structure (Mikami et al., 1989). The greatest benefit was brought by these compounds to the understanding of the electrophysiology of the Ca-channel, and this deeper insight into the function of the Ca-channel will certainly have some positive spin-off to pharmacology, again. The Ca-agonistic DHP's increase voltage-dependent Ca-influx into the cell and thus exert positive inotropic effects.

They may add a new mechanism of positive inotropic drug action to our therapeutic armentarium.

STRUCTURES

Five Ca-agonistic dihydropyridines have been published so far: YC-170 (Takenaka and Maeno, 1982, Nakaya et al., 1986), CGP 28-392 (Erne et al., 1984), BAY K 8644 (Schramm et al., 1983 a), H 160/51 (Gjörstrup, 1985), and 202-791 (Hof et al., 1985) (Figure 1).

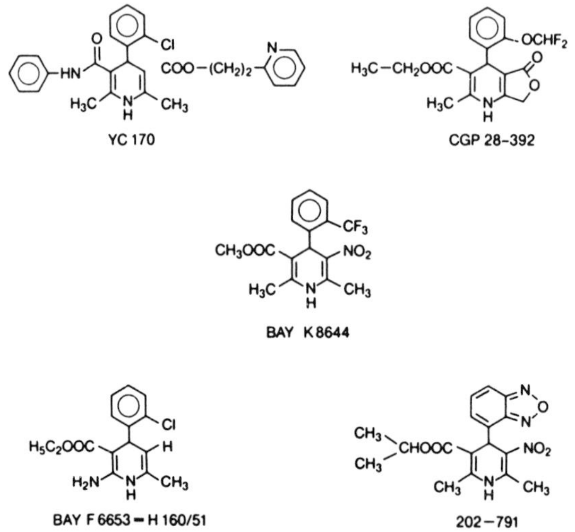

Figure 1. Chemical structure of Ca-agonists.

Only minor changes to the Ca-antagonistic DHP's resulted in a pharmacological profile just opposite to Ca-antagonism: the chemically common feature of the Ca-agonists is the combination of an ester moiety and a non-ester moiety at the DHP-ring (see Fig. 1). Due to this asymmetry there is a chiral center and therefore pairs of enantiomeres exist. For BAY K 8644 (Franckowiak et al., 1985), H160/51 (Gjörstrup et al., 1986), and for 202-791 (Hof et al., 1985) it has been shown, that the Ca-agonistic principle resides in only one of the enantiomeres, while the optical antipode is a Ca-antagonist.

For most of these compounds the potency of the Ca-antagonistic enantiomere is much lower; therefore, results obtained with the racemic mixture reflect the effect of the respective Ca-agonistic enantiomere; it is a "pure" Ca-agonistic response. For 202-791, however, similar potencies of both enantiomeres have been reported (Hof et al., 1985). The pharmacological profile of the racemic mixture therefore resembles the combined effects of a Ca-agonist and a Ca-antagonist. It can easily be shown that such a combination overcomes the main drawback of the Ca-agonists known so far, i.e. the unwanted vasoconstriction which is appearently inherent in this mechanism. Future development will show whether on this basis drug developments are possible or whether vasoconstriction can also be overcome by an increased cardioselectivity. There are distinct graduations between different Ca-agonists with respect to the extent of positive inotropic versus vasoconstrictive effects. The one extreme is represented by YC-170 which exhibits almost no inotropic but only vasoconstrictive effects (Nakaya et al., 1986).

HEMODYNAMIC EFFECTS

In the isolated perfused guinea pig heart, perfused according to Langendorff (Opie, 1965), the inotropic and vascular effects can be read from left ventricular peak pressure and coronary perfusion pressure at constant flow: while with 202-791, due to the combination effect mentioned above, there is a drop in coronary perfusion pressure - as seen with Ca-antagonists - concomitant with a rise of left ventricular isovolumic peak pressure (see Figure 2).

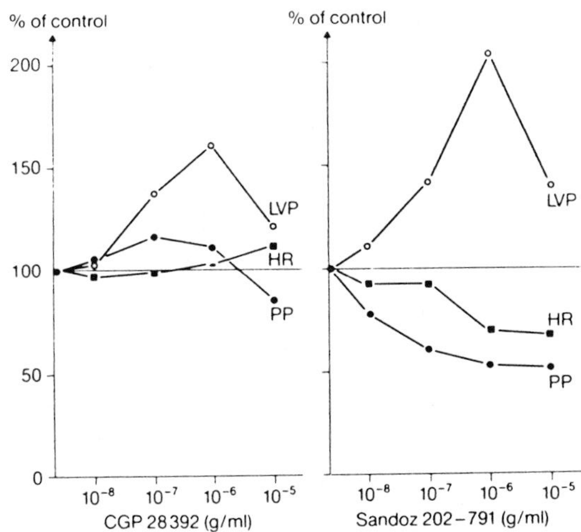

Figure 2. Effect of two Ca-agonists on left ventricular peak pressure (LVP), coronary perfusion pressure (PP), and heart rate (HR) in isolated perfused guinea-pig hearts.

The positive inotropic response is accompanied by a clear-cut vasoconstriction under a "pure" Ca-agonist like CGP 28-392.

Comparing the amount of positive inotropic response and vasoconstrictive properties of BAY K 8644 and CGP 28-392, studies in chronically instrumented, conscious dogs, revealed a marked difference also in vivo, (Preuss et al., 1984, Preuss et al., 1985): while CGP 28-392 exerts at comparable increases of mean aortic blood pressure no evident positive inotropic effects (as estimated from $LV(dP/dt)_{max}$), BAY K 8644 elicits a positive inotropic cardiac response of more than double the relative amount as compared to the pressure increase. In order to test whether this accelerated left ventricular pressure development increases cardiac performance, left ventricular function curves were obtained from left ventricular stroke work over left ventricular enddiastolic pressure from experiments in conscious, chronically instrumented dogs, paced at a constant heart rate of 150 bpm (see Figure 3).

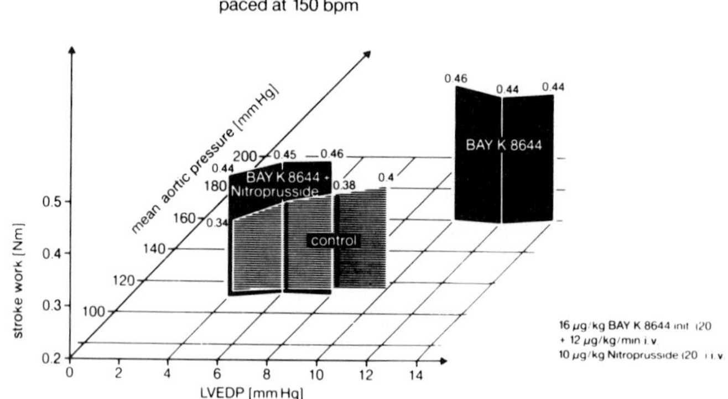

Figure 3. Ventricular function curves obtained in conscious dogs under the influence of BAY K 8644 with and without sodium-nitroprusside. (From Gross et al., 1985 by permission)

BAY K 8644 increases left vertricular stroke work, but due to the simultaneous vasoconstriction, the ventricular function curve is shifted to the upper right, i.e. to higher aortic and higher filling pressures. When the vasoconstriction was then antagonized by injection of an appropriate dose of the pure vasodilator sodium-nitroprusside, the left ventricular function curve could be shifted back to its control position. Under these conditions, the left ventricular stroke work was higher than for control, indicating that the heart performed a higher stroke work against the same aortic pressure from any filling pressure (Gross et al., 1985).

ECONOMY OF CONTRACTION

The therapeutic value of a positive inotropic principle substantially depends on the economy of the increased cardiac performance. We therefore measured the energy costs for the inotropic effect calculated from the external mechanical cardiac work (pressure-stroke volume product) and oxygen consumption In the isolated perfused working guinea-pig heart (Neely et al., 1967) (see Figure 4).

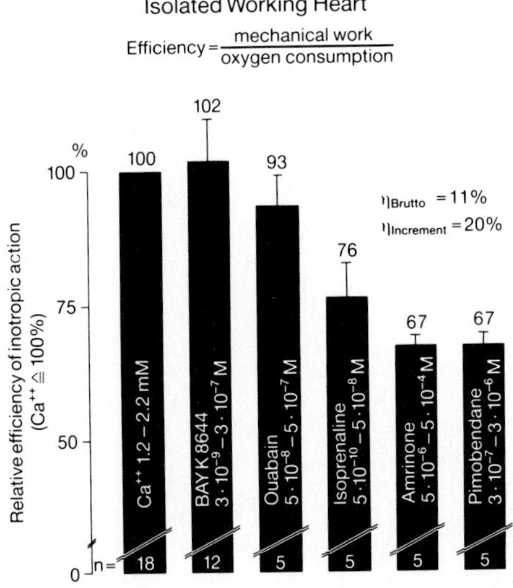

Figure 4. Net efficiency of the increase of cardiac performance under stimulation with various inotropic drugs. (given are means and S.E.M., n = number of exp.).

Gross efficiency under control conditions was 11%, close to the values reported in literature (Siess et al., 1981). Perfusion with increasing Ca-concentrations (1.2 to

2.2 mM) yielded about a doubled cardiac performance. Net efficiency for this inotropic response was 20%. Different inotropes were measured at concentrations eliciting similar inotropic effects ($5 \cdot 10^{-8}$ to $5 \cdot 10^{-7}$ M ouabain, $3 \cdot 10^{-9}$ to $3 \cdot 10^{-7}$ M BAY K 8644, $5 \cdot 10^{-10}$ to $5 \cdot 10^{-8}$ M isoprenaline, $5 \cdot 10^{-8}$ to $5 \cdot 10^{-4}$ M amrinone, and $3 \cdot 10^{-7}$ to $3 \cdot 10^{-6}$ M pimobendane). Optimal net efficiency could only be achieved under stimulation with Ca^{++}, ouabain or BAY K 8644, while under stimulation with cAMP-dependent drugs net efficiency markedly declined: with isoprenaline only 3/4 and with amrinone or with pimobendane only 2/3 of this net efficiency were achieved. This indicates that the Ca-agonist stimulates the heart more economically than cAMP-dependent drugs, which induce an over-proportionate increase of energy consumption. In this study the isolated hearts were paced at a constant heart rate; therefore, under in vivo conditions this difference should be even more pronounced due to the positive chronotropic effect of the latter drugs.

MODE OF ACTION

Pharmacological studies have revealed that Ca-agonists do not interact with intracellular structures,

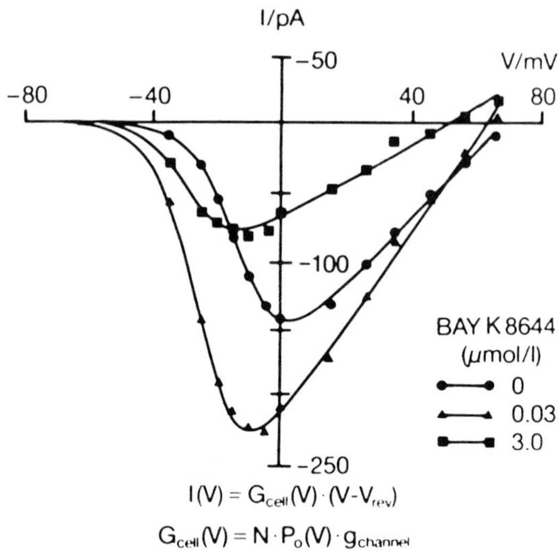

Figure 5. Effect of BAY K 8644 on peak Ca-current-voltage relationship. Holding potential was -70mV. (From Bechem and Schramm, 1988 by permission).

they exert no so-called "Ca-sensitizing" effect as evidenced by experiments with skinned fibers (Thomas et al., 1985 b). The contractile effect is caused by an enhanced Ca-influx into the cell, as could be demonstrated by measurements of ^{45}Ca-fluxes (Schramm et al., 1985). The concentration-dependent increase of action-potential duration (Thomas et al., 1985 a) plausibly explains the inotropic response by an effect on the sarcolemmal Ca-channels. Voltage-clamp studies using the whole cell recording patch clamp technique with BAY K 8644 elucidated a lot of the nature of the interaction between the Ca-agonist and the Ca-channel (see Figure 5): At concentrations exerting positive inotropic effects (30 nM; (-A-)), BAY K 8644 increases Ca-current in a voltage-dependent manner. Mainly in a voltage range between -60 mV and +10 mV the Ca-current is increased.

The peak of the current-voltage-relationship (IV-curve) is shifted to more negative potentials. However, there is no change in the reversal potential nor any significant change of the "ohmic" part of the IV-curve at positive clamp potentials. These findings give evidence against an influence of Ca-agonists on either single Ca-channel conductance or on the number of available Ca-channels but rather for a strong influence on the open

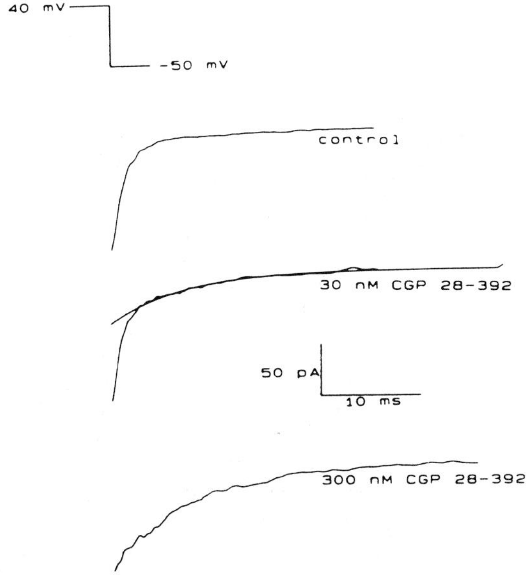

Figure 6. Effect of CGP 28-392 on "tail-currents". (From Bechem and Schramm, 1988, by permission).

probability of single Ca-channels. Besides the increase of Ca-current during the clamp-pulse, Ca-agonists slow the relaxation of Ca-current after repolarization ("tail-current") (see Figure 6) even at positive membrane-potentials where no steady-state elevation of Ca-current is seen. The time constant of the exponential decay increases more than sixfold (Bechem and Schramm, 1988).

The influence of Ca-agonists on the open and closed times of a single channel can be calculated from the activation and deactivation (tail current) kinetics of the Ca-current. While the closed times are not modified by BAY K 8644, the open times of the single Ca-channel are markedly prolonged. This has also been shown in single channel recordings (Hess et al., 1984). In Figure 7 the calculated open and closed times are given in a semilogarithmic plot for BAY K 8644 and CPG 28-392. The exponential voltage dependence of the open times is shifted by about 40 mV to more negative potentials by both compounds. This shift does not depend on drug concentration suggesting that the drug-bound Ca-channel has a fixed open time.

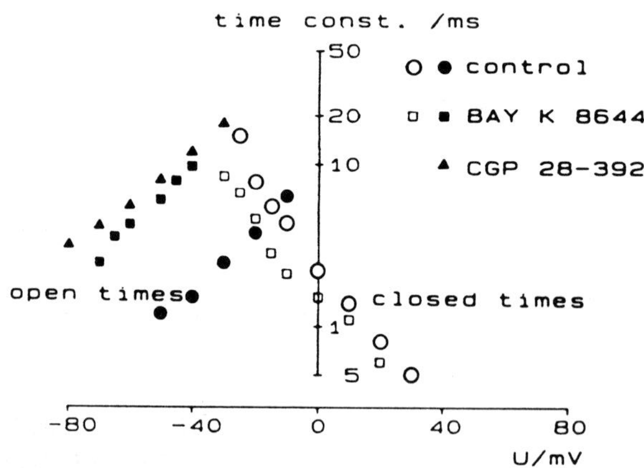

Figure 7. Single channel open and closed times as influenced by 3μM BAY K 8644 and CGP 28-392. (From Bechem and Schramm, 1988, by permission).

In experiments with low concentrations (Figure 6) therefore two components of the tail-current can be identified, a slow component representing the drug-bound channels and another one identical to the drug-free condition. Dose-dependence thus reflects fractional receptor occupancy. K_D-values (5 nM for BAY K 8644 and 100 nM for CGP 28-392) calculated from such experiments fit with those obtained from receptor binding studies (Janis et al., 1984, Laurent et al., 1985).

ANTAGONISTIC EFFECTS OF CA-AGONISTS

Functional experiments in isolated organs (see Thomas et al., 1984) have revealed that at high concentrations the Ca-agonistic effect of BAY K 8644 is diminished. This can also be read from the IV-curves in single myocardial cells (see Figure 5): Under 3 µmol/l Bay K 8644 the IV-curve is depressed below the control curve, but still peaks at more negative membrane potentials. This Ca-antagonistic effect of Ca-agonists at high drug-concentrations is not caused by a reduced single channel open probability but rather by a decrease of the number of available Ca-channels.

Similarly, if the cell is partly depolarized, the Ca-agonistic effect of BAY K 8644 is diminished or even reversed to Ca-antagonism: Sanguinetti and Kass (1984) have shown in voltage-clamp studies BAY K 8644 to increase Ca-current (500 nM) when the test pulse started from a well polarized holding potential (-62 mV). If, however, holding potential was only -45 mV the effect of BAY K 8644 resembles Ca-antagonism, since the IV-curve is depressed below its control. Such findings may help to understand tissue selectivity on the basis of different resting potentials. In terms of therapeutic use, they can be taken as an in-built safety mechanism, limiting Ca-influx and protecting the cell from Ca-overload.

THE UNIFYING MODEL

The electrophysiological studies with Ca-agonists have lead to models of drug-action and of the function of the Ca-channel itself (see Figure 8). Ca-agonists lead to a dose-dependent shift of the open-probability of the Ca-channel to the left, i.e. to more negative membrane-potentials. At low drug concentrations or during an action potential elicited in well polarized cells therefore Ca-influx is increased by a Ca-agonist. Since the steady-state inactivation curve is also shifted to the left, the number of available channels is dose-dependently decreased if resting membrane potential is low. Under these conditions Ca-agonists act like Ca-antagonists. Since for Ca-antagonistic DHP's also Ca-agonistic effects at very low concentrations have been found (see Thomas et al., 1984), the proposed mode of action might be common for both Ca-antagonists and Ca-agonists.

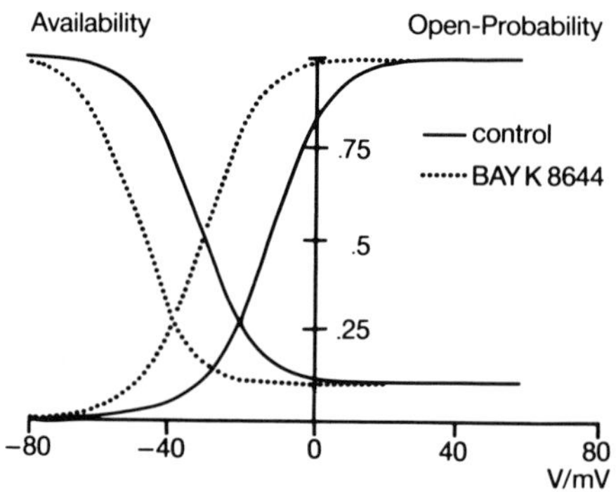

Figure 8. Shift of the open-probability and the steady-state inactivation curve by BAY K 8644, schematic.

Such detailed knowledge of the molecular mode of drug action may lead to the development of cardiospecific Ca-agonists for treatment of congestive heart failure.

REFERENCES

Bechem M, Schramm M (1987). Calcium Agonists. J Mol Cell Cardiol 19(Supl.II):63-75.

Bechem M, Schramm M (1988). Electrophysiology of Dihydropyridine Ca-Agonists. In Morad M, Nayler W, Kazda S, Schramm M (eds): "The Calcium channel: Structure, Function and Implications", Berlin, Heidelberg, New York, London, Paris, Tokyo. Springer Verlag, pp 63-70.

Bellemann P, Ferry D, Lübbecke F, Glossmann H (1981). ^{3}H-Nitrendipine, a potent calcium channel antagonist binds with high affinitiy to cardiac membranes. Arzneim Forsch/Drug Res 31:2064-2067.

Curtis BM and Catterall WA (1984). Purification of the calcium antagonist receptor of the voltage-sensitive calcium channel from skeletal muscle transverse tubulus. Biochemistry 23:2113-2118.

Erne P, Burgisser E, Bühler FR, Dubach B, Kühnis H, Meier M, Rogg N (1984). Enhancement of calcium influx in human platelets by CGP 28392, a novel dihydropyridine. Biochem Biophys Res Comm 118:842-847.

Franckowiak G, Bechem M, Schramm M, Thomas G (1985). The optical isomers of the 1,4-dihydropyridine BAY K 8644 show opposite effects on Ca-channels. Europ J Pharmacol 114:223-226.

Gjörstrup P (1985). Effects of H160/51, a new Ca-agonist, and its interaction with felodipine on cardiac and vascular tissue in vitro. Proc Cardiovascular Pharmacotherapy, Intern Symposium Geneva April 22-25, Abstr. 127.

Gjörstrup P, Harding H, Isaksson R, Westerlund C (1986). The enantiomers of the dihydropyridine derivative H160/51 show opposite effects of stimulation and inhibition. Eur J Pharmacol 122:357-361.

Glossmann H, Ferry DR, Goll A, Striessnig J, Schober M (1985). Calcium channels: basic properties as revealed by radioligand binding studies. J Cardiovasc Pharmacol - 7 Suppl 6:S20-S30.

Gross R, Bechem M, Kayser M, Schramm M, Taniel R, Thomas G (1985). Effects of the calcium agonistic dihydropyridine on the heart. In Fleckenstein A, VanBreemen C, Gross R, Hoffmeister F (eds): "Cardiovascular effects of dihydropyridine-type calcium antagonists and agonists", Berlin, Heidelberg, New York, Tokyo; Springer Verlag, pp 218-235.

Haas H and Härtfelder G (1962). C-Isopropyl-c-(N-methyl-homoveratryl)-y-aminopropyl-3,4-dimethoxy-phenylacetonitril, eine Substanz mit coronargefäßerweiternden Eigenschaften. Arzneim Forsch 12:549-558.

Hess P, Lansmann JB, Tsien RW (1984). Different modes of calcium channel gating behaviour favoured by dihydropyridine Ca-agonists and antagonists. Nature 311:538-544.

Hof RP, Rüegg UT, Hof A, Vogel A (1985). Stereoselectivity at the calcium channel: opposite action of the enantiomeres of a 1,4-dihydropyridine. J Cardiovasc Pharmacol 7:689-693.

Janis RA, Rampe D, Sarmiento JG, Triggle DJ (1984). Specific binding of a Ca channel activator BAY K 8644 to membranes from cardiac muscle and brain. Biochem Biophys Res Comm 121:317-323.

Laurent S, Kim D, Smith TW, Marsch JD (1985). Inotropic effect, binding properties and calcium flux effects of the calcium channel agonist CGP 28392 in intact cultured embryonic chick ventricular cells. Circ Res 56:676-682.

Mikami A, Imoto K, Tanabe T, Niidome T, Mori Y, Takeshima H, Narumiya S and Numa S (1989). Primary structure and functional expression of the cardiac dihydropyridine-sensitive calcium channel. Nature 340:230-233.

Nakaya H, Hattori Y, Tohse N, Kanno M (1986). Voltage-dependent effects of YC-170, a dihydropyridine calcium channel modulator in cardiovascular tissue. Naunyn-Schmiedeberg's Arch Pharmacol 333:421-430.

Neely JR, Liebermeister H, Battersby EJ and Morgan HE (1967). Effect of pressure development on oxygen consumption by isolated rat hearts. Am J Physiol 212(4):804-814.

Opie LH (1965). Coronary flow rate and perfusion pressure as determinants of mechanical function and oxydative metabolism of isolated perfused rat hearts. J Physiol (London) 180:529-541.

Preuss KC, Brooks HL, Gross GJ, Warltier DC (1985). Positive inotropic actions of the calcium channel stimulator, BAY K 8644, in the awake, unsedated dog. Bas Res Cardiol 80:326-332.

Preuss KC, Chung NL, Brooks HL, Warltier DC (1984). Cardiovascular effects of the nifedipine analog CGP 28392 in the conscious dog. J Cardiovasc Pharmacol 6:949-953.

Sanguinetti MC and Kass RS (1984). Regulation of cardiac calcium channel current and contractile activity by the dihydropyridine BAY K 8644 is voltage dependent. J Mol Cell Cardiol 16:667-670.

Schramm M, Thomas G, Towart R, Franckowiak G (1983 a). Novel dihydropyridines with positive inotropic action through activation of Ca^{2+} channels. Nature 303:535-537.

Schramm M, Thomas G, Towart R, Franckowiak G (1983 b). Activation of calcium channels by novel 1,4-dihydropyridines. Arzneim Forsch 33:1268-1272.

Schramm M, Towart R, Lamp B and Thomas G (1985). Modulation of calcium ion influx by the 1,4-dihydropyridines nifedipine and BAY K 8644. J Cardiovasc Pharmacol 7:493-496.

Siess M, Stieler K, Seifart HJ (1981). Zur Wirkung von ARL-115 BS auf Funktion und Sauerstoffverbrauch isolierter Meerschweinchenherzvorhöfe im Vergleich zu g-Strophantin und Theophyllin. Arzneim Forsch/Drug Res 31:165-170.

Takenaka T and Maeno H (1982). A new vasoconstrictor 1,4-dihydropyridine derivative, YC-170. Jap J Pharmacol 32:139P.

Thomas G, Groß R, Schramm M (1984). Calcium channel modulation: ability to inhibit or promote calcium influx resides in the same dihydropyridine molecule. J Cardiovasc Pharmacol 6:1170-1176.

Thomas G, Chung M, Cohen CJ (1985 a). A dihydropyridine (BAY K 8644) that enhances calcium currents in guinea-pig and calf myocardial cells. Circ Res 56:87-96.

Thomas G, Groß R, Pfitzer G, Rüegg JC (1985 b). The positive inotropic dihydropyridine BAY K 8644 does not affect calcium sensitivity or calcium release of skinned cardiac fibres. Naunyn-Schmiedeberg's Arch Pharmacol 328:378-381.

Vater W, Kroneberg G, Hoffmeister F, Kaller H, Meng K, Oberdorf A, Puls W, Schloßmann K, Stoepel K (1972). Zur Pharmakologie von 4-(2'Nitrophenyl)-2,6-dimethyl-1,4-dihydropyridin-3.5-dicarbonsäuredimethylester (Nifedipin, BAY A 1040). Arzneim Forsch 22:1-14.

Calcium Channel Modulators in Heart and Smooth Muscle: Basic Mechanisms and Pharmacological Aspects
S. Abraham and G. Amitai, editors
© 1990, VCH, Weinheim/Deerfield Beach, FL and Balaban, Rehovot/Philadelphia

Design of calcium channel modulators

MICHAEL W. WOLOWYK AND EDWARD E. KNAUS

Faculty of Pharmacy and Pharmaceutical Sciences, University of Alberta, Edmonton, Alberta, Canada T6G 2N8

INTRODUCTION

Heart disease is one of the leading causes of death in North America accounting for 30 to 40% of all deaths. Hypertension, coronary artery disease (angina) and congestive heart failure are the most prevalent cardiovascular diseases in people between 35 and 70 years of age. At the present time such cardiovascular disorders continue to present a major challenge to the medicinal chemist, pharmacologist and clinician to discover adequate therapies. The most dramatic development in the treatment of such diseases involves the use of drugs such as the calcium antagonists nifedipine, verapamil and diltiazem (Fleckenstein, 1983). Among these, the dihydropyridine (DHP) calcium channel antagonists (eg. nifedipine), have received the most attention because they are most amenable to structural modification. Our research has been directed to designing and testing novel DHP drugs which act as calcium channel antagonists as well as agonists, with a focus on developing and identifying compounds with a higher tissue selectivity, longer duration of action and fewer side effects.

A number of studies (Janis and Triggle, 1983; Triggle and Janis, 1983; Bossert et al., 1981; Rodenkirchen et al., 1979) have identified relevant structural features of the DHPs that are important for calcium channel antagonist activity. Many second generation compounds with differential activity between heart, vascular smooth muscle and even some with selectivity for specific blood vessels in different locations of the body have been synthesized. Similar compounds which open the calcium channel and have effects diametrically opposite to those of the antagonists have also been identified (eg. Bay K 8644). Such calcium channel agonists are very effective in stimulating the heart to beat more efficiently and with greater force. Unfortunately the vasoconstrictor and arrhythmogenic properties preclude their clinical use.

RESULTS AND DISCUSSION

Our approach has involved the synthesis and evaluation of novel C-4(pyridyl) derivatives of the 1,4-DHP class of compounds. By carefully delineating the structural and conformational features important for activity, new compounds have been designed with therapeutic potential. Such compounds also provide useful probes for characterizing differences in receptor site interactions and calcium channel function in different tissues.

The first question addressed was whether a pyridyl (pyr) ring substituent at the C-4 position of the 1,4-DHP structure would be acceptable for calcium channel antagonist activity? Most second generation derivatives of nifedipine possessed a 2-nitrophenyl ring substituent at C-4. The relative calcium channel antagonist potencies of DHP derivatives in which the C-4 substituent was either a 2-pyr, 3-pyr or 4-pyr ring are compared in Figure 1. Among these the relative potency order for isomeric C-4 pyr analogues was 2-pyr > 3-pyr > 4-pyr. When considering the effects of different ester substituents at positions C-3 and C-5, increasing the size of the alkyl ester substituents enhanced activity. Those compounds which had asymmetric ester substituents were more potent than those having identical substituents. In the latter examples the C-4 position is a chiral center, and thus the possibility exists that one of the enantiomers may be even more active.

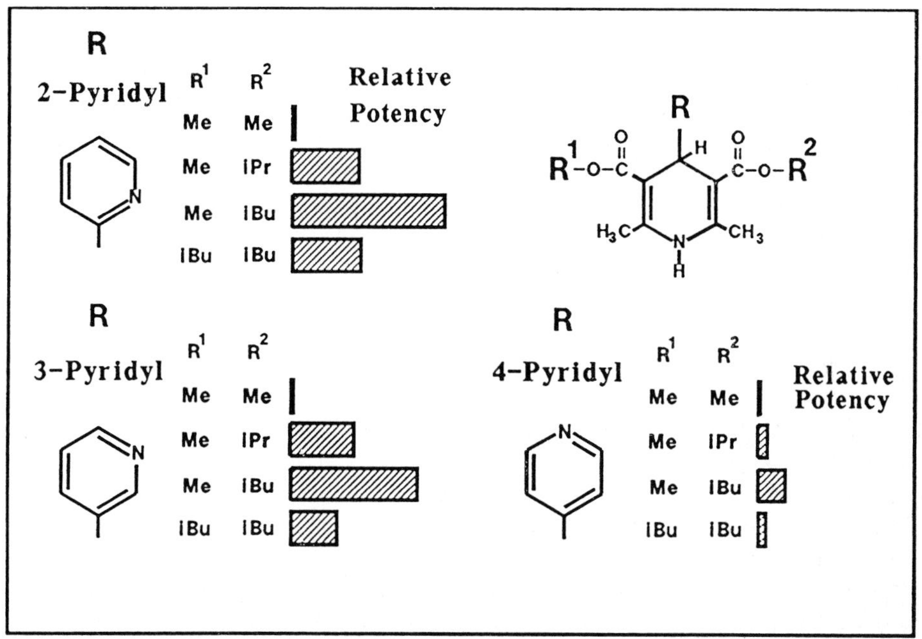

Figure 1. Relative potencies of Dialkyl 1,4-dihydro-2,6-dimethyl-4-(pyridinyl)-3,5-pyridinedicarboxylates. From data of Dagnino et al. 1986.

Conformational changes of the DHP structure, which affect the ability of these compounds to interact with the DHP receptor sites on the calcium channel, have also been thoroughly investigated. Dagnino et al. (1987a; 1987b) have shown that nifedipine and related analogues exist in a boat conformation with the C-4 ring system positioned perpendicular to the 1,4-DHP ring (Figure 2A-D). Replacement of the C-4 pyridyl ring with a dihydropyridyl or tetrahydropyridyl ring resulted in a sequential change in the steric conformation of the C-4 substituents

(Figure 2C, 2D and 2E) to the point where the 2′-tetrahydropyridyl ring (Figure 2E) was no longer perpendicular to the DHP ring. This was accompanied with a loss of activity.

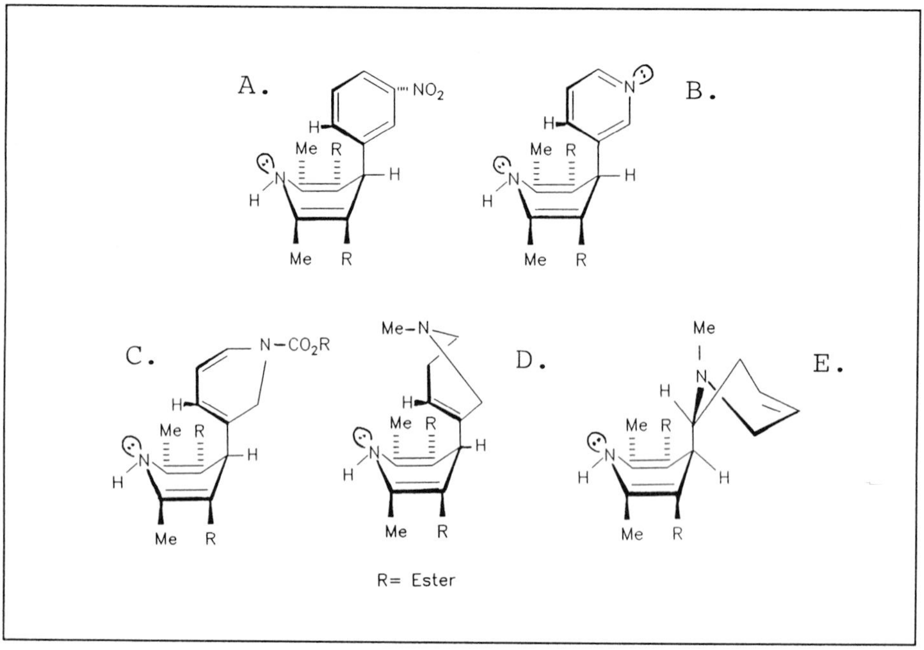

Figure 2. Three dimensional conformations of 1,4-DHP's with modified C-4 substituents: A. nitrophenyl, B. pyr, C. dihydropyridyl, D. 3′-tetrahydropyridyl and E. 2′-tetrahydropyridyl.

The steric effect of replacing the 1,4-DHP ring with a 1,2-DHP ring led to a dramatic reduction in calcium channel antagonist activity (Soboleski et al., 1988; Wynn et al., 1988). This reduction in activity is attributed to: 1. the fact that the diene moiety of the 1,2-DHP ring system is planar relative to the boat-shaped 1.4-DHP ring; 2. a major change in the relative orientation of the C-4 ring system, which is perpendicular in the 1,4-DHP system but co-planar in the 1,2-DHP system; and 3. a concomitant change in the non-bonded interactions between the C-3, C-4 and C-5 substituents. Fossheim (1986) also provided evidence showing that the most active 1,4-DHP compounds exhibited the smallest degree of puckered ring distortion from planarity.

The ester substituents at C-3 and C-5 also have been shown to have a subtle influence on the activity and/or conformation of DHP calcium antagonists (Ramesh et al., 1987). Although increasing the size and bulk of the esters tended to increase the lipophilic properties and activity, very large bulky esters (eg. tert-butyl) are believed to flatten the 1,4-DHP ring to a less puckered conformation. The increased steric interactions between the C-3, C-4 and C-5 substituents would alter the position of the perpendicular C-4 ring system and also tend to flatten the boat-shaped 1,4-DHP ring system. In the latter class of compounds the rank order of activity changed from 2-pryridinyl > 3-pryridinyl > 4-pryridinyl to 3-pryridinyl > 4-pryridinyl > 2-pryridinyl in that series.

In light of the useful structural and conformational information obtained from such studies, and in particular, the fact that pyridyl ring substituents were bioisosteric with a nitro phenyl ring at position C-4, provided an opportunity to design and synthesize a number of novel DHP compounds with unique activity.

Matowe et al. (1989) recently described the pharmacological properties of AK-2-38, a novel C-4 2-pyridyl 1,4-DHP analog, which exhibited twice the potency of nifedipine on smooth muscle. The dose related effects of AK-2-38 on smooth and cardiac muscle are compared in Figure 3. AK-2-38 caused a potent inhibition of smooth muscle contractility mediated by the voltage sensitive calcium channel (receptor agonist or high KCl induced) with an $IC_{50} = 6.7 \times 10^{-9}$ M. In the dosage range which inhibited smooth muscle, AK-2-38 was a partial agonist on cardiac muscle. This high differential activity could be therapeutically beneficial as an antihypertensive agent, especially in patients with compromised myocardial function. In attempts to further characterise A-2-38, consideration must be given to the fact that it is a racemate. Therefore, the possibility exists that one of the enantiomers may be even more potent or selective.

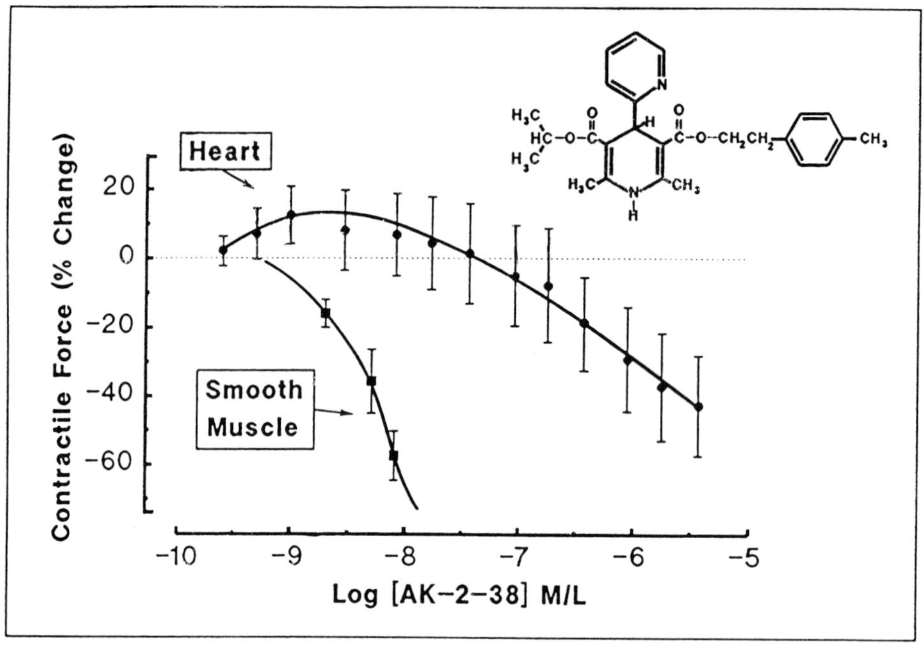

Figure 3. Effects of AK-2-38 on cardiac (guinea pig atria) and guinea pig ileal longitudinal smooth muscle. Values are means ± SEM (n=3).

In another series of DHP derivatives structural features, which are associated with a prolonged duration of action, were identified (Dagnino et al., 1987a; Wolowyk et al., 1987). Two of the 37 compounds tested are depicted in figure 4. Structurally, the difference between these two compounds is that D4DHP7 has identical isobutoxycarbonyl esters at C-3 and C-5 of the DHP ring and D4DHP8 has one of these large ester groups replaced by a smaller methoxycarbonyl ester. Functionally this difference was observed to cause a prolonged degree of smooth muscle calcium channel antagonist activity in the case of D4DHP7.

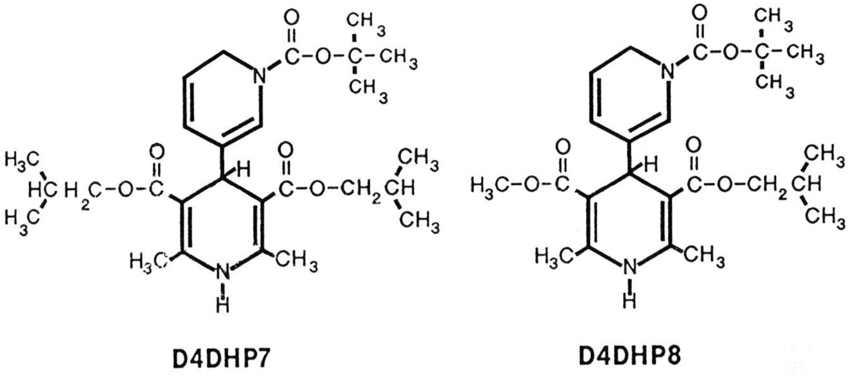

Figure 4. 3,5-Disubsituted 4-dihydro-[1-alkyl (phenyl)oxycarbonyl]pyridinyl-1,4-dihydro-2,6-dimethylpyridines.

The property of relative ligand irreversibility was a common feature of seven derivatives which all possessed bulky lipophilic identical isopropoxycarbonyl or isobutoxycarbonyl ester substituents at the C-3 and C-5 positions. Compounds having smaller identical esters (methoxycarbonyl or ethoxycarbonyl) had shorter durations of action. This effect was independent of the nature of the C-4 dihydropyridinyl nitrogen substituent, and was considered to be due, in part, to the increased hydrophobic hydrogen bonding interactions and possibly conformational changes in the DHP ring induced by these bulky substituents.

Such compounds could also be used to further investigate the nature of the dihydropyridine receptor(s) which control voltage operated calcium channels. Furthermore, such compounds could be clinically useful as well.

In addition to the above mentioned examples, recent preliminary work in our laboratory has led to the discovery of a unique class of DHP compounds with cardioselective calcium channel agonist and smooth muscle calcium channel antagonist activity. This could mean that at least two subtypes of calcium channels may exist, one predominantly in cardiac tissue and another subtype predominantly in smooth muscle. Thus, the investigation of novel DHP analogues continues to provide more information and useful probes which could shed more light on the heterogeneity of voltage sensitive calcium channels.

ACKNOWLEDGEMENTS

The work presented has been supported by operating grants from the Canadian Medical Research Council and equipment, fellowships and studentships from the Alberta Heritage Foundation for Medical Research.

REFERENCES

Bossert F, Meyer H, Wehinger F (1981). 4-Aryl dihydropyridines, a new class of highly active calcium antagonists. Angew Chem Int Ed Engl 20:762-769.

Dagnino L, Li-Kwong-Ken MC, Wynn H, Triggle CR, Wolowyk MW, Knaus EE (1986). Synthesis and calcium channel antagonist activity of dialkyl-1,4-dihydro-2,6-dimethyl-4-(pyridinyl)-3,5-pyridine dicarboxylates. J Med Chem 29:2524-2529.

Dagnino L, Li-Kwong-Ken MC, Wynn H, Wolowyk MW, Triggle CR, Knaus EE (1987a). Synthesis and calcium channel antagonist activity of dialkyl 4-(dihydropyridinyl)-1,4-dihydro-2,6-dimethyl-3,5-pyridine dicarboxylates. J Med Chem 30:640-646.

Dagnino L, Li-Kwong-Ken MC, Wolowyk MW, Triggle CR, Knaus EE (1987b). Synthesis and calcium channel antagonist activity of dialkyl hexahydro-1,2',6'-trimethyl(bipyridine)-3',5'-dicarboxylates. Eur J Med Chem 22:499-503.

Fleckenstein A (1983). "Calcium and antagonism in heart and smooth muscle." New York: J. Wiley & Sons, pp 286-329.

Fossheim R (1986). Crystal structure of the DHP calcium antagonist felodipine. DHP binding prerequisites assessed from crystallographic data. J Med Chem 29:305-307.

Janis RA, Triggle DJ (1983). New developments in calcium channel antagonists. J Med Chem 26:775-785.

Matowe WC, Akula M, Knaus EE, Wolowyk MW (1989). AK-2-38, a nifedipine analogue with potent smooth muscle calcium antagonist action and partial agonist effects on isolated guinea pig left atrium. Proc West Pharmacol Soc 32:305-307.

Ramesh M, Matowe WC, Wolowyk MW, Knaus EE (1987). Synthesis and calcium channel antagonist activity of alkyl t-butyl esters of nifedipine analogues containing pyridinyl substituents. Drg Des Del 2:79-89.

Rodenkirchen R, Bayer R, Steiner R, Bossert F, Meyer H, Moller E (1979). Structure-activity studies on nifedipine in isolated cardiac muscle. Naunyn-Schmiedeberg's Arch Pharmacol 310:69-78.

Soboleski DA, Li-Kwong-Ken MC, Wynn H, Triggle CR, Wolowyk MW, Knaus EE (1988). Synthesis and calcium channel antagonist activity of nifedipine analogues containing 1,2-dihydro pyridyl in place of the 1,4-dihydropyridyl moiety. Drg Des Del 2:177-189.

Triggle DJ, Janis RA (1983). Calcium antagonists and ionophores. In Grover AK, Daniel EE (eds): "Calcium and contractility." Clifton NJ: Humana, pp 37-60.

Wolowyk MW, Li-Kwong-Ken MC, Dagnino L, Wynn H, Knaus EE (1987). Novel dihydropyridine calcium channel antagonists: Structural features which extend duration of action. Proc West Pharmacol Soc 30:97-100.

Wynn H, Ramesh M, Matowe W, Wolowyk MW, Knaus EE (1988). Synthesis and calcium channel antagonist activity of 1,2-dihydropyridyl analogues of nifedipine. Drg Des Del 3:245-256.

Calcium Channel Modulators in Heart and Smooth Muscle: Basic Mechanisms and Pharmacological Aspects
S. Abraham and G. Amitai, editors
© 1990, VCH, Weinheim/Deerfield Beach, FL and Balaban, Rehovot/Philadelphia

Structure-activity and regulatory perspectives for calcium channel activators and antagonists

J. FERRANTE, Y.W. KWON, M. GOPALAKRISHNAN, R. BANGALORE, X.-Y. WEI, W. ZHENG, M. HAWTHORN, D.A. LANGS AND D.J. TRIGGLE

School of Pharmacy State University of New York Buffalo, NY and Medical Foundation of Buffalo, Buffalo, NY, USA

INTRODUCTION

Voltage-dependent Ca2+ channels may be viewed as a family of pharmacologic receptors. They may be distinguished by a number of permeation, electrophysiologic and pharmacologic characteristics [Table 1].

TABLE 1. Properties of Plasmalemmal Ca2+ channels

	Channel Class		
	L	T	N
Activation voltage	-10mV	-70mV	-30mV
Inactivation rate	slow	fast	moderate
Conductance	25pS	8pS	13pS
Permeation	Ba2+>Ca2+	Ba2+=Ca2+	Ba2+>Ca2+
1,4-DHP sensitivity	sensitive	insensitive	insensitive
w-CTX sensitivity	sensitive [neurons]	insensitive	sensitive

1 Data compiled from several sources and is not intended to indicate that channels have uniquely the indicated properties.

As pharmacologic receptors these Ca2+ channels may be expected to have certain properties including the presence of specific drug binding sites, the existence of both activator and antagonist ligands,coupling to regulatory proteins that are associated with other classes of pharmacologic receptor, regulation by both homologous and heterologous influences and the altered expression of channel number and function in pathological conditions.

All of these anticipated characteristics have been demonstrated for the L class of Ca2+ channel: it is likely that similar developments will occur for the other channel classes. The existence of specific binding sites that accommodate both channel activators and antagonists raises the issue of whether such agents mimic the function of endogenous regulators (Triggle, 1988). Definitive answers are not available, but peptide candidate molecules have been described (Callewaert et al. 1988; Janis et al. 1988).

STRUCTURE-ACTIVITY RELATIONSHIPS

There is now general agreement that several discrete ligand binding sites are associated with the Ca2+ channel (reviewed in Janis et al. 1987) These sites are linked allosterically one to the other and to the permeation and gating machinery of the channel (Figure 1) and mediate the

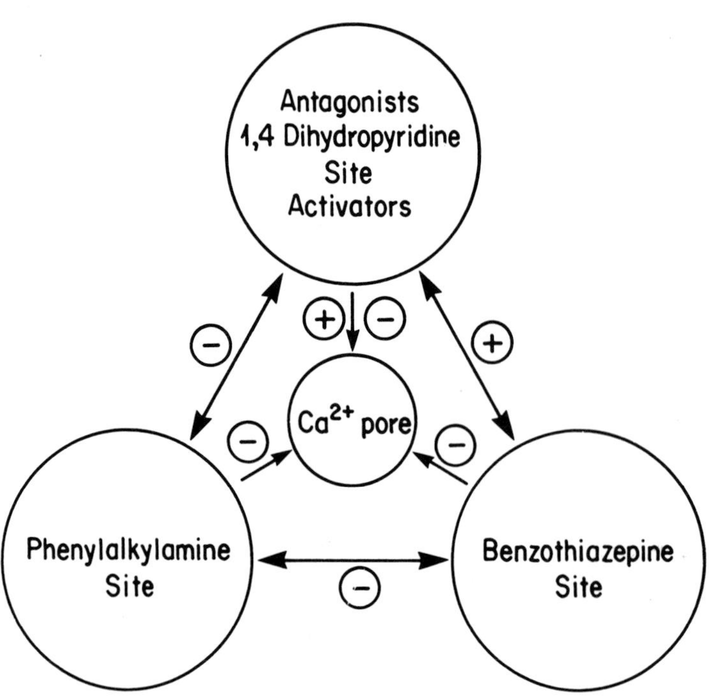

Figure 1. Schematic representation of the major drug binding sites of the L class of voltage-dependent Ca2+ channel. The

sites are depicted as linked allosterically one to the other and to the permeation and gating machinery of the channel.

dominantly cardiovascular pharmacology and therapeutics of the phenylalkylamine (verapamil), 1,4-dihydropyridine (nifedipine) and benzothiazepine (diltiazem) classes of agent. It is likely that other structural classes represented by pimozide, MDL 12,330A and HOE 166 may occupy other specific binding sites. The 1,4-dihydropyridine structure has attracted much attention because it includes the most potent agents and because this structure embraces both potent antagonists and activators (Figure 2).

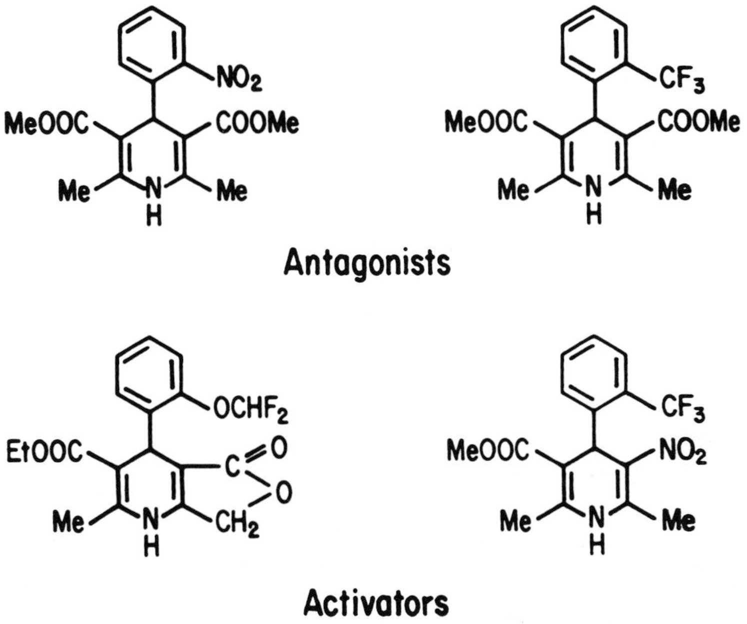

Antagonists

Activators

Figure 2. The structures of 1,4-dihydropyridine activators and antagonists.

The general structural requirements for activity in the 1,4-dihydropyridines are outlined in Figure 3 and have been derived from pharmacologic, in vivo and in vitro, and radioligand binding studies (reviewed in Janis et al. 1987; Triggle et al. 1989b).

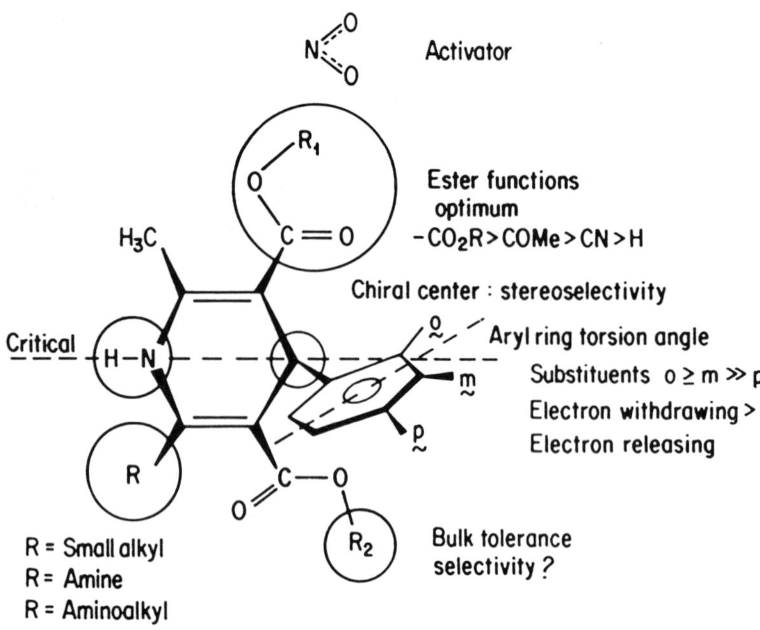

Figure 3. Representation of the major structural features determining activity in the 1,4-dihydropyridines.

There are several features of interest. It was early realized that these agents were more potent in smooth than in cardiac muscle, although the same qualitative structure-activity relationship was expressed. Furthermore, even in a single tissue different structure-activity expressions may be obtained according to the stimulus mode (Bolger et al. 1983; Triggle et al. 1989a; Figure 4). However, radioligand binding studies reveal the same high affinity binding sites in smooth and cardiac muscle membranes as well as in other cell types [reviewed in Janis et al. 1987; Triggle et al. 1984). Additionally, activators and antagonist of the 1,4-dihydropyridine series show remarkably similar structural requirements, and although the most potent activators possess a C5-NO2 group (Triggle et al. 1989b; Figure 2), activator properties may be detected, albeit transiently, even in agents of potent antagonist character.

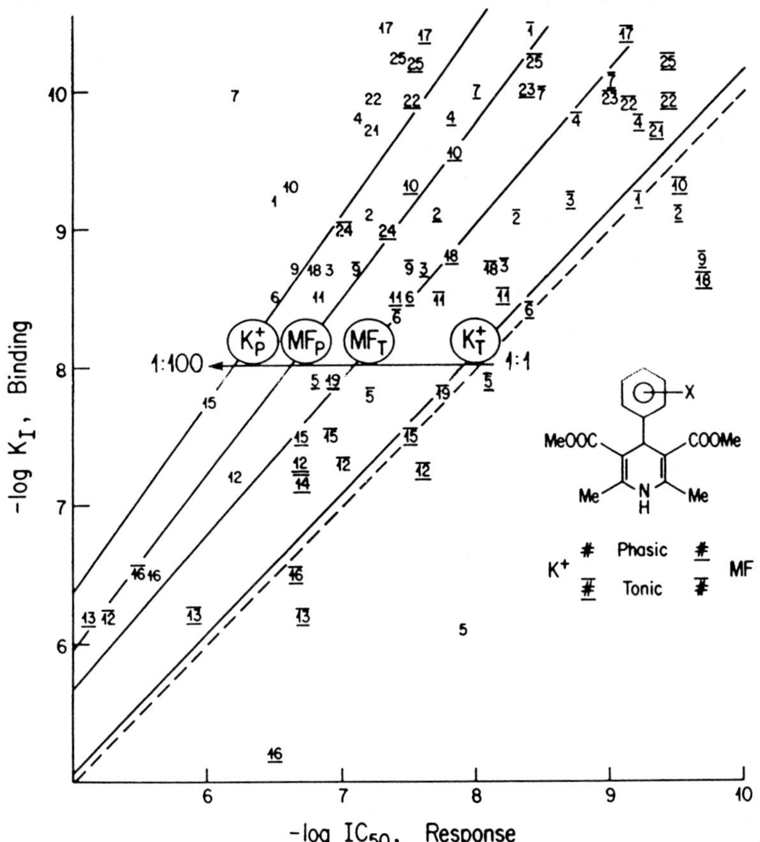

Figure 4. Correlations between the abilities of a series of 1,4-dihydropyridines of the nifedipine class to inhibit specific [3H]nitrendipine binding to guinea pig ileal longitudinal smooth muscle membranes and to inhibit the phasic and tonic components of tension response to K+ depolarization and to methylfurmethide [MF] activation of muscarinic receptors in this smooth muscle. Reproduced with permission of The New York Academy of Sciences from Triggle et al. (1989a).

Thus a number of important questions may be raised for the 1,4- dihydropyridine class of ligands (Triggle 1989). These include:

1. What are the structural demands for 1,4-dihydropyridine interaction at the Ca2+ channel?
2. What is the molecular basis for differentiation between activator and antagonist 1,4-dihydropyridines?
3. What is the basis for the observed tissue selectivity of the 1,4-dihydropyridines?

State-Dependent Drug-Channel Interactions

Structure-activity expressions of channel active ligands can be interpreted only in terms of the several conformations adopted by the channel proteins as the channel cycles through resting, open and inactivated states. The Ca2+ channel undergoes voltage-dependent, ligand-modulated transitions between several discrete states (Figure 5).

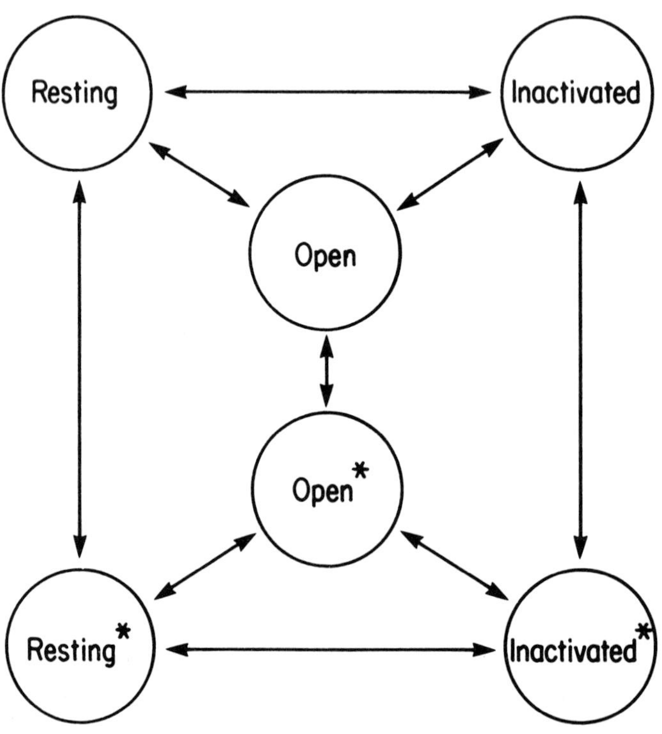

Figure 5. The channel depicted as cycling through resting, open and inactivated states. Each of these states may exhibit preferential affinity or access to specific ligands.

Ligands may exhibit preferential affinity for or access to one or other of these states (Hondeghem et al. 1985; Triggle, 1989). In principle, qualitatively or quantitatively different structure-activity relationships may be expressed for each of these several states. The 1,4-dihydropyridines exhibit these properties.

Electrophysiologic studies have demonstrated the voltage-dependence of interaction of the 1,4-dihydropyridine antagonists, apparent affinity increasing with increasing depolarization (Bean et al. 1984; Sanguinetti et al. 1984). This voltage-dependence can be demonstrated directly through radioligand binding studies (Table 2).

TABLE 2. Voltage-dependent Interactions of 1,4-dihydropyridines in Cardiac Cells (Wei et al. 1989).

	Polarized		Depolarized	
	K_D $\times 10^{-9}$ M	B_{Max} fmoles/mg	K_D $\times 10^{-9}$ M	B_{Max} fmoles/mg
[3H]PN 200 110	3.57	50.1	0.06	47.2
[3H]Bay K 8644	5.15	63.1	5.56	62.3

The binding of the antagonist [3H]PN 200 110 is increased by approximately 100 fold by depolarization in intact ventricular myocytes. Additionally, comparison of a 1,4-dihydropyridine series in depolarized cells and in membrane fragments indicates identical affinities in the two preparations (Wei et al. 1989; 13; Figure 6a). Thus, consistent with several electrophysiologic studies binding to membrane fragments likely represents interaction with the inactivated state of the channel. This may underlie the numerous observations that 1,4-dihydropyridine binding affinities are very similar in membrane preparations from many cell types (reviewed in Triggle et al. 1984; Triggle et al. 1989b), despite the diversity of expression of harmacologic activity. 1,4-Dihydropyridine activators and antagonists reveal significant differences in their voltage-dependent interactions (Figure 6b).

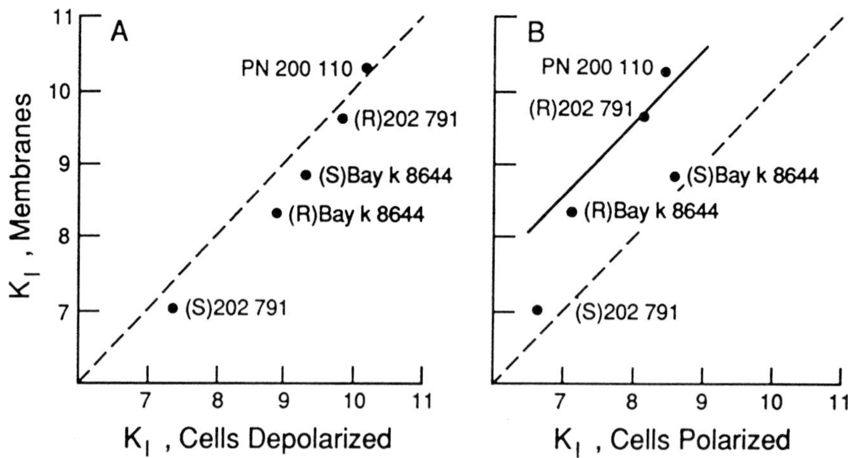

Figure 6. Correlations between binding affinities of 1,4-dihydropyridines in cardiac cells and membranes.
a: Correlation for a series of substituted nifedipine derivatives between binding in depolarized neonatal rat ventricular cells and membrane preparations from the same cells.
b: Correlation between binding to polarized and depolarized neonatal rat ventricular cells (Wei et al. 1989).

The virtual independence of activator affinity from membrane potential, observed in both direct and indirect radioligand binding approaches, probably accommodates the observations of the activator/antagonist discrimination observed in enantiomeric pairs of 1,4-dihydropyridines and the observations that the activator properties of 1,4-dihydropyridines become antagonist with decreasing membrane potential (Kass 1987; Wei et al. 1986). Thus, the stereochemical demands for interaction with the inactivated [antagonist] and open [activator] state of the channel are different.

There are insufficient data available to permit a detailed comparison of the structural requirements for 1,4-dihydropyridine activators and antagonists. For a limited series, however, comparison of activity in cardiac and smooth muscle preparations does not reveal the major differences in activity, approximately > 100 fold, seen with antagonists (Kwon et al. 1989; Figure 7). This observation also is accommodated by the state-dependent hypothesis since it is likely that the extent of open channel availability is more similar than the extent of inactivated channel state. Additional evidence consistent with the thesis of differential structure-activity expression according to channel state is provided by the data of Table 3 and 4.

TABLE 3. Pharmacological Activities of 1,4-Dihydropyridine Activator Analogs of Bay K 8644 in Cardiac and Smooth Muscle (Kwon et al. 1989).

Substituent	Papillary muscle	Femoral artery
	EC_{50} M	
2-CF_3	6.7×10^{-8}	4.2×10^{-9}
3-NO_2	4.2×10^{-8}	4.9×10^{-8}
3-CF_3	3.5×10^{-7}	9.6×10^{-8}
2-NO_2	1.4×10^{-7}	1.4×10^{-7}
4-NO_2	2.3×10^{-6}	3.3×10^{-7}

Bay K 8644 = 2,6-dimethyl-3-carbomethoxy-5-nitro-4-substitutedphenyl-1,4-dihydropyridine.

TABLE 4. Pharmacologic Activities of 1,4-Dihydropyridine Activators and Antagonists (Kwon et al. 1989).

	Antagonists		Activators	
	$-\log IC_{50}$ M		$-\log EC_{50}$ M	
Substituent	ileum	heart	ileum	heart
$2-NO_2$	9.52	6.43	6.68	6.85
$3-NO_2$	9.05	6.24	7.06	6.38
$4-NO_2$	6.72	4.50	6.35	5.64

Compounds based on the nifedipine and Bay K 8644 structures with substituent variation in the phenyl ring.

This reveals that the role of aromatic substituents in the 4-phenyl ring of 1,4-dihydropyridine activators and antagonists is different (Kwon et al.1989).

Most 1,4-dihydropyridines are neutral hydrophobic species and the available structure-activity relationships are derived from these agents. New 1,4-dihydropyridines possessing basic functions, including amlodipine and related compounds (Figure 7), suggest additional subtleties of action.

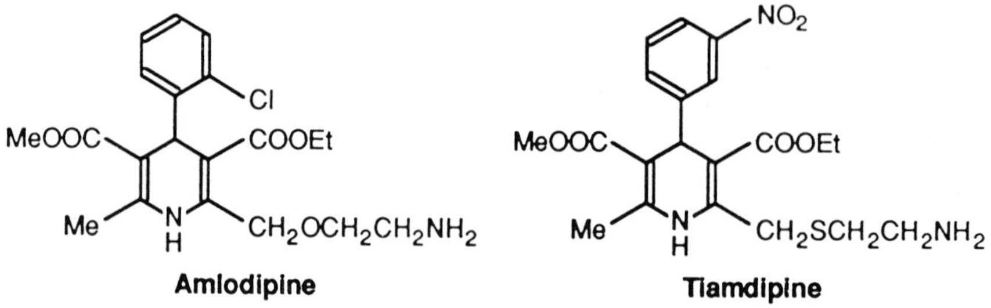

Figure 7. The structures of amlodipine and tiamdipine.

These compounds exhibit very high stereoselectivity, > 1000 fold, slow onset and offset kinetics as the protonated species, and a structure-activity relationship which suggests that they adopt a binding mode that may be different from the neutral 1,4-dihydropyridines of the nifedipine class (Kwon et al. 1989a). As the neutral species, however, these agents show onset and offset kinetics that are basically very similar to those of nifedipine and other rapid-acting 1,4-dihydropyridines (Kass et al. 1989). This suggests that the access of charged and uncharged 1,4-dihydropyridine derivatives may be different (Figure 8).

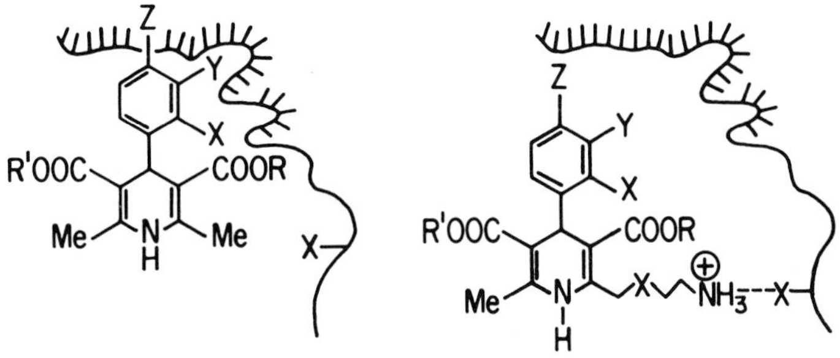

Z
Y
X
R'OOC
COOR
Me
N
H
Me
X
Z
Y
X
R'OOC
COOR
Me
N
H
X
NH_3^+
X

Figure 8. Alternative binding modes of nifedipine and amlodipine series of 1,4-dihydropyridines.

THE REGULATION OF CALCIUM CHANNELS

Consistent with their designation as pharmacologic receptors Ca2+ channels are regulated both in number and function by homologous and heterologous influences and under several experimental and pathological conditions (reviewed in Ferrante and Triggle, Submitted for publication). Of particular interest to a focus on structure-activity relationships is channel regulation mediated by chronic ligand occupancy or chronic channel activation. Additionally, such phenomena are of immediate importance to issues of tolerance and/or withdrawal during cardiovascular drug therapy.

The effects of chronic Ca2+ channel ligand administration have been examined both in vivo and in vitro [reviewed in Ferrante and Triggle, Submitted for publication). Chronic nifedipine and verapamil administration in vivo resulted in a down-regulation of high affinity binding sites in both brain and heart (Gengo et al. 1988); Panza et al. 1985). In vivo studies are complicated because of the possible reflex effects on sympathetic discharge and other regulatory pathways brought about by the vasodilating properties of the antagonists. Chronic treatment of PC 12 cells with nifedipine or Bay K 8644 produced up-and down-regulation of both channel number and function, as measured by $^{45}Ca^{2+}$ uptake and whole cell current (Skatteboli et al. 1989); Table 5)]. These latter results are very similar to the chronic ligand administration studies documented for other receptor types.

TABLE 5. Homologous Regulation In PC 12 Cells Following Chronic (5 day)) Exposure To Nifedipine and Bay K 8644 (Panza et al. 1985).

		Nifedipine 5×10^{-8} M/5 days	Bay K 8644 5×10^{-7} M/5 days
B_{Max}	control	46.1 fmoles/mg	43.8 fmoles/mg
B_{Max}	experimental	59.6 fmoles/mg[a]	33.4 fmoles/mg[a]
K_D	control	100.8 pM	75.1 pM
K_D	experimental	112.8 pM[b]	96.2 pM[b]

a Significantly different from control
b Not significantly different from control

Both membrane potential and Ca2+ may be viewed as "endogenous signals" of the Ca2+ channel. Accordingly, chronic activation of the channel by depolarization or elevation of intracellular Ca2+ might be expected to exert regulatory influence. Chick neural retina cells show a time- and concentration-dependent down regulation of both channel number and function upon chronic exposure to elevated K+ (Figure 9).

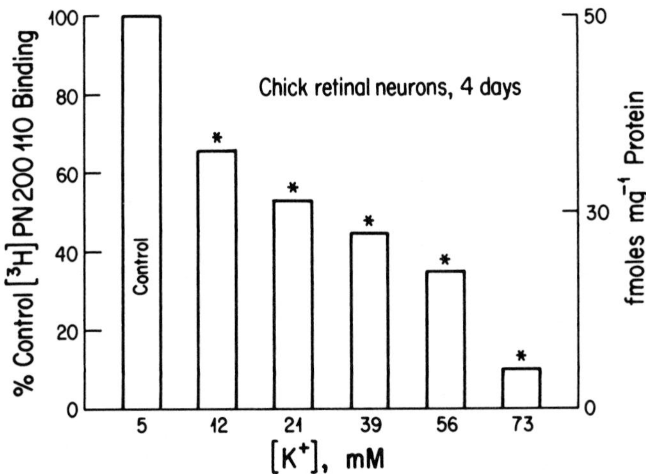

Figure 9. The down regulation of [3H]1,4-dihydropyridine binding sites in chick neural retina cells by chronic K+ depolarization.

The effects of elevated K+ are mimicked by the Ca2+ ionophore ionomycin. These studies parallel those previously reported by Delorme et al. (1989) for the PC 12 cell. It is probable, however, that not all cell types are equally sensitive to such regulatory influences. Thus cardiac cells do not appear to down-regulate their 1,4-dihydropyridine binding sites on chronic depolarization or chronic modulation by isoproterenol.

SUMMARY

The Ca^{2+} channel antagonists have proved to be major players on the cardiovascular drug field. Simultaneously, these agents and the corresponding activator 1,4-dihydropyridines have proved to be invaluable molecular tools with which to classify and characterize Ca^{2+} channels. It is certain that continued investigation of the structural and regulatory requirements for the actions of these drugs will lead to a better understanding of their clinical roles but also to the more facile development of new and improved agents.

REFERENCES

Bean B.P. Nitrendipine block of cardiac calcium channels: high affinity binding to the inactivated state. Proc. Nat. Acad. Sci. USA, 81: 6388-6392, 1984.

Bolger G.T. Gengo P. Klockowski R. Luchowski E. Siegel H. Janis R.A. Triggle A.M. and Triggle, D. J. characterization of the binding of the Ca^{2+} channel antagonist, [3H]nitrendipine, to guinea pig ileal smooth muscle. J. Pharmacol. Exp. Therap., 225: 291-309, 1983.

Callewaert G., Hanbauer I.and Morad M. Modulation of calcium channel in cardiac and neuronal cells by an endogenous peptide. Science, 243: 663-668, 1989.

DeLorme E.M. Rabe C.S. and McGee R. Regulation of the number of functional voltage-sensitive Ca^{2+} channels on PC 12 cells by chronic changes in membrane potential. J. Pharmacol. Exp. Therap., 244: 838-843, 1988.

Gengo P. Skattebol A. Moran J.F. Gallant S. Hawthorn M. and Triggle D.J. Regulation by chronic drug administration of neuronal and cardiac calcium channel, beta adrenoceptor and muscarinic receptor levels. Biochem. Pharmacol., 37: 627-633, 1988.

Hondeghem L.M. and Katzung B.G. Antiarrhythmic agents: the modulated receptor mechanism of action of sodium and calcium channel-blocking drugs. Ann. Rev. Pharmacol. Toxicol., 24: 387-423, 1985.

Janis R.A., Silver P. and Triggle D.J. Drug action and cellular calcium regulation. Adv. Drug. Res., 16: 309-591, 1987.

Janis R.A. Shrikhande A.V., Johnson D.E., McCarthy R.T., Howard A.D., Greguski R.and Scriabine A. Isolation and characterization of a fraction from brain that inhibits 1,4[3H]dihydropyridine binding and L-type calcium channel current. FEBS Lett., 239: 233-236, 1988.

Kass R.S. Voltage-dependent modification of cardiac calcium channel current by optical isomers of Bay K 8644: implications for channel gating. Circ. Res., 61 [Suppl. I], 1-5, 1987.

Kass R.S. and Arena J.P. Influence of pH on calcium channel block by amlodipine a charged 1,4-dihydro-pyridine compound: implications for location of the dihydropyridine receptor. J. Gen. Physiol., in press, 1989.

Kwon Y.W. Franckowiak G. Langs D.A. Hawthorn M. Joslyn A. and Triggle D.J. Pharmacologic and radioligand binding analysis of the actions of 1,4-dihydropyridine activators related to Bay K 8644 in smooth muscle, cardiac muscle and neuronal preparations, Naunyn-Schmied. Arch. Pharmacol., in press, 1989.

Kwon Y.W. Zhong Q. Wei X.-Y. Zheng W. and Triggle D.J. The interactions of 1,4-dihydropyridines bearing a 2-(2-aminoethylthio)methyl substituent at voltage-dependent Ca2+ channels of smooth muscle, cardiac muscle and neuronal tissues. Naunyn-Schmied. Arch. Pharmacol., submitted for publication, 1989.

Panza G. Grebb J.A. Sanna E. Wright A.G. and Hanbauer I. Evidence for down-regulation of [3H]nitrendipine recognition sites in mouse brain after long term treatment with nifedipine or verapamil. Neuropharmacol., 24: 1113-1117, 1985.

Sanguinetti M.C. and Kass R.S. Voltage-dependent block of calcium channel current in calf cardiac Purkinje fibers by dihydropyridine calcium channel antagonists. Circ. Res., 55: 336-348, 1984.

Skattebol A. Brown A.M. and Triggle D.J. Homologous regulation of voltage-dependent calcium channels by 1,4-dihydropyridines, submitted for publication, 1989.

Triggle D. J. Endogenous ligands for the calcium channel: myths and realities, In, The Calcium Channel: Structure, Function and Implications, Eds., Morad M., Nayler W., Kazda S. and Schramm M. Springer-Verlag, Berlin, 1988.

Triggle D.J. and Janis R.A. Calcium channel antagonists: New perspectives from the radioligand binding assay. Mod. Methods Pharmacol., 2: 1-28, 1984.

Triggle D.J. Structure-function correlations of 1,4-dihydropyridine calcium channel antagonists and activators, In, Molecular and Cellular Mechanisms of Antiarrhythmic Agents, Eds., L. Hondeghem and B. G. Katzung, Futura Press, 1989.

Triggle D.J. Zheng W. Hawthorn W.M. Kwon Y. W. Wei X.-Y. Joslyn A. Ferrante J. and Triggle D.J. Calcium channels in smooth muscle: properties and regulation. Ann. New York Acad. Sci., in press, 1989a.

Triggle D.J., Langs D.A. and Janis R.A. Ca2+ channel ligands: structure-function relationships of the 1,4-dihydropyridines. Med. Res. Revs., in press, 1989b.

Wei X.-Y. Luchowski E. Rutledge A. Su C.M. and Triggle D.J. Pharmacologic and radioligand binding analysis of the actions of 1,4-dihydropyridine activator-antagonist pairs in smooth muscle. J. Pharmacol. Exp. Therap., 239: 144-152, 1986.

Wei X.-Y. Rutledge and Triggle D.J. Voltage-dependent binding of 1,4-dihydropyridine Ca2+ channel antagonists and activators in cultured neonatal rat ventricular myocytes. Mol. Pharmacol., in press, 1989.

Author Index

The OHOLO Conferences

The Israel Institute for Biological Research (IIBR), founded in 1952, is oriented towards applied research, development and production, mainly in the fields of biology, chemistry, ecology and public health. Basic research studies related to and evolving out of the applied projects are also part of the Institute's program. The Institute is divided into three scientific divisions: chemistry, biology and environmental sciences.

Since 1956, IIBR has been organizing international meetings on selected topics in biology, chemistry and environmental science. Topics for the conference are selected on the basis of growing awareness and importance in which the Israeli scientific community has special interest. The purpose of these gatherings is to encourage the dialogue between Israeli scientists, and their colleagues from abroad.

The Conferences were at first held at OHOLO, a seminar center on the shores of the Sea of Galilee. But the OHOLO Conferences, as they came to be called, soon outgrew their first home, and have been held at various places. The 33th Conference was held in Eilat.

OHOLO Conferences Since 1956

1956 Bacterial Genetics
1957 Tissue Cultures in Virilogical Research
1958 Inborn and Acquired Resistance to Infection in Animals
1959 Experimental Approach to Mental Diseases
1960 Cryptobiotic Stages in Biological Systems
1961 Virus-Cell Relationships
1962 Biological Synthesis and Function of Nucleic Acids
1963 Cellular Control Mechanism of Macromolecular Synthesis
1964 Molecular Aspects of Immunology
1965 Cell Surfaces
1966 Chemistry and Biology of Psychotropic Agents
1967 Structure and Mode of Action of Enzymes
1968 Growth and Differentiation of Cells In Vitro
1969 Behavior in Animal Cells in Culture
1970 Microbial Toxins
1971 Interaction of Chemical Agents with Cholinergic Mechanisms
1972 New Concepts in Immunity in Viral and Rickesttsial Diseases
1973 Strategies for the Control of Gene Expression
1974 Sensory Physiology and Behavior
1975 Air Pollution and the Lung
1976 Host-Parasite Relationships in Systemic Mycoses
1977 Skin: Drug Application and Evaluation of Environmental Hazards
1978 Extrachromosomal Inheritance in Bacteria
1979 Neuractive Compounds and Their Cell Receptors
1980 New Developments with Human and Veterinary Vaccines
1981 Biomimetic Chemistry and Transition-State Analogs: Approaches to Understanding Enzyme Catalysis
1982 Behavioural Models and the Analysis of Drug Action
1983 Mechanisms of Viral Pathogenesis (From Gene to Pathogen)
1984 Boundary Layer Structure: Modelling and Application to Air Pollution and Wind Energy
1985 Basic and Therapeutic Strategies in Alzheimer's and Other Age-Related Neuropsychiatric Disorders
1986 Model Systems in Neurotoxicology: Alternative Approaches to Animal Testing
1987 Modern Approaches to Animal Cell Technology (A Joint Meeting with ESACT)
1988 Computer-Assisted Modeling of Receptor Ligand Interactions: Theoretical Aspects and Applications to Drug Design
1989 Calcium channel modulators in ischemic heart disease and other cardiovascular disorders: Basic and therapeutic approach.
1990 Novel Strategies in Production and Recovery of Biologicals from Recombinant Microorganisms and Animal Cells.

VANDERBILT UNIVERSITY
3 0081 021 541 879